AF564565

Recent Advances in Pineapple Production, Postharvest and Processing Technology

NIPA® GENX ELECTRONIC RESOURCES & SOLUTIONS P. LTD.

New Delhi-110 034

About the Editors

Dr. Prahlad Deb was born on 9th January, 1981 in Coochbehar, West Bengal. He completed his Masters degree in Horticulture from Uttar Banga Krishi Vishwavidyalaya (UBKV), Pundibari, Coochbehar, West Bengal. Thereafter, he qualified NET in Horticulture conducted by ASRB (ICAR) in 2008 as well as 2009. Subsequently, he completed his doctoral degree in Fruit Science from UBKV in 2010. Since last 12 years, he has been involved in teaching as Assistant Professor at the Department of Horticulture and Post-Harvest Technology, Visva-Bharati University, Santiniketan, West Bengal. Apart from teaching, he has been actively involved in various research and extension activities. Dr. Deb has handled major research projects funded by BARC, GOI and National Horticulture Board, GOI. Till now five students have been conferred the Ph.D. degree under his supervision. He has published more than 70 research papers in different journals of National and International repute; around 20 book chapters and one book. Beside these, he has also published a considerable number of newspaper articles, technical booklets, practical manuals etc. Dr. Deb has delivered lectures in around 50 National and International seminars and symposiums conducted within India and abroad. He is life member as well as an Executive Committee member of eight professional International societies including ISHS.

Dr. Nilesh Bhowmick has completed his B.Sc. (Hort.), M.Sc. (Hort.) in Fruits & Orchard Management and Ph.D. in Horticulture & Post Harvest Technology from Uttar Banga Krishi Viswavidyalaya, Bidhan Chandra Krishi Viswavidyalaya, and Visva-Bharatai, respectively. He has served in the rank of Assistant Director of Horticulture under the Dept. of Food Processing Industries and Horticulture, Govt. of WB around for two years. After that he had started his career as Assistant Professor and at present he is working as Professor, Dept. of Pomology & Post Harvest Technology, Uttar Banga Krishi Viswavidyalaya, Pundibari, Cooch Behar, West Bengal. He has 4 edited books, 70 full length papers and 15 book chapters in his credit. He is handling a numbers of UG & PG courses, as well as research projects sponsored by ICAR, Department of Science & Technology (Govt. of India), Department of Science & Technology & Biotechnology (Govt. of WB), and RKVY. He has supervised 3 doctoral, 11 master degree students and 2 doctoral, 3 master degree students are working under his supervision at present. He is the Life Members of several national, international professional societies, and has given various television, radio talks, as well as acted as resource person in HRD training for farmers. Dr. Bhowmick has awarded as Young Scientist of the Year 2017 by the International Foundation for Environment & Ecology, Kolkata, Best Young Scientist Award (2019) by the Society of Tropical Agriculture, New Delhi and got the Best Publication Award (2013) by Society for Advancement of Human and Nature (SADHNA), Dr. Y S Parmar University of Horticulture & Forestry, Himachal Pradesh. He has the credit for the several Best Oral and Poster Presentation Awards in different International and National seminars. His area of interest in teaching and research is production technology, stress physiology, crop improvement of fruit crops, minor and underutilized fruits.

Dedication

This book is dedicated to our teacher Late (Dr.) Swapan Kumar Ghosh, Ex-Professor & Ex-Dean, Faculty of Horticulture, Uttar Banga Krishi Viswavidyalaya, Pundibari, Cooch Behar, West Bengal for his continuous moral support and inspiration. Dr. Ghosh had devoted his life for the teaching of horticultural subjects particularly for Fruit Science & Post Harvest Technology. His research interest was on standardizing of package of practices of fruit crops, evaluation of different fruit crops, and storage life, postharvest improvement of several horticultural crops. He was a true teacher, mentor, motivator and icon of a numbers of his students and fellow teachers. He contributed a lot in overall development of horticulture through policy making, technological development, handling research projects, authoring of books, research papers, popular articles, guiding of many M.Sc. and Ph.D. students and catering a number of courses. After immense contribution he had retired from the service as the Professor, Dept of Pomology & Post Harvest Technology, UBKV on 30.11.2016 and breathed his last on 1st August, 2020.

Recent Advances in Pineapple Production, Postharvest and Processing Technology

Prahlad Deb
Assistant Professor
Department of Horticulture and Post-Harvest Technology
Visva-Bharati University
Santiniketan, West Bengal

Nilesh Bhowmick
Professor
Department of Pomology and Post-Harvest Technology
Faculty of Horticulture
Uttar Banga Krishi Viswavidyalaya
Pundibari, Cooch Behar, West Bengal

NIPA® GENX ELECTRONIC RESOURCES & SOLUTIONS P. LTD.
New Delhi-110 034

NIPA. GENX ELECTRONIC RESOURCES & SOLUTIONS P. LTD.

101,103, Vikas Surya Plaza, CU Block
L.S.C. Market, Pitam Pura, New Delhi-110 034
Ph : +91 11 27341616, 27341717, 27341718
E-mail: newindiapublishingagency@gmail.com
www: www.nipabooks.com

For customer assistance, please contact
Phone: + 91-11-27 34 17 17
Fax: + 91-11-27 34 16 16
E-Mail: feedbacks@nipabooks.com

Print ISBN: 978-93-58872-76-7

ebook ISBN: 978-93-58877-00-7

Composed and Designed by NIPA®.

উত্তরবঙ্গ কৃষি বিশ্ববিদ্যালয়
পুন্ডিবাড়ী, কোচবিহার, পশ্চিমবঙ্গ-৭৩৬১৬৫
UTTAR BANGA KRISHI VISWAVIDYALAYA
P.O. PUNDIBARI, DIST. COOCH BEHAR, WEST BENGAL- 736165

প্রফেসর দেবব্রত বসু
উপাচার্য
Prof. Debabrata Basu
M.Sc. (Ag.), Ph.D., DDE
Vice-Chancellor

মো°/Mob: +91-9434748016
ইমেল/E-mail: vc@ubkv.ac.in
vcubkvv@gmail.com
ওয়েবসাইট/ Website: www.ubkv.ac.in

Foreword

In the world of tropical agriculture, few fruits evoke the same sense of delight and fascination as the pineapple. With its crown of spiky leaves and sweet, tangy flavor, the pineapple holds a special place in culinary traditions around the globe. From its origins in South America to its widespread cultivation in tropical regions worldwide, the pineapple has captivated people for centuries with its unique combination of taste, aroma, and nutritional benefits.

This book, "Recent Advances in Pineapple Production, Postharvest and Processing Technology", curated by Dr. Prahlad Deb of Visva-Bharati and Prof. Nilesh Bhowmick of Uttar Banga Krishi Viswavidyalaya. serves as a comprehensive compendium on all facets of pineapple cultivation. It offers a deep dive into its history, taxonomy, advanced propagation techniques, crop improvement, scientific nutrient management, crop protection, processing, utilization, and more.

Readers will embark on a journey through the rich history of this iconic fruit, tracing its humble beginnings in the tropical forests of Paraguay and Brazil to its cultivation in countries spanning the equatorial belt. From the discovery o f the pineapple by European explorers to its introduction to new lands by colonial powers, the book unravels fascinating narratives that have shaped the pineapple's extraordinary legacy.

Diving into the world of pineapple cultivation, readers will discover a wealth of knowledge on selecting optimal cultivars, soil preparation, planting techniques, irrigation, and pest management. The book also explores biotechnological interventions, flower regulation, and year-round production practices, providing invaluable insights for scientists and growers. Innovative approaches such as high-density planting and organic farming techniques are scientifically highlighted to meet the increasing demand for this beloved fruit while prioritizing sustainability.

Beyond cultivation, the book explores the diverse array of pineapple varieties and their culinary applications, from fresh consumption to industrial uses. Emphasizing sustainable practices, it advocates for environmentally responsible approaches to pineapple cultivation, essential for meeting the rising global demand while conserving natural resources and promoting biodiversity. This book is an indispensable resource for pineapple enthusiasts, whether farmers, researchers, or avid consumers. With its blend of scientific rigor and practical wisdom, it promises to inform, inspire, and delight, serving as a beacon for all those who share a passion for this extraordinary fruit.

I extend my congratulations to the editors, Dr. Prahlad Deb and Prof. Nilesh Bhowmick, for compiling this comprehensive guide. I am confident that it will serve as a valuable resource for the scientific and academic community, growers, traders, and lovers of pineapple alike.

14/5/24

Prof. Debabrata Basu
Vice-Chancellor

Preface

In the world of tropical agriculture, few fruits evoke the same sense of delight and fascination as the pineapple. With its crown of spiky leaves and sweet, tangy flavour, the pineapple holds a special place in culinary traditions around the globe. From its origins in South America to its widespread cultivation in tropical regions worldwide, the pineapple has captivated people for centuries with its unique combination of taste, aroma, and nutritional benefits.

This book is a comprehensive guide to all aspects of pineapple particularly, history, taxonomy, advanced propagation techniques, crop improvement, scientific nutrient management, crop protection, processing, utilization etc. It delves into the rich history of this iconic fruit, tracing its journey from its humble beginnings in the tropical forests of Paraguay and Brazil to its cultivation in countries spanning the equatorial belt. Readers will discover the fascinating stories behind the discovery of the pineapple by European explorers, its introduction to new lands by colonial powers, and its eventual spread to virtually every corner of the globe. Various pineapple varieties cultivated worldwide from the popular Smooth Cayenne and Queen Victoria to lesser-known varieties such as Red Spanish and Sugarloaf, each type offers its own unique flavor profile, texture, and suitability for different growing conditions. Readers will learn how to identify and propagate their preferred varieties, enabling them to tailor their cultivation practices to maximize quality and productivity.

At the heart of this book lies a thorough exploration of pineapple cultivation practices. From selecting the optimal cultivars and preparing the soil to planting, irrigation, and pest management techniques, readers will gain invaluable insights into the art and science of growing pineapples. This holistic note on pineapple offers practical advice and expert tips to help the pineapple growers looking to diversify the crops or a seasoned agriculturalist seeking to optimize pineapple yield. Furthermore, this book provides an in-depth examination of the biotechnological interventions, flower regulation, year round production practices which will be immense help for the scientists as well as growers of pineapple. To meet the increasing demand of the fruit the high density pineapple cultivation of pineapple are widely practiced and the advanced technique of high density planting of pineapple are scientifically highlighted.

In addition to cultivation, this book explores the myriad ways in which pineapples can be processed and utilized. From fresh consumption and juicing to canning, drying, and fermentation, pineapples lend themselves to a wide range of culinary applications. Moreover, pineapple by-products such as juice, pulp, fibre and other products are valuable commodities with diverse industrial uses, further enhancing the economic potential of pineapple cultivation. Throughout this book, the importance of sustainable and environmentally responsible practices is emphasized. As the global demand for pineapples continues to rise, it is imperative that growers prioritize methods that minimize environmental impact, conserve natural resources, and promote biodiversity. From organic farming techniques to integrated pest management strategies, readers will discover innovative approaches to pineapple cultivation that prioritize both productivity and sustainability. Overall, combining scientific rigor with practical wisdom this book is a comprehensive resource for anyone with an interest in pineapples, whether as a hobbyist, farmer, researcher, academician or enthusiast. Anyone seeking to enhance knowledge, improve practices, or simply indulge love for this extraordinary fruit, this book is sure to inform, inspire, and delight.

Editors are thankful to all the authors of chapters for their worth contribution, New India Publishing Agency for publication and allied support and Mr. Avisek Mondal, MANAGE, Hyderabad for designing the book.

Prahlad Deb
Nilesh Bhowmick

Contents

Foreword ix
Preface xi

1. **Importance and Uses of Pineapple** **1**
Prahlad Deb, Pradipto Kumar Mukherjee and Payel Das

2. **Industrial Uses of Pineapple** **17**
Ishani Joardar and Subhasish Dutta

3. **Global Scenario of Pineapple Production and Export** **27**
Trina Adhikary and Satpal Singh

4. **Status of Pineapple Production in North Eastern India** **41**
Abhijit Debnath, H.S. Datta, J. Uchoi, Harish G.D R. Dangmei, R.K. Chaurasia, D. Pasweth, C. A. Pariat S.P. Bhutia, L. Sanga, S. Borang and E. Reang

5. **Status of Pineapple Production in Eastern India** **81**
Agniva Halder

6. **Status of Pineapple Production in Southern India** **91**
Kosgi Mounika, Peddaveeri Pravalika Reddy and Kanne Sushmitha

7. **History, Taxonomy and Genetic Resources of Pineapple** **99**
Polu Parameshwar, V.S.S.V. Prasanna, Susan Subba Diya Ghosal, Nilesh Bhowmick

8. **Reproductive Biology of Pineapple** **111**
Santhoshkumar G M, Bidyarani Devi Senjam Polu Parameshwar, Novin Chamling, Dinabandhu Samanta and Nilesh Bhowmick

9. **Recent Advances in Pineapple Improvement** **119**
Nilesh Bhowmick, Santhoshkumar G.M. and Dinabandhu Samanta

10. Biotechnological Intervention in Pineapple Improvement..... 145
Santhoshkumar G.M., Bidyarani Devi Senjam Novin Chamling and Nilesh Bhowmick

11. Edaphic and Climatic Requirement of Pineapple.................. 159
M. Premi Devi, Prahlad Deb and U.K. Behera

12. Propagation and Quality Planting Material Production of Pineapple.. 167
Ranjit Pal

13. Planting System of Pineapple... 181
Tanmoy Mondal and Fatik Kumar Bauri

14. Nutrient Management in Pineapple 191
Prerna Baraily

15. Status of Organic Pineapple Cultivation in India 213
Mahabub Alam and Agniva Halder

16. Cultural Strategies Adopted in Pineapple Cultivation 219
Arkendu Ghosh and Koyel Dey

17. Uses of Growth Regulators and Flower Regulation in Pineapple.. 229
Usha Kumari, Prahlad Deb

18. Harvesting and Post-Harvest Operations of Pineapple.......... 245
Rocky Thokchom, Nikita Khoisnam and Goutam Mandal

19. Physiological Disorders of Pineapple 259
Sananda Mondal and Debasish Panda

20. Important Pests of Pineapple and Management 265
Panchal Bhattacharjee, Sayan Sau, Prahlad Deb and Pallab Datta

21. Ecofriendly Management of Insect Pests Infesting Pineapple.. 275
Manasij Das and Swarnali Bhattacharya

22. Diseases of Pineapple and Management.......................................291
Pradipto Kumar Mukherjee

23. Existing Marketing System of Pineapple in West Bengal - A Look 305

Banani Das and Kalyan Kanti Das

24. Post Harvest Management, Processing and Value Addition of Pineapple 313

Arghya Mani, V.S.S.V. Prasanna and Surajit Mitra

Index 343

1

Importance and Uses of Pineapple

Prahlad Deb, Pradipto Kumar Mukherjee and Payel Das

Department of Horticulture and Postharvest Technology
Institute of Agriculture, Visva-Bharati, Sriniketan, West Bengal, India

Pineapple [*Ananas comosus* (L.) Merr.] is an important commercial tropical fruit crop under the family Bromeliaceae. Centre of origin of this fruit can be traced back to Brazil, South America. It is eulogized as "Queen of Fruit" or "Friendship Fruit" or "Haven Fruit" or "Golden Queen". The word 'Pineapple' came from the Spanish word "Pina" based on appearance of fruit which resembles a pine cone. Now, pineapple is termed as "Boon for dry land agriculture". Pineapple is an introduced crop in India and was first introduced in Goa by Portuguese. The largest producer of pineapple is Thailand and India ranks 6th in the world pineapple production. In India, Tripura, Assam, West Bengal, Kerala are the leading pineapple producers. Pineapple is a CAM (Crassulacean Acid Metabolism), monocarpic (each stem flowers only once and dies after fruiting), herbaceous perennial plant (Baruwa, 2013) and is suitable for cultivation in xerophytic conditions (Baraily and Deb, 2018).

The earliest written references to pineapple are found in the reference of Christopher Columbus, Gonzalo Fernández de Oviedo y Valdés, and Sir Walter Raleigh. They found pineapple growing in the West Indies mainly for for food and wine making. The Portuguese sailors were responsible for early dissemination of the pineapple and introduced it to Saint Helena in 1502. They also carried it to Africa and India by about 1550. Cultivation of pineapple had spread over most of the tropical countries including some of the islands of the South Pacific before the end of the 16th century. Costa Rica, Brazil, China, India, and Thailand are now major growers of pineapple.

Pineapple was considered as stable component of the crop complex and the diet of native Americans in the lowland tropics before the discovery of the fruit by Cristobal Colón (Christopher Columbus) on 4 November 1493 The European explorers have been impressed by the beauty and delicious taste of the fruit and often mentioned it in their chronicles. The references indicate that domesticated pineapple was widely distributed in Orinoco, Amazon, coastal Brazil around Rio de Janeiro and in the Caribbean provinces prior to the arrival of Columbus (Levvit, 2014). In few cases, the Europeans themselves dispersed the cultivation of pineapple in their continent. The names 'nanas' and 'ananas' has been extensively used throughout South America and the Caribbean provinces.

European explorers have been found acute domestication and selection exhibited in pineapples. The Americans have distinguished landraces from the wild types and had developed scientific crop growing practices, including the production cycle. Landraces (e.g. the Andean 'Perolera' and 'Manzana') were found with variation in fruit yield and quality during that time. Apart from the fresh fruit, the native Americans used to prepare alcoholic beverages (pineapple wine, chicha and guarapo), fibre, and medicine too (as an emmenagogue, abortifacient, antiamoebic and vermifuge and for the correction of stomach disorders). They have also domesticated a smooth-leaved type pineapple (curagua) with a higher yield of long and strong fibres. In Brazil and Philippines, there exists many traditional industry based on pineapple fibre and as 'piña cloth' mentioned as early as 1571 (Alvin and Kozlowski, 1977).

Pineapple was among the choicest food among the Europeans from the early 1500s and they introduced and grew it in greenhouses. The first successful greenhouse cultivation of pineapple has been done by Le Cour, or La Court at the end of the 17th century near Leyden. He published a monograph on horticultural practices of pineapple including flower 'forcing'. Pineapple was introduced to English gardeners in 1719 and France in 1730 rom The Netherlands to. Perrotet introduced pineapple variety Cayenne Lisse ('Smooth Cayenne') in 1819 from French Guiana. Most of these early varieties like 'Smooth Cayenne' and 'Queen' have been declined from commercial cultivation in Europe and pineapple fruit was imported from the West Indies. These two varieties have been spreaded to all tropical and subtropical provinces from Europe. The Spaniards and Portuguese sailors introduced 'Singapore Spanish' to Africa and Asia during 16th and 17th centuries.

Pineapple is considered as native to southern Brazil and Paraguay where the wild relatives are abundantly present. After introduction of pineapple by Portuguese to India and Java, the fruit become very much suitable with the edaphic and climatic condition and further spread throughout the Far East. The Indians traders and sailors dispersed the pineapple to the West Indies before the arrival of Columbus. The Spanish carried the pineapple into the Philippines, Hawaii and Guam during early 16th century. The fruit has been reached to England in 1660 and has started to grow under greenhouses around 1720 (Samson, 1980).

After introduction of pineapple in Hawaii by John Kidwell its large-scale pineapple cultivation has began during early 19the century by the US companies. James Dole as the most popular pineapple industrialist has been moved to Hawaii in 1899 and started pineapple cultivation in 1900 in company name as Dole and Del Monte in the island of Oahu in 1901 and 1917, respectively as the US's largest growers of pineapples.

After dissolving Pineapple Research Institute (USA) in 1986, Del Monte took variety 73 -114 and dubbed it into MD-2 to its plantations in Costa Rica which was popular as Gold Extra Sweet pineapple, known internally as MD-2. MD-2 is a hybrid developed from Pineapple Research Institute in Hawaii, which conducted research on behalf of Del Monte, Maui Land and Pineapple Company, and Dole.

Pineapple is an important tropical fruit and global trade on fresh pineapple has increased to 100 % during the last one decade. Although, India is the fifth largest producer of pineapple in the world, it has only 0.1% global pineapple market share. The pineapple market of different Asian countries and the countries around the Indian ocean can be exploited by Indian pineapple industry.

Nutritional importance: Pineapple is known as a queen of fruit due to its excellent taste and flavour. It is consumed or served as fresh, juiced, jam, jelly, candy and many other forms in day to day life. Pineapple is very familiar with our modern health-conscious diet plan now a day. The pineapple, with its sweet taste and distinctively tropical aroma, isn't just a delicious treat, it also packs a punch when it comes to nutritional value. It is a beautiful tropical fruit which have exceptional juiciness, vibrant tropical flavour and uncountable health benefits. Pineapple is primarily composed of water, which makes up approximately 86% of its weight and 13-19% total solids, where glucose sucrose and fructose are the main components. (Dull, 1971) The remaining portion consists of carbohydrates, dietary fibre, vitamins, minerals, and trace elements. Pineapple is a rich source of carbohydrates, mainly in the form of natural sugars such as sucrose, glucose, and fructose. Carbohydrate is an important source of energy in our body. In 100g fresh ripe pineapple, 12.39g of carbohydrate is present and in pineapple syrup the amount goes up to 20.20g (Debnath *et al.*, 2012). Pineapple is indeed a good source of dietary fibre, which is essential for a healthy digestive system. Hundred gram of fresh ripe pineapple can provide 1.2-1.4 grams of fibre in our body (Debnath *et al.*, 2012;

Hossain *et al.*, 2015). Dietary fibre helps regulate digestion, promotes satiety, and can contribute to maintaining a healthy weight.

It also plays a role in lowering cholesterol levels and reducing the risk of certain diseases, such as heart disease and type II diabetes. Including pineapple in your diet as part of a balanced meal plan can be a tasty way to increase your fibre intake (Sabahelkhier *et al.*, 2010). Pineapple is an excellent source of ascorbic acid (Vitamin C), a powerful antioxidant that boosts immune function, promotes skin health, and enhances iron absorption. This property of pineapple fights with different bacterial and viral disease-causing organism in our body. Vitamin C content of pineapple ranges from 15-20mg per 100g of fruit pulp (Hossain *et al.*, 2015).While not as abundant as vitamin C, pineapple contain significant amount of vitamin A. It also contains small amounts of vitamin A (23-130 I.U). Vitamin A is essential for maintaining healthy vision specially in low light conditions. It also supports immune function, reproduction, and cellular communication. Beta-carotene the precursor of vitamin A is found in pineapple contributing to it vitamin A content. Vitamin A is present in pineapple in smaller amounts but still contributes to overall nutrition. (Hemalatha and Anbuselvi, 2013)

Pineapple contains vitamin B6, albeit in smaller amounts compared to vitamin C and vitamin A (Hossain *et al.*, 2015; Joy, 2010). Vitamin B6, also known as pyridoxine, is involved in over 100 enzyme reactions in the body. It plays a crucial role in metabolism, neurotransmitter synthesis, and immune function. Adequate intake of vitamin B6 is necessary for optimal brain health and the production of neurotransmitters like serotonin and dopamine.

Pineapple also contains trace amounts of other B vitamins, including thiamine (vitamin B1) (0.079-0.092mg/100g), riboflavin (B2), niacin (B3), pantothenic acid (B5), and folate (B9), although in smaller quantities compared to vitamin C.

Beside the different essential vitamins several essential minerals exist in pineapple including that potassium, manganese, Copper, Magnesium, Iron, Zinc, Calcium, Phosphorus and sodium (Ali *et al.*, 2020).Potassium, an electrolyte crucial for maintaining fluid balance, nerve function, and muscle contractions, is abundant in pineapple. A 100-gram serving of fresh pineapple provides approximately 113-150 mg of potassium, contributing to the recommended daily intake of this vital mineral (Debnath *et al.*, 2012; Hossain *et al.* 2015).Manganese, often overshadowed by other minerals, plays a pivotal role in numerous physiological processes, including bone formation, wound healing, and metabolism (Debnath *et al.*, 2012). Pineapple boasts a notable manganese content, with around 1.3 mg per 100 grams, making it a valuable dietary source of this essential trace mineral. Copper, though required in small amounts, is indispensable for various enzymatic reactions, iron metabolism, and the synthesis of connective tissues. Copper is present in Pineapple aiding in the maintenance of optimal health and functionality. Magnesium, renowned for its multifaceted roles in energy production, muscle function, and nerve transmission, is present in moderate amounts in pineapple. With roughly 12-14 mg per 100 grams, pineapple contributes to the daily magnesium requirements, supporting overall well-being. Iron, a fundamental component of haemoglobin essential for oxygen transport, is found in pineapple, albeit in modest quantities. A 100-gram serving provides approximately 0.28-0.37 mg of iron, contributing to the dietary iron intake and promoting haematological health. Zinc, an off-overlooked mineral critical for immune function, wound healing, and DNA synthesis, is present in pineapple in small but notable amounts. With approximately 0.12 mg per 100 grams, pineapple aids in meeting the body's zinc requirements, bolstering immune resilience and cellular integrity. Calcium, renowned for its pivotal role in bone health, muscle contractions, and nerve signalling, is relatively low in pineapple compared to other minerals. However, with approximately 7-16 mg per 100 grams, pineapple still offers a modest contribution to the daily calcium intake, supporting skeletal integrity and cellular functions. Phosphorus, integral for bone mineralization, energy metabolism, and cell membrane structure, is present in pineapple, albeit in limited quantities. With roughly 7-11 mg per 100 grams, pineapple contributes modestly to the overall phosphorus intake, supporting various physiological processes. Sodium, crucial for maintaining fluid balance, nerve function, and muscle contractions, is present in pineapple, albeit in negligible amounts. Its sodium content is minimal, with approximately 1 mg per 100 grams, making pineapple a suitable choice for individuals monitoring their sodium intake. (Debnath *et al.*, 2012; Joy, 2010)

Pineapple contains a good number of organic acids too. Citric acid is the most abundant in it. Ash content, nitrogenous compound and lipids (0.1%) are very low in the pulp of pineapple. Proteins are developed from the nitrogenous compound and 25-30% of nitrogenous compounds are true proteins. One of the unique components of pineapple is bromelain, a mixture of proteolytic enzymes known for their anti-inflammatory properties and digestive benefits. Bromelain is an enzyme found in pineapple juice and in the stem of pineapple plants (Rowan *et al.*, 1990). It has various potential health benefits and uses. Bromelain helps break down proteins in the digestive system, aiding in digestion (Santos *et al.*, 2021). It's often used as a natural digestive enzyme supplement, particularly for individuals with pancreatic insufficiency or other digestive disorders (Maurer, 2001). Pineapple is relatively low in calories, with approximately 50-55 calories per 100 grams, making it a suitable option for calorie-conscious individuals.

Table 1: Nutritional composition of pineapple per 100g

Nutrient	Amount/100g
Energy	50-55 cal
Carbohydrate	12.39 g
Fibre	1.2-1.6 g
Protein	0.39-0.54 g
Fat	0.43 g
Vitamin C	15-20 mg
Vitamin A	23-130 IU
Vitamin B_1	0.079-0.092 mg
Vitamin B_2	0.031-0.036 mg
Vitamin B_3	0.489 mg
Vitamin B_6	0.110 mg
Potassium	113-150 mg
Manganese	1.3 mg
Magnesium	12-14 mg
Phosphorus	7-11 mg
Zinc	0.012 mg
Iron	0.28-0.37 mg
Calcium	7-16 mg
Sodium	1 mg

(Debnath *et al.*, 2012; Hossain *et al.*, 2015).

Medicinal Value: Pineapple fruits provide a rich array of vitamins, minerals, dietary fibre, antioxidants, and many other medicinal compounds. Pineapple can be used as an alternative source of nutrition for personal healthy life. Pineapples are rich in vitamin C, which plays a crucial role in supporting the immune system. Adequate intake of vitamin C can help boost the body's defences against infections and illnesses. Ascorbic acid or vitamin C is a powerful antioxidant which fights against bacterial and viral infection in human body (Carr and Magginii, 2017). One ripe, healthy pineapple fruit can provide roughly 16.2% of the daily required amount of vitamin C (Hemalatha and Anbuselvi, 2013). Vitamin C is a potent antioxidant that aids in the synthesis of collagen in blood vessels, bones, muscle, and cartilage as well as the absorption of iron. Moreover, vitamin C lowers the incidence of some cancers, such as stomach, oesophageal, and colon cancer, and delays the onset of urinary tract infections in pregnant mothers (Chen *et al.,* 2022). As per the 2005 Dietary Guidelines for Americans released by the U.S. Department of

Health and Human Services, half a cup of raw pineapple has just 37 calories and 28 mg of vitamin C. Pineapple juice is a very good blend of sugar and acidity which is served as an option for drinking plain or mixed with other fruit juices.

The root and fruit of pineapple are either eaten as an anti-inflammatory and as aproteolytic agent. In Tripura, it is used as anthelmintic agent (Dey *et al.,* 2013). A root extract used for diarrhoea in rural tribal area. Pineapple is a digestive aid and a natural anti-inflammatory fruit. Pineapples include vitamin B1, or thiamine, which is necessary for the body's cells to turn carbohydrates into energy. Vitamin B is necessary for the heart, muscles, and neurological system to function. It is also essential for the creation of red blood cells. Pineapples are a good source of several important minerals, such as manganese, a trace mineral that is important for bone building and the production and activation of specific enzymes (Debnath *et al.*, 2012; Hossain *et al.,* 2015). Additionally, pineapples contain the trace mineral copper. It controls heart rate and blood pressure and aids in the absorption of iron. Thirteen percent of the acidity in pineapple juice comes from malic acid. Malic acid has additional health benefits. It strengthens the immune system, encourages firm, smooth skin, supports good dental health, and lowers the danger of toxic metal poisoning (Niklander *et al.,* 2018).

Dietary fibre, including that found in pineapple, is associated with lower cholesterol levels and improved heart health (Hossain *et al.,* 2015). Consuming pineapple fibre as part of a balanced diet may help reduce the risk of heart disease and stroke. Fibre helps slow down the absorption of sugar into the bloodstream, which can help stabilize blood sugar levels and promote a feeling of fullness and aid in weight management. This property is particularly beneficial for individuals with diabetes or those at risk of developing diabetes The fibre in pineapple can aid in the removal of toxins and waste products from the body, promoting overall detoxification and cleansing of the digestive system. Pineapple fibre, being a natural source of dietary fibre, can help promote regular bowel movements and alleviate constipation. As dietary fibre levels rise, there appears to be a corresponding decrease in blood cholesterol, which lowers the risk of heart disease. Due to its high manganese level, pineapple juice is a fantastic option for increasing sperm quality and fertility. Manganese content in one cup of this juice is 1.3 mg, whereas the daily recommended dose for males is 2.3 mg.

Bromelain has been shown to accelerate wound healing by reducing inflammation and promoting the growth of new tissue. It is sometimes used topically in creams or ointments for this purpose (Rowan *et al.*, 1990). This enzyme mixture has been used traditionally in various cultures as a remedy for a range of health issues. Some of the potential health benefits of bromelain include aiding digestion, reducing inflammation, relieving muscle soreness, promoting wound healing, and supporting immune function. Incorporating pineapple into your diet can offer numerous health benefits, but it's essential to consume it as part of a balanced diet for optimal results. Bromelain offers numerous therapeutic advantages, including the management of conditions like angina pectoris, bronchitis, sinusitis, post-surgery recovery, and thrombophlebitis (Pavan *et al.,* 2012). It aids in wound healing, improves drug absorption especially antibiotics and alleviates symptoms of osteoarthritis, diarrhoea, and certain cardiovascular issues.

Uses of Pineapple

Pineapple offers a wide range of nutritional benefits due to its high vitamin, enzyme, and antioxidant content. Pineapples are low in calories, despite their sweetness. They are high in thiamine and ascorbic acid, which aid in energy production, immune system support, and macular degeneration prevention. Manganese, which is necessary for the development of strong bones and connective tissue, is found in nearly 75% of the daily recommended value in the fruit. Due to the presence of a large amount of dietary fibre and bromelain, an enzyme that breaks down protein, it also helps digestion (Szalay, 2018).

Food: Pineapple fruits are commonly consumed fresh or in the form of pineapple juice. The crown, rind, eyes, and core of a pineapple ripened in plants are best eaten fresh, and only the crown, rind, eyes, and core must be removed. Pineapple adds a sweet and tangy flavour to fruit chat (also known as fruit chaat), a popular dish in South Asian cuisine. It's often mixed with other fruits like apples, bananas, grapes, and pomegranate seeds, along with a blend of spices and sauces for a flavorful salad. Pineapple is a common ingredient in smoothies and juices, adding sweetness and a tropical flair to drinks. You can blend pineapple with other fruits like mango, banana, or strawberries for a refreshing beverage. Grilling pineapple brings out its natural sweetness and adds a smoky flavour. Grilled pineapple slices can be served as a side dish or used as a topping for burgers, tacos, or salads. Pineapple can be used in desserts like cakes, tarts, and pies. It pairs well with coconut in tropical desserts or can be caramelized and served with ice cream for a decadent treat. Pineapple salsa or chutney adds a burst of flavour to several dishes like grilled fish or chicken. It combines the sweetness of pineapple with spicy peppers, onions, and herbs for a delicious accompaniment. Pineapple chunks can be skewered with other fruits or meats and grilled for tasty kebabs. Pineapple is a popular ingredient in tropical cocktails like piña coladas, maitais, and rum punches. You can also use it to garnish drinks for an extra touch of tropical flair. Pineapple can be used as a natural sweetener in recipes. Its natural sugars can enhance the flavour of dishes like sauces, marinades, and even savory dishes like stir-fries.

The fermented pulp is used to make nata de pina, a popular sweetmeat in the Philippines. Salads made with young, tender shoots are popular in Africa. Raw or boiled, the terminal bud, also known as "cabbage," and the inflorescences are eaten. In Guatemala, young shoots known as "hijos de pina" are sold in vegetable markets.

Pineapple apples have high moisture content, high sugar content, high vitamin C content, and low crude fibre content. Pineapple waste includes non-reducing sugars, carbohydrates, and protein, all of which are used as a nutrient medium for microbe growth and yeast fermentation to produce ethanol and single-cell protein (Hemalatha and Anbuselvi, 2013).

Medicine: Pineapple has long been used in folk medicine for medical purposes in addition to its nutritive properties. It stimulates appetite, acts as a diuretic, is contraceptive and aids in the removal of intestinal worms. It's been used to treat ulcers and improve fat excretion, among other aspects. Bromelain, a proteolytic enzyme found in commercial meat tenderizers and used as a soft tissue anti-inflammatory and topical debridement agentis derived from pineapple (Omojasola *et al* 2008).

Fibre: Pineapple leaves produce a solid, white, silky fibre, and some cultivars are bred specifically for fibre production, with the young fruits removed to preserve the plant's vitality. Because of its long, wide, and stiff leaves, the 'Perolera' cultivar is ideal for fibre extraction. The retted material is thoroughly cleaned, sun-dried, and combed (Rohrbach *et al* 2014). The "Perolera" has been used to make thread in Malacca and Borneo for many years. Shoemakers in India prize the thread, which was once used in the Celebes. Pineapple fibre has long been used to string jewellery and weave tribal chiefs' caps in West Africa. Pineapple fibre is used to weave coarse textiles that resemble grass cloth by the people of Kwangtung province and Hainan Island. This fibre is used to wrap or sew cigars in Guam, and it is hand twisted into casting net. 'Pina' cloth is highly valued in Panay, a Philippine island, and in Taiwan. Pineapple fibre is also used in Taiwan to weave coarse cloth for farmers' underwear (Morton 1987).

Bromelain: Of the various similar compounds found in pineapples, bromelain stands out as the primary and most extensively studied among these proteolytic enzymes (Varilla *et al.* 2021). Bromelain is primarily composed of proteases, which are enzymes that break down proteins. Bromelain stands as a distinctive plant protease, extractable from diverse plant parts like the fruit pulp, stem, peel, and leaves. Naturally derived from pineapple, this substance can be extracted from parts of the fruit that are typically discarded as waste in industry, making efficient use of the plant's resources. Its concentration notably surpasses that in the fruit when found in the stem, rendering it a particularly accessible and plentiful source of bromelain (Tochi *et al.* 2008). This versatile product holds significant value with its broad range of uses across industries such as food, cosmetics, and pharmaceuticals. Research indicates that bromelain plays a vital role in various capacities, acting as an anti-inflammatory, antibacterial, anticoagulant, antithrombotic, antioedematous, fibrinolytic, and potentially even as an anticancer agent (Pavan *et al.* 2012). Bromelain consists of a diverse blend of proteases and non-proteases, comprising mainly stem bromelain (80%), along with fruit bromelain (10%), ananain (5%), and additional components (Nadzirah *et al.* 2013).

Pineapple wastes: Pineapple peel solid waste can be successfully converted into methane gas with an attractive concentration of 48 percent after 20 days using an indigenous microorganism. The ability of anaerobic digestion to treat waste was proven by the biodegradation of pineapple peel solid waste to methane gas. Biogas production from pineapple industrial waste not only provides an alternative source of green energy, but it also helps to maintain a pollution-free environment (Chulalaksananukul *et al* 2012).

Feed for animal: The outer peel, skin, and core from pineapple canning industries, known as bran, as well as the leaves, are used as ruminant feed. Fresh pineapple waste was preferred by cattle over fermented pineapple waste with higher acidity, and pineapple waste from the field or the cannery is used as dairy feed. Dried and ensiled pineapple waste can be used as additional roughage for dairy cattle, replacing up to 50% of the roughage in the total mixed ration (Hossain and Bepary, 2015).

Aesthetic value: The use of fruit-bearing plants for ornamental purposes has been increasing in recent years, making the development of such varieties an interesting alternative for the cut flower and ornamental plant segments (Souza *et al.*, 2012a; 2012b).The ornamental fruit crops has been emerged as an interesting alternative with differentiated and original products for use as cut flowers, potted plants, foliage and mini fruits (Souza *et al.*, 2012c; 2014; Santos *et al.*, 2015; Colombo *et al.*, 2017; Nóbrega *et al.*, 2017). Pineapple is a versatile fruit that can be used in various dishes and presentations, culinary uses, health care, crafts and decorative arrangements.

(*Source*: Pereira *et al.*, 2018)

Pineapple can be used in fruit bouquets or arrangements, where it's cut into decorative shapes like stars, hearts, or flowers. These bouquets are often used as centre pieces at events or as gifts for special occasions. Pineapples are a popular motif in home decor due to their tropical and welcoming vibe. You can find pineapple-themed decorations like artwork, throw pillows, and kitchen accessories to add a touch of whimsy to your living space. Pineapple leaves can be used in floral arrangements to add texture and visual interest. They're often used in tropical-themed bouquets or as accents in centre pieces.

Pineapple-scented candles are popular for their sweet and tropical fragrance. You can find pineapple-scented candle wax or essential oils to create your own candles at home. Tropical decoration with pineapple is the new trend in the world of interior design. Well, because of it is apart from its exotic and refreshing appearance, pineapple has a deeper meaning that makes it a symbol of good luck and hospitality. In many cultures, pineapples are considered a welcome gift, making them an ideal choice for decorating living spaces. Its characteristic appearance, with its prickly skin and sweet, juicy fruits, gives any room a cheerful and welcoming atmosphere. In addition, the pineapple is versatile and can be integrated into the decoration in various ways. pineapple adds a touch of holiday spirit and brings the feeling of sun, beach and sea directly into your home. With their vibrant colour and exotic charm, pineapples are definitely the perfect companion for tropical decorations. Pineapple flowers are an edible garnish, but ones that taste as good as they look. The thin edges crisp lightly in the oven, while the middle becomes chewy and the pineapple's tart-sweetness concentrates to a candy-like confection.

(*Source*: https://www.bespoke-bride.com/, https://www.shemakesandbakes.com/)

Ornamental pineapple species: There are many wild relatives and related species of pineapple having ornamental or aesthetic value for their characteristic leaves, inflorescence etc. There are also many striking variegated pineapple cultivars. The leaves can have central or marginal longitudinal stripes of white, cream, yellow, pink, brown or ochre depending on the cultivar. These plants can be very striking in the garden. *Ananas macrodontes*, *Ananas comosus* var. *ananassoides* 'Nanus', *Ananas comosus* var. *bracteatus*, *Ananas comosus* var. *erectifolius* 'Chocolat', *Ananas comosus* var. *bracteatus* 'Tricolor', *Ananas comosus* var. *bracteatus* 'Candystripe', *Ananas comosus* 'Golden Rocket', *Ananas comosus* 'Ivory' etc. are some popular types. *Ananas comosus* var. *bracteatus* has distinctive red bracts. *Ananas comosus* var. *bracteatus* 'Tricolor' looks great when mass planted. *Ananas comosus* var. *bracteatus* 'Candystripe' has a central stripe. *Ananas comosus* 'Golden Rocket' has spineless leaves with central variegation. This can be cream in the shade but is a warm pink to almost red in full sun. *Ananas comosus* 'Ivory' has ornamental white banded leaves.

Ananas comosus var. *erectifolius* 'Chocolat', *A. comosus* var. *bracteatus* 'Tricolor', *A. comosus* var. bracteatus 'Candystripe', *A. comosus* 'Ivory'
(*Source*: https://gardendrum.com/2014/07/16/)

References

Ali, M. M., Hashim, N., Abd Aziz, S. and Lasekan, O. 2020. Pineapple (Ananas comosus): A comprehensive review of nutritional values, volatile compounds, health benefits, and potential food products. Food Research International, 137: 109675.

Alvin, P. T. and Kozlowski, A. 1977 Eco physiology of Tropical Crops, Academic Press, London, pp, 144

Baraily, P. and Deb, P. 2018. Effect of integrated nutrient management on growth and yield of pineapple (cv. Kew), International Journal of Chemical Studies, 6(5): 1691-1695.

Baruwa, O. I. 2013. Profitability and constraints of pineapple production in Osun State, Negeria. Journal of Horticultural Research, 21(2):59-64. DOI: 10.2478/johr2013-0022.

Carr, A. C. and Maggini, S. 2017. Vitamin C and Immune Function. Nutrients, 9(11): 1211. https://doi.org/10.3390/nu9111211

Chen, Z., Huang, Y., Cao, D., Qiu, S., Chen, B., Li, J., Bao, Y., Wei, Q., Han, P. and Liu, L. 2022. Vitamin C Intake and Cancers: An Umbrella Review. Frontiers in nutrition, 8: 812394. https://doi.org/10.3389/fnut.2021.812394

Colombo, R. C., Favetta, V., Cruz, M. A., Carvalho, D. U., Roberto, S. R.and Faria, R. T. 2017. Acclimatization and growth of ornamental pineapple seedlings under organic substrates. Ornamental Horticulture, 23(3): 257-262. DOI: https://doi. org/10.14295/oh.v23i3.1040

Debnath, P., Dey, P., Chanda, A. and Bhakta, T. 2012. A Survey on Pineapple and its medicinal value. Scholars Academic Journal of Pharmacy, 1(1): 24-29.

Dey, P., Debnath, P., and Bhakta, T. 2013. Evaluation of anthelmintic activity of pineapple fruit extract using Indian earthworm (Pheritima posthuma). Mintage Journal of Pharmaceutical and Medical Sciences, 2, 26-27.

Dull, G. G. 1971. The pineapple: general. In: A. C. Hulme (Ed.). The Biochemistry of Fruits and their Products, Academic Press, New York. 2: 303-324

Hemalatha, R. and Anbuselvi, S. 2013. Physicochemical constituents of pineapple pulp and waste. Journal of Chemical and Pharmaceutical Research, 5(2):240-242.

Hossain, M. F., Akhtar, S. and Anwar, M. 2015. Nutritional value and medicinal benefits of pineapple. International Journal of Nutrition and Food Sciences, 4(1): 84.

Joy, P. P. 2010. Benefits and uses of pineapple. Pineapple Research Station (Kerala Agricultural University), Vazhakulam, pp, 670-686.

Levvit, R. (2014) A noble present of fruit': A transatlantic history of pineapple cultivation, In: Garden History, Garden Trust, 42(1): 106-119. (https://www.jstor.org/stable/24636289)

Maurer, H. R. 2001. Bromelain: Biochemistry, pharmacology and medical use. Cellular and Molecular Life Sciences, 58: 1234-1245.

Nadzirah, K. Z., Zainal, S., Noriham, A. and Normah, I. 2013. Efficacy of selected purification techniques for bromelain. International Food Research, 20, 43-46.

Niklander, S., Fuentes, F., Sanchez, D., Araya, V., Chiappini, G., Martinez, R. and Marshall, M. 2018. Impact of 1% malic acid spray on the oral health-related quality of life of patients with xerostomia. Journal of oral science, 60(2): 278–284. https://doi.org/10.2334/josnusd.17-0164

Nóbrega, D. S., Peixoto, J. R., Vilela, M. S., Faleiro, F. G., Gomes, K. P. S., Sousa, R. M.,D. and Nogueira, I. 2017. Agronomic descriptors and ornamental potential of passion fruit species. Ornamental Horticulture, 23(3): 357-362, DOI: https://doi. org/10.14295/oh.v23i3.1053

Omojasola, P. F., Jilani, O. P. and Ibiyemi, S. A 2008. Cellulase Production by some Fungi Cultured on Pineapple Waste. Nature and Science, 6(2): 64-79.

Pavan, R., Jain, S., Shraddha and Kumar, A. 2012. Properties and therapeutic application of bromelain: a review. Biotechnology Research International, 976203. https://doi. org/10.1155/2012/976203

Pereira, G. N. D., DeSouza, E. H., Souza, J. D. S. and Souza, F. V. D. 2018. Public perception and acceptance of ornamental pineapple hybrids. Ornamental Horticulture (Campinus), 24(2): 116-124.

Rowan, A. D., Buttle, D. J. and Barrett, A. J. 1990. The cysteine proteinases of the pineapple plant. Biochemical Journal, 266(3): 869.

Sabahelkhier, K. M., Hussain, A. S. and Ishag, K. E. A. 2010. Effect of maturity stage on protein fractionation, in vitro protein digestibility and anti-nutrition factors in pineapple (Ananas comosus) fruit grown in Southern Sudan. African Journal of Food Science, 4(8): 550 -552.

Samson, J. A. 1980. 'Pineapple' in Tropical Fruits, London, pp, 171

Santos, A. R. A., Souza, E. H. Souza, F. V. D. Fadini, M., Girardi, E. A., Soares Filho, W. S. 2015. Genetic variation of Citrus and related genera with ornamental potential. Euphytica, 205 (2): 503-520. DOI: https://doi.org/10.1007/s10681-015-1423-2

Santos, D. I., Martins, C. F., Amaral, R. A., Brito, L., Saraiva, J. A., Vicente, A. A. and Moldão-Martins, M. 2021. Pineapple (Ananas comosus L.) By-Products Valorization: Novel Bio Ingredients for Functional Foods. Molecules (Basel, Switzerland), 26(11): 3216.

Souza, E. H., Costa, M. A. P. C., Costa Júnior, D. S. Santos-Serejo, J. A. and Souza, F. V. D. 2014. Selection and use recommendation in hybrids of ornamental pineapple. Revista Ciências Agronômicas, 45(2): 409-416, DOI: http://dx.doi.org/10.1590/S1806-66902014000200024

Souza, E. H., Costa, M. A. P. C., Souza, F. V. D., Costa Junior, D. S., Amorim, E. P., Silva, S.O. and Santos-Serejo, J. A. 2012. Genetic variability of banana with ornamental potential. Euphytica, 184 (3): 355-367. Doi: 10.1007/s10681-011-0553-4

Souza, E. H., Souza, F. D., Costa, M. A. P. C., Costa Júnior, D. S., Santos-Serejo, J. A., Amorim, E. P. and Ledo, C. A. S. 2012a. Genetic variation of the Ananas genus with ornamental potential. Genetic Resources and Crop Evolution, 59(7): 1357-1376. DOI: https://doi.org/10.1007/s10722-011-9763-9

Souza, F. V. D., Carvalho, A. C. P. P. and Souza, E. H., 2012b. Abacaxi Ornamental. In: Paiva, P. D. O.; Almeida, E. F. A. (Org.). Produção de Flores de Corte, 1: 19-39.

Szalay, J. 2018. Pineapple: Health Benefits, Risks & Nutrition Facts. Retrieved from Live Science : https://www.livescience.com/45487-pineapple-nutrition.html

Tochi, B. N., Wang, Z., Xu, S. Y. and Zhang, W. 2008. Therapeutic application of pineapple protease (bromelain): a review. Pakistan journal of nutrition, 7(4), 513-520.

Varilla, C., Marcone, M., Paiva, L. and Baptista, J. 2021. Bromelain, a Group of Pineapple Proteolytic Complex Enzymes (Ananas comosus) and Their Possible Therapeutic and Clinical Effects. A Summary. Foods (Basel, Switzerland), 10(10): 2249. https://doi.org/10.3390/foods10102249.

2

Industrial Uses of Pineapple

Ishani Joardar and Subhasish Dutta

Departmentof Biotechnology, Haldia Institute of Technology (HIT)
ICARE Complex, Hatiberia, Haldia, West Bengal, India

Considered as one of the leading members of the family *Bromeliaceae*, pineapple is one of the most extensively consumed fruits in the world. Fruit juices, extracts, concentrates and processed edibles are consumed worldwide. Besides, plant fibres, enzymes obtained from the stems and leaves find many applications across industries. Pineapple residue mainly consists of pulp, stem, leaves and peels and the huge wastes produced as a result of improper handling, harsh environmental conditions, decaying and deterioration. Thesecan be successfully processed and extracted for further inexpensive and sustainable utilization instead of discarding. Agro-based and chemical industries have been switching to using plant-based substrates from various resources. Above all, all the natural substrates obtained are highly renewable, won't contribute to the global waste production, require less expenditures in processing, less toxic and has considerably better biochemical properties than the synthetic alternatives.

Industrial Importance of Pineapple

Pineapple Leaf Fibre (PALF)

Back in the days, pineapple leaf fibre was considered to be an agricultural residue until it's potential capabilities as an organic fibre and its commercial prominence heaved into view as a consequence of pioneering researches (Uddin *et al.*2017). According to Technical Association of Pulp and Paper Industry (TAPPI) standards, many chemical constituents like holocellulose, α-cellulose, lignin, and ash content of PALF were analyzed from a different source of fibres, with varying environmental conditions, cost, etc. (TAPPI,1991). Mainly because of its highest tensile strength (413 to 1627 MNm^{-2}), high cellulose content, as compared to other organic fibres with its Young's modulus varying from 34.5 to 82.51 GNm^{-2}. Furthermore, it has also been ascertained that upon its conversion to yarn, a 13% rise in its strength is observed and a bundle of saturated PALF adheres a reduced strength by 50% (Asim *et al.,* 2015). Various researches and aided projects of SITRA (South India Textile Research) under UNDP/UNIDO have demonstrated the production of 100% PALF yarn by blending with other synthetic or natural fibres like cotton or silk. The threads, thus produced, are used to make fabrics, fancy carpets, mops, curtains, etc. where pineapple-jute blended yarns are used by the textile and fabric industries (Mishra *et al.*, 2004; Debnath, 2016). It is also used in furnishings, decorations, sports items, baggage, quilts due to its aesthetic and elegant fabric (Anon, 1992). Decorticated and excoriated fibres are enzymatically altered to make paper (Asim *et al.,* 2015). Other uses of the PALF include the manufacture of, ridged reinforced roofing sheets, medicines, cosmetics, biopolymer coating for various chemicals (Asim *et al.,* 2015; Natarajan *et al.,* 1997; Jayanta, 2003; Geethamma *et al.*, 1998). PALF is used in the industries after surface modifications for the productions of V- belt cord, air-bag tying cords, conveyor belt cord, lightweight duck cloth, thermal insulator, etc. (Sapuan *et al.*, 2011; Basu and Chellamani, 2003). Moreover, its high mechanical and tensile strength has made it suitable for the manufacture of construction and building supplies, aerospace and automobile industries. Hence, it also has the potency of being used as reinforcing fillers and fibre reinforced plastics with the help of vinyl-ester, polyethene, polyester, polycarbonate, polypropylene (Asim *et al.,* 2015; Mishra *et al.*, 2004; Sapuan *et al.*, 2011; Basu and Chellamani, 2003). The reinforced composites of PALF are also considered as an economical substitute against the expensive, unsustainable synthetic fibres with glass fibres costing around1200–1800US$/tonnes and the natural fibres priced between 200 and 1000 US$/tonnes (Asim *et al.,* 2015). In India, white, raw pineapple leaf fibre is priced around Rs. 580 per kg in India-Mart.

Bromelain

Properties: Mainly obtained from the stems of pineapple, the aqueous, concentrate or enzyme, bromelain is extracted from its leaves and can also be found in the fully-ripened fruit (Asim *et al.,* 2015; Golden and Smith, 2012). It's a source of sulfhydryl proteolytic enzyme (Golden and Smith, 2012) and consists of various thiol endopeptidases, peroxidases, cellulase, glucosidase, carbohydrates, glycoproteins and many protease inhibitors (Bhattacharya, 2008).

Importance & applications: There are numerous therapeutic uses of this low-cost, economic by-product mainly due to its less toxicity and capabilities of being orally administered, compared to other plant enzymes is what makes thiscrude extract of pineapple thrive commercially (Pavan*et al.*, 2012). Some of the advantages of using bromelain are as follows, i.e., **(i)** According to various researches, bromelain having a composition of two proteases helps in the disintegration of collagen or meat tenderization (beef, chicken etc.) more inexpensively. It is also used in brewing and baking industries (Ketnawa *et al.*, 2012).as it helps in the relaxation of dough by working rapidly on its glutens and in the clarification of alcoholic beverages respectively. Being a proteolytic enzyme, it also aids in digestion, break-down of proteins into simpler amino acids and prevents constipation (Debnath *et al.*, 2012). **(ii)** Protects crops from phytopathogens because cysteine protease shows antifungal activities, helps in curing dermatological problems and is preferred over surgical debridement by helping in the rapid healing of wounds (Garcia *et al.*, 2012; Hu *et al.*, 2011). **(iii)** Other medicinal attributes include minimization of cardiovascular disorders, diabetes, brain-stroke, thrombosis, etc. due to its property as an inhibitor of the aggregation of blood platelets. Bromelain has anti-inflammatory and analgesic characteristics which help in relieving Osteoarthritis and in its pathogenesis (Mojcik and Shevach, 1997). **(iv)** Bromelain supplements are recommended as an adjuvant medicinal alternativefor treating chronic inflammatory and autoimmune diseases (Barth *et al.*, 2005)It interacts with cell signalling cascades and pathways to withstand gastroenteric pathogens like *Escherichia coli* and *Vibrio cholera* to prevent diarrhoea. Clinical studies have also revealed its outstanding anticancer activities which involve a reduction in tumour formation, apoptosis of tumour cells by the inhibition of regulators that promote cell survival like Akt and Erk. It prevents tumour progression and angiogenesis by suppressing NF-κB, Cox-2, and PGE2 activity. These promising traits might make bromelain suitable for oral enzyme therapies of cancer patients in the long run (Chobotova *et al.*, 2010).

Pectin from Pineapple Wastes

Application and Uses: (i) One of the primary significances and known uses of pectin is as a food stabilizer, thickening agents in jams, pickles and marmalades. However, there are many other biomedical and industrial applications of this simple, inexpensive fruit derivative. **(ii)** High-methoxyl pectin is used mostly as a thickening agent for increasing the viscosity in fruit juices, and the low-methoxyl pectin, on the other hand, promotes a constant texture in canned food items (Wicker *et al.*, 2003). **(iii)** It aids in rapid digestion, acts as a cure to diarrhoea and constipation, used in biomedical instruments as adhesives for wound healing and is one of the most versatile food stabilizers. It also considered as a healthier substitute of fat in many baked items, for reducing acidity levels in yoghurt. Powdered pectin has been used as a source of making jam at homes since a long time. Moreover, a level of 0.5-1% pectin is considered optimal as a food additive (Sundeen, 2009).

Extraction Technologies and Cost: First extracted and discovered in the 1800s, pectin is a complex polysaccharide which is abundantly found in the peels of citrus fruits, apple pomaces and terrestrial plants. One of the most convenient sources for its Extraction involves boiling for two hours or microwaved heating for fifteen minutes after proper pre-treatments. Extraction agents like CDTA and EDTA are used. It is then separated by filtration, precipitation and centrifugation processes. After drying, it is converted into a powdered texture for commercial purposes (Sundeen, 2009; Stephen, 1995). According to the 2016 IMR quarterly reviews on food hydrocolloids, HM and LM pectin costs around $9 per lb and 10.60$ per lb respectively and in India, it is priced at around Rs. 1500 per kg of pectin powder.

Vinegar

Application and uses: Vinegar is produced extensively throughout the world using varying techniques and sources; however, the overall benefits and functional properties stand unaltered. Besides a critical ingredient for enhancing the flavour of food or acting as a preservative for increasing shelf life, the health benefits associated are unprecedented. According to clinical studies, it has antidiabetic (Salbe *et al.,* 2009), anti-inflammatory, antimicrobial properties (Budak *et al*., 2014), helps in lowering cholesterol levels thus decreasing the chances of CVD (Budak *et al*., 2014), prevents hypertension (Kando *et al.* 2001) and increases the cognitive functions of the brain in humans. Researches also proved its efficacy over synthetic ones in reversing hepatotoxicity in rats (Dougnon*et al.,* 2009). Also, being a rich source of phenolic acids is said to show antineoplastic traits (Roleira *et al.,* 2015), thus preventing tumour metastasis and inducing an immune response in breast cancer (Mohamad *et al*., 2019).

Extraction: Vinegar obtained from pineapple peels possesses the same potency as any other commercial alternative, although the cost-effectiveness of this process is much more sustainable. Before processing, all the pre-treatments must be thoroughly done, which include washing, peeling, cutting etc. Following which, the fermentation of the raw material is carried outusing *Saccharomyces cerevisiae and Acetobacter pasteurianus.* Finally, it is filtered, packed, and the bottles are stored in a dark and cold place. In another experiment, vinegar production from pineapple peels was done with the help of saccharification procedures using various enzymes which proved that enzymatic treatments on pineapple wastes have a more significant impact in increasing the net yield (Roda *et al*., 2014).

Ethanol (from Pineapple wastes)

Application and uses: Deriving bioethanol from plant-based residues like pineapple wastes are considered to be a more sustainable means without leaving unfavourable carbon footprints or causing complete destruction of food crops as the only pineapple remains, or decayed pulp is used as a substrate. Researches have also proven that ethanol obtained from pineapple wastes are also safe for use in petrol or diesel engines (Hossain and Fazliny, 2010).

Extraction: Alcohol production is done from pineapple wastes with the help of sugars, enzyme treatments like hemicellulase and cellulase, followed by fermentation procedures using commercial yeasts like *Saccharomyces cerevisiae* and *Zymomonas mobilis*(Koffi and Han,1990). Besides, ethanol was also produced in large amounts without any pre-treatments using deficient respiration strain *Saccharomyces cerevisiae* ATCC 24553 with pineapple cannery wastes and juice of the discarded pulp as the low-cost substrates (Nigam,2000). Furthermore, using a morepurified carbohydrate binding molecule from *Trichoderma harzianum* is said to increase the yield of fermentable sugars in the peels and pulp ofpineapple(Anbuselvi and Muthumani, 2014). It is the high levels of sugar, protein, acid content and crude fibres that make it an efficient substrate in fermentation for obtaining ethanol with the help of yeasts (Azhar *et al*., 2017).

Citric Acid Obtained from Pineapple Wastes

Application & uses: Disclosed in the 1900s by Krebs that citric acid helps in the metabolism of all the aerobic organisms, this tricarboxylic acid aids as a food preservative, flavour enhancer, buffering agent. As an emulsifier, it is considered as a key industrial chemical thus highly emerging in the economy, which is widely used across various industries like cosmetic, food, beverage etc. (Thauer, 1988; Verhoff, 2005). Its uses involve applications in detergents and dishwashing industries, as a crosslinker in biomedical and pharmaceutical industries (Reddy *et al.,* 2015), polyols for sustainable sources of packaging, in tissue engineering (Sun *et al.,* 2014) and as an extracting agent. Researches have proved that citric acid is a safe disinfectant against many viruses, especially the human norovirus (Koromyslova *et al.,* 2015). Its outstanding metal chelation properties makes it suitable for cleaning nuclear sites which are highly contaminated by heavy metals; also, unlike phosphate, citric acid doesn't contribute to marine eutrophication (Kantar and Honeyman, 2006).

Extraction technique: With the increase in global demand for citric acid, Pineapple wastes have been used as a potential low-cost substrate for the production of citric acid by submerged and solid-state fermentation using

Aspergillus niger and *Saccharomyces cerevisiae*. The former was shown to give a larger yield of citric acid under suitable conditions (Ola *et al.,* 2009).

Soluble Sugar

In the early development of fruit, the predominant sugars were glucose and fructose. Only after the sixth week, sucrose concentration seemed to increase before its harvest. It has been observed that the activity of sucrose synthase was considerably high in the un-ripened fruit, and its level declined with its development (Chen and Paull, 2000). Also, the content of sugars is considered essential for determining the quality of the fruit (Py *et al.*, 1987). Although sufficient studies have been offered to understand the knowledge regarding the content of sugar present in pineapple, not enough research or technological advancements have been done to realize its industrial utilization and extractions of the same.

Pineapple Waste-based Animal Feed

Pineapple residues mixed with rice straw are used as an efficient substitute for roughage in ruminant animals (Sruamsiri, 2007). Moreover, it was seen that including these wastes in proper ratios for dairy cows gave promising yield at a considerably less feed cost without compromising on the milk production (Chinvaroj *et al.*, 2001). Instead, providing ensiled pineapple waste to the cattle was shown to increase the net output of milk per day by 3.1 litres (Gowda *et al.*, 2015). Although bran and pineapple wastes are hard to ensile, it can be diversified with withered grass, hay, rice straw etc.to decrease its acidic content and to make it easier to digest (Sruamsiri, 2007). Ensiled pineapple residues are thus economic as it replaces 70% of the supplements without increasing the cost of feeding or affecting daily weight gains and fertility; instead, it also overcame the disposal issues (Prado *et al.*, 2003). However, the inclusion of the same diet in non-ruminant animals did not prove to be equally affluent.

Further researches also proved that the bagasse obtained from pineapple fruit pulp and its constituent residual levels are fit for animal feedstock consumption, as far as insecticides are concerned (Cabrera *et al.,* 2000). Combined with maize, wheat bran and molasses, it is used as a fodder crop for cattle. The by-products obtained from pineapple and its bran are appropriately dried and used as a feedstock for farm animals, mainly because of its sugar content. It is considered a high energy crop;however, it needs to be supplemented with a nutritious diet rich in protein alongside (Henke,1931).Researches have also deduced the efficiency of including wheat offal alongside pineapple bran (Makinde *et al.*, 2011). It has also been found that the residues will act as an efficient carbon source substitute for single-cell protein (SCP) production using yeast.It will help in increasing the overall protein content in the diet, and the conversion efficiencies were found to be equally promising (Dhanasekaran *et al.*, 2011; Guerra*et al.*, 1986).

Conclusion

The crude residues obtained from pineapple wastes are considered as a novel, economic and feasible substitutes mainly becauseofitssustainable and high dietary values. The industrial prospects of bromelain, pineapple leaf fibre and vinegar extracted from the pineapple stem, leaves and pulp are colossal. They have multiple applications across various industries, be it biomedical or as animal fodder. Switching to such inexpensive substrates has not only proven to produce better yield but, are considered sustainable as well. Thus the prospects of pineapple wastes and similar fruit/vegetable residues are considered highly promising in this regard. However, it has been thoroughly concluded that only abundant and thorough scientific investigation and awareness will help to generate a lucrative market for mass production.

References

Anbuselvi, S. and Muthumani, S. 2014. Extraction and physicochemical characteristics of ethanol from pineapple pulp and waste. International Journal of ChemTech Research, 6(4): 2374-2376.

Anon, A.1992. A guide to unusual natural fibres: pineapple leaf fibre (PALF), Textiles, 21(3).

Asim, M., Abdan, K., Jawaid, M., Nasir, M., Dashtizadeh, Z., Ishak, M. R. and Hoque, M. E. 2015. A review on pineapple leaves fibre and its composites. International Journal of Polymer Science.https://doi.org/10.1155/2015/950567.

Asim, M.,KhalinaAbdan, M.,Jawaid, M., Nasir, Dashtizadeh, Z., Ishak, M. R. and EnamulHoque, M. 2015.A Review on Pineapple Leaves Fibre and Its Composites, International Journal of Polymer Science, Article ID 950567.

Azhar, S. H. M., Abdulla, R., Jambo, S. A., Marbawi, H., Gansau, J. A., Faik, A. A. M. and Rodrigues, K. F. 2017. Yeasts in sustainable bioethanol production: A review. Biochemistry and Biophysics Reports, 10: 52-61.

Ban-Koffi, L. and Han, Y. 1990. Alcohol production from pineapple waste. World Journal of Microbiology and Biotechnology, 6(3): 281-284.

Barth, H., Guseo, A. and Klein, R.2005. In vitro study on the immunological effect of bromelain and trypsin on mononuclear cells from humans,European Journal of Medical Research,10(8): 325–331.

Basu, P. R. K. A. and Chellamani,K. P. 2003.Jute and pineapple leaf fibres for the manufacture of technical textiles,Asian Textile Journal, 12: 94–96.

Bhattacharyya, B. K. (2008). Bromelain: an overview.Natural Product Radiance, 7(4):359-363.

Budak, N. H., Aykin, E.,Seydim, A. C., Greene, A. K. and Guzel□Seydim, Z. B. 2014. Functional properties of vinegar. Journal of food science, 79(5):R757-R764.

Budak, N.H., Doguc, D.K., Savas, M. C., Seydim, A.C., Tas, T.K. andCiris, M.I.2011.Effects of apple cider vinegars produced with different techniques on blood lipids in high-cholesterol-fed rats. Journal of Agricultural and Food Chemistry.59:6638–44.

Cabrera, H. A. P., Menezes, H. C., Oliveira, J. V. and Batista, R. F. S. 2000. Evaluation of residual levels of benomyl, methyl parathion, diuron, and vamidothion in pineapple pulp and bagasse (Smooth cayenne). Journal of Agricultural and Food Chemistry, 48(11), 5750-5753.

Chen, C. C.and Paull, R. E. 2000. Sugar metabolism and pineapple flesh translucency. Journal of the American Society for Horticultural Science, 125(5): 558-562.

Chinvaroj, S., Poathong, S. andWanasitchaiwat, V. Use of pineapple residues in Total Mixed Ration for lactating cows. Annual Report on Animal Nutrition Division 2001, Department of Livestock Development.Ministry of Agriculture and Cooperatives, Thailand.

Chobotova, K., Vernallis, A. B. and Majid,F. A. A. 2010. Bromelain's activity and potential as an anti-cancer agent: current evidence and perspectives,Cancer Letters, 290(2): 148–156.

Debnath, P., Dey. P., Chanda, A. and Bhakta, T. 2012. A Survey on Pineapple and its medicinal value. Scholars Academic Journal of Pharmacy.1(1).

Debnath, S. 2016. Pineapple Leaf Fibre—A Sustainable Luxury and Industrial Textiles. In Handbook of Sustainable Luxury Textiles and Fashion. Springer Singapore. 35-49.

Dhanasekaran, D., Lawanya, S., Saha, S., Thajuddin, N. and Panneerselvam, A. 2011. Production of single cell protein from Pineapple waste using yeast. Innovative Romanian food Biotechnology, 8: 26.

Dougnon, T.J., Kpodekon, T.M., Ahissou, H., Gbenou, J., Loko, F. andLaleye, L.2009. Protective effect of stem pineapple on Wistar rat poisoning with paracetamol. International Journal of Chemical and Biological Sciences, 3:688–93.

Femi-Ola, T. O., Oluyege, J. O. and Gbadebo, A. O. 2009. Citric acid production from pineapple waste. Continental Journal of Microbiology , 3:1-5.

Geethamma, V. G., Mathew,K. T.,Lakshminarayanan, R. and Thomas,S.1998. Composite of short coir fibres and natural rubber: effect of chemical modification, loading and orientation of fibre, Polymer, 39(6-7):1483–1491.

Golden, K. D. and Smith-Marshall, J. 2012. Characterization of bromelain from Morindacitrifolia (Noni). Journal of Scientific Research, 4(2): 445-445.

Gowda, N.K.S., Vallesha, N.C., Awachat, V.B., Anandan, S., Pal, D.T. and Prasad, C.S. 2015. Study on evaluation of silage from pineapple (Ananascomosus) fruit residue as livestock feed. Tropical Animal Health and Production, 47:557–61.

Guerra, N. B., Stamford, T. M. M., Medeiros, R. B., Freitas, C. P., Maia, S. R. and Cavalcante, M. L. 1986. Food Science: Protein Enrichment of Pineapple Waste for Animal Feeds. Food and Nutrition Bulletin, 8(1): 1-4.

Henke, L. A. 1931. Pineapple bran as a feed for livestock.University of Hawaii. Makinde, O. A., Odeyinka, S. M. and Ayandiran, S. K. 2011. Simple and quick method for recycling pineapple waste into animal feed. Livestock Research for Rural Development, 23(9).

Hossain, A. B. M. S. and Fazliny, A. R. 2010. Creation of alternative energy by bio-ethanol production from pineapple waste and the usage of its properties for engine. African Journal of Microbiology Research, 4(9): 813-819.

Hu, W., Wang, A. M. and Wu, S. Y. 2011. Debriding e ffect of bromelain on firearm wounds in pigs,The Journal of Trauma ,71(4):966–972.

Jayanta, B.2003.Plastics filled with agrowastes,Popular Plastics and Packaging, 48(11): 63–74.

Kantar, C., Honeyman, B.D. 2006. Citric acid enhanced remediation of soils contaminated with uranium by soil flushing and soil washing. Journal of Environmental and Chemical Engineering, 132:247–255.

Ketnawa, S., Chaiwut, P. and Rawdkuen, S. 2012. Pineapple wastes: A potential source for bromelain extraction. Food and Bioproducts Processing, 90(3): 385-391.

Kondo, S., Tayama, K., Tsukamoto, Y., Ikeda, K. andYamori, Y. 2001. Antihypertensive effects of acetic acid and vinegar on spontaneously hypertensive rats. Bioscience, Biotechnology, and Biochemistry, 65:2690–94.

Koromyslova, A.D., White, P.A. and Hansman, G.S. 2015. Treatment of norovirus particles with citrate. Virology, 485:199–204.

López García, B., Hernández, M. and Segundo, B. S. 2012. Bromelain, a cysteine protease from pineapple (Ananascomosus) stem, is an inhibitor of fungal plant pathogens. Letters in Applied Microbiology, 55(1): 62-67.

Mishra, S., Misra, M., Tripathy, S. S., Nayak, S. K. andMohanty, A. K. 2001. Potentiality of pineapple leaf fibre as reinforcement in PALF-polyester composite: Surface modification and mechanical performance. Journal of Reinforced Plastics and Composites, 20(4): 321-334.

Mishra, S., Mohanty, A. K., Drzal, L. T., Misra, M. and Hinrichsen, G. 2004. A review on pineapple leaf fibers, sisal fibers and their biocomposites.Macromolecular Materials and Engineering, 289(11): 955-974.

Mohamad, N. E., Abu, N., Yeap, S. K., Lim, K. L., Romli, M. F., Sharifuddin, S. A. and Alitheen, N. B. 2019. Apoptosis and metastasis inhibitory potential of pineapple vinegar against mouse mammary gland cells in vitro and in vivo. Nutrition & Metabolism, 16(1): 49.

Mojcik, C. F. and Shevach,E. M. 1997Adhesion molecules: a rheumatologic perspective,Arthritis and Rheumatism, 40(6): 991–1004.

Nasir Uddin, N., Miah, S., Jalil, M.A.,Islam, M. and Siddika,A. 2017.A review on extraction, characterization and application of pineapple leaf fiber (half) in textiles and other fields. International Journal of Advanced Research.5: 112-116. (ISSN 2320-5407). www.journalijar.com

Natarajan, V., Thangamani, K. and Thilagavathi, G.1997.Unconventional Natural Fibres and Their Textile Applications, AICTE-ISTE Summer School.

Nigam, J. N. 2000. Continuous ethanol production from pineapple cannery waste using immobilized yeast cells. Journal of Biotechnology, 80(2): 189-193.

Pavan, R., Jain, S., and Kumar, A. 2012. Properties and therapeutic application of bromelain: a review. Biotechnology Research International.976203.doi: 10.1155/2012/976203.

Prado, I.N., Lallo, F.H., Zeoula, L.M., Caldas Neto, S.F., Nascimento, W.G. and Marques, J. A.2003. Bulls performance in feedlot with levels of substituting corn silage by pineapple by-products silage. Brazilian Journal of Animal Science, 32:737–44.

Py, C., Lacoecilhe, J.J. and Teisson, C. 1987. The pineapple: Cultivation and uses. G.-P. Maisonneuve et Larose, Paris

Reddy, N., Reddy, R. and Jiang, Q. 2015. Crosslinking biopolymers for biomedical applications. Trends Biotechnol,33:362–369.

Roda, A., De Faveri, D. M., Dordoni, R. and Lambri, M. 2014. Vinegar production from pineapple wastes–Preliminary saccharification trials. Chemical Engingineering, 37.

Roleira, F. M., Tavares-da-Silva, E. J., Varela, C. L., Costa, S. C., Silva, T., Garrido, J. and Borges, F. 2015.Plant derived and dietary phenolic antioxidants: Anticancer properties. Food Chemistry, 183: 235-258.

Salbe, A. D., Johnston, C. S., Buyukbese, M. A., Tsitouras, P. D. and Harman, S. M. 2009. Vinegar lacks antiglycemic action on enteral carbohydrate absorption in human subjects. Nutrition Research, 29(12):846-849.

Sapuan, S. M., Mohamed, A. R., Siregar, J. P. and Ishak, M. R. 2011. Pineapple LeafFibers and PALF-Reinforced Polymer Composites. In Cellulose Fibers: Bio-and Nano-Polymer Composites, Springer Berlin Heidelberg. 325-343.

Singh, T. A., Sarangi, P. K. and Singh, N. J. 2018. Tenderisation of meat by bromelain enzyme extracted from pineapple wastes. International Journal of Current Microbiology and Applied Sciences, 7(9):3256-3264.

Sruamsiri, S. 2007. Agricultural wastes as dairy feed in Chiang Mai. Animal Science Journal,78:335–41.

Stephen, A. M. and Phillips, G. O. 2016. Food polysaccharides and their applications.CRC press.

Sun, D., Chen, Y., Tran, R.T., Xu, S., Xie, D., Jia, C., Wang, Y., Guo, Y., Zhang, Z., Guo, J., Yang, J., Jin, D. and Bai, X. 2014. Citric acid-based hydroxyapatite composite scaffolds enhance calvarial regeneration. Scientific Report, 4:6912.

Sundeen, K. A. 2009. Pectin The Miracle Molecule (Doctoral dissertation, University of Pennsylvania).

Technical Association of the Pulp and Paper Industry, TAPPI Standard and Suggested Methods, Technical Association of the Pulp and Paper Industry, New York, NY, USA, 1991.

Thauer, R.K.1988. Citric-acid cycle, 50 years on. European Journal of Biochemistry, 176:497–508.

Verhoff, F.H. 2005. Citric acid, Ullmann's encyclopedia of industrial chemistry. Wiley-VCH, Weinheim

Wicker, L., Ackerley, J. L., and Hunter, J. L. 2003. Modification of pectin by pectinmethylesterase and the role in stability of juice beverages.Food Hydrocolloids, 17: 809-814.

Yogesh M., Hari Rao A. N. 2015. Study on Pineapple Leaves Fibre and its Polymer based Composite: A Review, International Journal of Science and Research, 2319-7064.

3

Global Scenario of Pineapple Production and Export

Trina Adhikary and Satpal Singh

Department of Fruit Science, College of Horticulture & Forestry
Punjab Agricultural University, Ludhiana, Punjab, India

Pineapple is a wonderful tropical fruit having exceptional juiciness, vibrant tropical flavour and immense health benefits. Pineapple contains considerable calcium, potassium, fibre, and vitamin C. It is low in fat and cholesterol. Vitamin C is the body's primary water soluble antioxidant, against free radicals that attack and damage normal cells. It is also a good source of vitamin B1, vitamin B6, copper and dietary fibre. Pineapple is a digestive aid and a natural Anti-Inflammatory fruit. A group of sulfur-containing proteolytic (protein digesting) enzyme (bromelain) in pineapple aids digestion. Fresh pineapples are rich in bromelain used for tenderising meat. Bromelain has demonstrated significant anti-inflammatory effects, reducing swelling in inflammatory conditions such as acute sinusitis, sore throat, arthritis and gout and speedy recovery from injuries and surgery. Pineapple enzymes have been used with success to treat rheumatoid arthritis and to speed tissue repair as a result of injuries, diabetic ulcers and general surgery. Pineapple reduces blood clotting and helps remove plaque from arterial walls. Studies suggest that pineapple enzymes may improve circulation in those with narrowed arteries, such as angina sufferers. Pineapples are used to help cure bronchitis and throat infections. It is efficient in the treatment of arterioscleroses and anaemia. Pineapple is an excellent cerebral toner; it combats loss of memory, sadness and melancholy. Pineapple fruits are primarily used in three segments, namely, fresh fruit, canning and juice concentrate with characteristic requirements of size, shape, colour, aroma and flavour.

Global Scenario

Pineapple exhibits increasing demand world-wide, over the years. The global trade is around 50% as fresh fruit, 30% as canned product and 20% as juice concentrate. World trade on fresh pineapple has shown 100% increase during the last one decade. Even though India is the sixth largest producer of pineapple in the world with a share of about 8% in production, its share in the world market is negligible. The main pineapple producers are Brazil, Thailand, Philippines, Costa Rica, China, India and Indonesia. The different Asian countries and the countries around the Indian Ocean are importing about two lakh tonnes of pineapple in a year, mostly coming from distant countries. The leading exporters are Costa Rica, Belgium, Coted'lvoire, Philippines, Ghana, Netherlands, USA and France. Major importers are USA, Belgium, France, Italy, Germany, Japan and UK. MD2 is the most popular variety in the international market because of its colour, flavour, shape, life span and ripeness being superior to other varieties.

Top Exporters and Importers

Countries export fresh and processed pineapple products

Fresh pineapple	Canned pineapple	Concentrated pineapple juice	Plain pineapple juice
Costa Rica	Thailand	Thailand	Costa Rica
Costa Rica	Thailand	Thailand	Costa Rica
Panama	Indonesia	Costa Rica	El Salvador
Ecuador	Kenya	Indonesia	Thailand
Honduras	China	South Africa	
Mexico	Vietnam		
Côte d'Ivoire	Malaysia		
Ghana			
Guatemala			
Malaysia			

Export Trend

Export of pineapples has significantly increased in last few years; however, the total quantity exported is negligible. Increase in exports has taken place from 717 tons in 2002-03 to 3785 tons in 2006-07. However, export, has decreased from 4407 tons in 2005-06 to 3785 tons in 2006-07.

World Production

According to FAO statistics, the pineapple is the eleventh most cultivated fruit, with just over 49million tonnes produced in 2019.As illustrated in Figure, world's production is climbing steadily and has risen by more than 8 million tonnes between 2000 and 2013.

Top 10 Producer Countries

The top ten producer countries, shown in Figure 2. Worldwide 26,289,762 tonnes of pineapple is produced per year.Costa Rica is the largest pineapple producer in the world with 2,930,661 tonnes production volume per year. Brazil comes second position production while India ranked 6.

Exports

The top ten exporter countries account for 97% of fresh fruits on the markets. Just 3 million tonnes of fresh fruit are traded. On the one hand, the huge populations of China or India, for example, consume a significant part of the locally produced fruit. On the other hand, material requirements for processing units are significant, estimated at between 6 and 6.5 million tonnes (fresh fruit equivalent). Hence the proportion of fresh pineapples exported and of fruits for processing amounts to between 9 and 9.5 million tonnes.

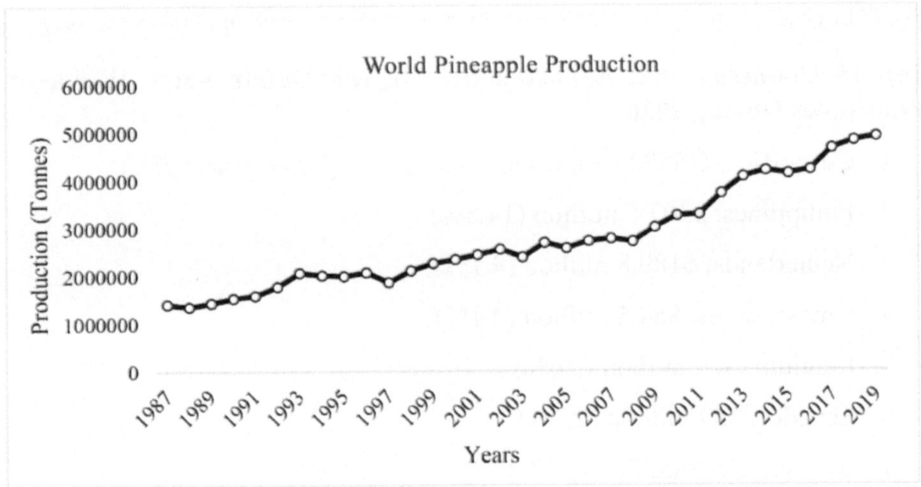

Fig. 1: World's pineapple production, thousand tonnes (1987-2019)
Source: FAOSTAT:http://www.fao.org/faostat/.

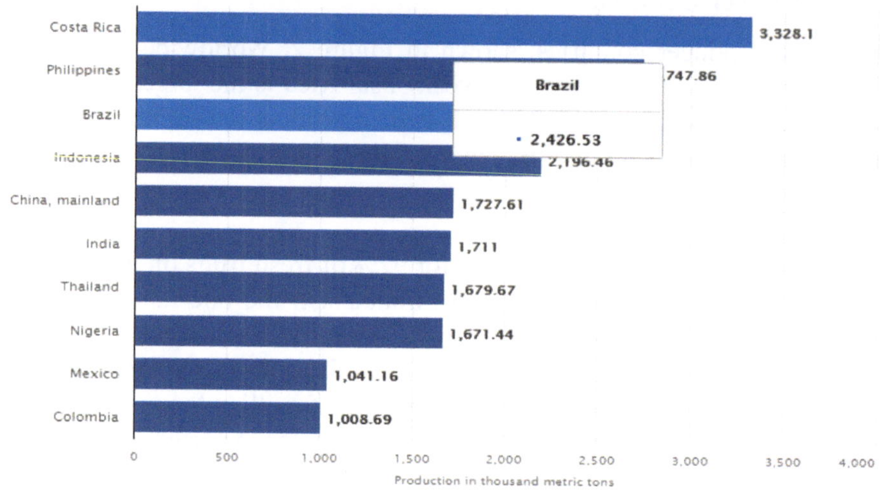

Fig. 2: Leading countries in pineapple production worldwide in 2019 *(in 1,000 metric tons)*

Top 15 Countries that Exported the Highest Dollar Value Worth of Pineapples During 2020

1. Costa Rica: US$922.7 million (44.4% of exported pineapples)
2. Philippines: $307.7 million (14.8%)
3. Netherlands: $189.8 million (9.1%)
4. United States: $84.5 million (4.1%)
5. Belgium: $75 million (3.6%)
6. Ecuador: $59 million (2.8%)
7. Taiwan: $55.8 million (2.7%)
8. Honduras: $49.7 million (2.4%)
9. Guatemala: $40.4 million (1.9%)
10. Ghana: $31 million (1.5%)
11. Ivory Coast: $28.7 million (1.4%)
12. Spain: $25.7 million (1.2%)
13. Mexico: $24.8 million (1.2%)
14. Portugal: $18.2 million (0.9%)
15. Germany: $16.9 million (0.8%)

The listed 15 countries shipped 92.9% of global pineapples exported in 2020 by value. Costa Rica stands out by virtue of its specialisation. The world number two producer is the leading exporter country since it sells nearly all of its production fresh, with the remaining aimed at the juice industry. Conversely, the Philippines only exports approximately 20% of its production fresh, the remaining 80 % is mainly aimed at processing. Mexico is in a similar situation, though for considerably lower quantities. This means that the main producer countries, apart from Costa Rica, send their production for local consumption, and above all processing. This is the case with Thailand, the Philippines, Indonesia and China, which have a powerful hold over the processed products niche (jam, plain or concentrated juice) which dominates the Asian supply. Conversely, the main fresh pineapple exporter countries are in Latin America, and to a lesser degree Africa with Côte d'Ivoire and Ghana.

National Scenario

India ranked sixth with a share of about 8 % of the world production of pineapples. The total area under pineapple cultivation in India is 108 thousand hectares with a production of 1777 thousand metric tonnes. India exports pineapple mainly to Nepal, Maldives, United Arab Emirates, Saudi Arabia, Kazakhstan, Oman, Bahrain, Bangladesh, Zambia, Pakistan and Qatar. 'Kew' and 'Mauritius' are the two varieties of pineapple grown in India. It is grown in Karnataka, Meghalaya, West Bengal, Kerala, Assam, Manipur, Tripura, Arunachal Pradesh, Mizoram, and Nagaland. It is also cultivated on limited areas in the coastal belt of Tamil Nadu, Goa and Orissa. Though Assam has the largest area under pineapple West Bengal is the largest producer. Karnataka, West Bengal and Bihar are the three states reporting high productivity. Overall, Indian productivity of approximately 16.00 t/ha poorly compares with the world average of 22.58 t/ha.

Table 1: Area, Production and Productivity of Pineapple in India (1987-1988, 1991-1992 to 2020-2021-2nd Advance Estimate)

Year	Area	% age of Total	Production	%age of Total	Productivity
	(' 000 Hectare)	Area	(' 000 MT)	Production	(MT/Hectare)
1987-88	45.7	1.6	578	2.1	12.6
1991-92	57.1	2	768.5	2.7	13.5
1992-93	59.4	1.9	859	2.6	14.5
1993-94	62.2	2	1006.7	2.7	16.2
1994-95	66.4	1.5	1055.2	2.7	15.9
1995-96	71.2	2.1	1071.1	2.6	15
1996-97	68.7	1.9	924.6	2.3	13.5

Year	Area	% age of Total	Production	%age of Total	Productivity
	(' 000 Hectare)	Area	(' 000 MT)	Production	(MT/Hectare)
1997-98	69.3	1.9	937.1	2.2	13.5
1998-99	74.2	2	1006.4	2.3	13.6
1999-00	75.5	1.9	1025.4	2.2	13.6
2000-01	78.1	2	1221.1	2.8	15.5
2001-02	77.2	1.9	1182.3	2.7	15.3
2002-03	79.8	2.1	1171.7	2.6	14.7
2003-04	80.9	1.7	1234.2	2.7	15.3
2004-05	82.8	1.7	1278.9	2.6	15.4
2005-06	82.4	1.5	1262.6	2.3	15.3
2006-07	87	1.6	1362	2.3	15.7
2007-08	80	1.4	1245	1.9	15.6
2008-09	84	1.4	1341	2	16
2009-10	91.9	1.5	1386.8	1.9	15.1
2010-11	89	1.4	1415	1.9	15.9
2011-12	102.4	1.5	1500	2	14.7
2012-13	105.2	1.5	1570.6	1.9	14.9
2013-14	109.9	1.5	1736.7	2	15.8
2014-15	116	-	1984	-	17.1
2015-16	110	-	1924	-	17.5
2016-17	111	-	1861	-	16.8
2017-18	103	-	1706	-	16.6
2018-19	104	-	1711	-	16.5
2019-20	106	-	1732	-	16.3
2020-21	108	-	1777	-	16.5

Source: Ministry of Agriculture, Govt. of India (Retrieved-Indiastat.com)

State Scenario

Maximum area under pineapple cultivation is in Assam (16 thousand ha), where productivity is of medium scale. Total production is higher in West Bengal, where productivity is also high (30 MT/Ha). Productivity is much lower in Karnataka, Kerala and Meghalaya states.

Table 2: Selected State-wise Area, Production and Productivity of Pineapple in India (2015-2016 to 2017-2018)

States	Area (`000 Hectare)			Production (`000 MT)			Productivity (MT/Hectare)		
	2015-16	**2016-17**	**2017-18**	**2015-16**	**2016-17**	**2017-18**	**2015-16**	**2016-17**	**2017-18**
Andhra Pradesh	3.79	4.29	4.52	55.38	64.41	71.33	14.60	15.00	15.78
Arunachal Pradesh	6.96	4.00	3.09	37.33	26.97	23.11	5.36	6.75	7.49
Assam	16.20	16.27	16.30	285.17	268.92	296.52	17.61	16.53	18.19
Bihar	4.30	4.31	4.26	116.30	116.58	115.13	27.05	27.05	27.03
Karnataka	2.48	2.69	2.62	155.41	164.26	163.73	62.74	61.06	62.42
Kerala	10.97	17.29	8.22	305.67	310.32	69.72	27.86	17.95	8.49
Maharashtra	0.10	-	0.35	0.40	-	11.26	3.89	-	31.88
Manipur	13.66	13.36	14.16	128.51	127.03	134.11	9.41	9.51	9.47
Meghalaya	11.58	12.16	12.37	123.13	140.95	144.73	10.64	11.59	11.70
Mizoram	4.86	4.97	5.03	32.87	32.98	27.64	6.76	6.64	5.50
Nagaland	8.99	9.56	9.53	127.81	132.62	132.83	14.22	13.88	13.93
Odisha	0.95	0.94	0.97	11.68	11.55	11.53	12.31	12.29	11.89
Sikkim	0.07	0.15	0.15	0.00	0.00	0.47	0.00	0.00	3.21
Tamil Nadu	0.77	0.80	0.83	26.22	26.26	25.74	34.00	32.95	31.12
Telangana	-	-	0.00	-	-	0.05	-	-	50.00
Tripura	12.69	8.85	8.73	180.26	94.67	127.00	14.20	10.70	14.55
West Bengal	11.00	11.20	11.41	330.07	336.11	345.15	30.01	30.01	30.26
Others	0.46	0.46	0.43	8.04	7.45	5.71	17.40	16.26	13.26

Source: Ministry of Agriculture and Farmers Welfare, Govt. of India.

Domestic Strengths for Exporting Pineapple

Domestic strengths for exporting pineapple are given in the following points

- Pineapple is cultivated in India in North East, West Bengal, and Bihar in eastern side and in Karnataka, Goa, Maharashtra etc, to some extent, in western side.
- Western ports can be exploited for exporting to Gulf and European Union which will save on transport also.
- Entire technology for obtaining high yields and for harvesting fruits throughout the year is available for putting into practice right away.
- Internationally acceptable cultivars like Kew or Giant Kew which is also called Cayenne or Smooth Cayenne are grown extensively in India.
- Agri Export Zone for promoting exports of pineapple has already been established in North Bengal.
- From location point of view, India is better placed for exporting pineapple to Gulf countries as well as to European countries, compared to South East Asian countries like Philippines and Central and South American countries.
- There are good prospects for cultivation of organic pineapples in Kodagu district of Karnataka and Ratnagiri district in Maharashtra state.

Table 3: Major importing countries of India's pineapple

2018-19		2019-20		2020-21		Growth (%)	Share (2020-21)	Rs. (Lacs)
Country	**Qty (t)**	**Rs. (Lacs)**	**Qty (t)**	**Rs. (Lacs)**	**Qty (t)**			
U. Arab Emts	956.22	710.27	1,015.64	758.93	690.19	543.06	-28.44	27.36
Qatar	802.28	529.87	537.53	364.02	628.1	420.14	15.42	21.17
Maldives	660.96	503.69	689.37	535.92	486.22	406.29	-24.19	20.47
Nepal	3,507.70	736.62	3,758.34	493.92	2,379.91	356.5	-27.82	17.96
France	0.4	7.2	0.82	27.65	1.8	59.3	114.47	2.99
USA	0.02	0.01	0	0	3.22	42.39	∞	2.14
Russia	50.08	63.9	139.09	174.48	16.78	24.19	-86.14	1.22
Baharain Is	170.58	98.03	130.55	78.86	36.97	19.9	-74.77	1
Saudi Arab	313.59	191.81	96.34	54.89	31.89	19.07	-65.26	0.96
Oman	447.33	332.51	293.98	194.17	34.23	15.62	-91.96	0.79
Iran	0.07	0.04	0	0	45.49	13.68	∞	0.69
Bhutan	23.29	6.62	0	0	37.1	12.87	∞	0.65
Kuwait	5.22	2.71	10.64	7.61	24.85	11.06	45.34	0.56
Jordan	0	0	0	0	24	9.87	∞	0.5
Afghanistan	0	0	0	0	48	9.07	∞	0.46
Austria	0	0	0.01	0.01	0.4	6.69	66,800.00	0.34
Pakistan	0.04	0.03	0	0	16	4.4	∞	0.22
Canada	0.02	0.09	0.23	1.29	0.6	3.96	206.98	0.2
Turkey	0.06	0.03	0.1	0.04	37.89	3.22	7,950.00	0.16
Belgium	0	0	0	0	1.5	2.48	∞	0.12
Lithuania	0	0	0.04	0.36	0.04	0.5	38.89	0.03
U K	0	0	1.5	0.93	0.03	0.44	-52.69	0.02
Sri Lanka	0.13	0.07	0	0	0.05	0.25	∞	0.01
Indonesia	0.07	0.04	0	0	0.02	0.03	∞	0

2018-19		2019-20		2020-21		Growth (%)	Share (2020-21)	Rs. (Lacs)
Country	Qty (t)	Rs. (Lacs)	Qty (t)	Rs. (Lacs)	Qty (t)			
Sierra Leone	0	0	0	0	0.02	0.02	∞	0
Argentina	0.02	0.01	0	0	0	0	0	0
Bahamas	0.03	0.02	0	0	0	0	0	0
Brazil	0.02	0.01	0	0	0	0	0	0
Cayman Is	0.01	0	0	0	0	0	0	0
China P Rp	0.03	0.01	0	0	0	0	0	0
Egypt A Rp	0.4	0.18	0	0	0	0	0	0
Gambia	0.05	0.04	0	0	0	0	0	0
Germany	0.07	0.03	0.01	0.01	0	0	-100	0
Hong Kong	0.17	0.2	0.1	0.08	0	0	-100	0
Iraq	0	0	4.24	3.88	0	0	-100	0
Italy	0.4	12.61	0.4	12.42	0	0	-100	0
Japan	0	0	0.02	0.01	0	0	-100	0
Kenya	0.86	1.84	0	0	0	0	0	0
Latvia	0.02	0.02	0	0	0	0	0	0
Liberia	0.02	0.01	0	0	0	0	0	0
Madagascar	0.02	0.01	0	0	0	0	0	0
Malaysia	0.09	0.06	0	0	0	0	0	0
Mali	0	0	1.4	0.37	0	0	-100	0
Monaco	0	0	0.12	0.05	0	0	-100	0
Mongolia	0	0	2.46	0.37	0	0	-100	0
Namibia	0	0	0.04	0.01	0	0	-100	0
Netherland	0	0	0	0	0	0	0	0
Norway	0.02	0.02	0	0	0	0	0	0
Panama Republic	0.06	0.03	0	0	0	0	0	0

2018-19		2019-20		2020-21		Growth (%)	Share (2020-21)	Rs. (Lacs)
Country	Qty (t)	Rs. (Lacs)	Qty (t)	Rs. (Lacs)	Qty (t)			
Singapore	0.07	0.02	0	0	0.02	0	0	0
Switzerland	0	0	0	0	0	0	0	0
Thailand	0.02	0.01	0	0	0	0	0	0
Ukraine	0.02	0.01	0	0	0	0	0	0
Uzbekistan	1.71	0.93	0	0	0	0	0	0
Total	**6,942.17**	**3,199.61**	**6,682.97**	**2,710.28**	**4,545.32**	**1,985.00**	**-26.76**	**100**

Source: DGCIS Annual Export, APEDA

Factors Affecting the Supply

Meteorological phenomena (floods, droughts) and disease outbreaks can bring down the planned production. Yet these factors often remain limited to a region, with another one able to offset the shortfall. Speculation around pineapple production, in particular the one for transformation, leads to production varying considerably. Hence, a fall in purchase prices for the producer because of high production will lead to disaffection of producers, and therefore a fall in production the following year. A probable rise in rates will encourage producers to plant more. For example in Thailand, an abundant production in April/June 2012 lead the government to set up an emergency fund for thousands of planters affected by the fall in purchase price : 4 THB/kg, up to 8 tonnes per producer, a long way off the 5 to 6 THB/kg paid in 2010 and 2011. In 2013, the situation improved, prices climbed back above the 5 THB/kg mark due to a fall in production. These fluctuations are less perceptible for fresh pineapples. However, pressure from volumes of fresh fruit of unequal quality has driven importers, in recent years, to favour certain sources to the detriment of others.

Prospects

Steadily ascending graphs of world pineapple exports and world imports, herald further development potential over the coming years. Still under-developed markets such as Russia, Eastern Europe, the Near-East and Middle-East are all possible consumption reserves, provided that certain geopolitical constraints are lifted. World production is able to meet further growth in fresh pineapple demand, since only 12 to 13 % of production is exported. The main limiting factor remains the price for potential consumers counterbalanced by the revenue for producers/exporters. In this respect, the development prospects seem more mixed. Small or modestly sized producers/exporters will have more difficulty cutting their margins in order to remain competitive in a context of generally rising costs of agricultural inputs and energy. The bigger facilities will find it easy to adapt to these constraints by economies of scale, and by means of financial operations. The general international context is vital, since it partly influences the economic results. Thus, while certain Latin American sources are on the wane due to lack of competitiveness, Costa Rica, the leading exporter country, is continuing its rise, offsetting the more or less marked decline of the neighbouring sources. However, the rebalancing of the euro/dollar exchange rate in recent months could eventually halt Costa Rican expansion, reducing the revenue of shareholders of the big companies dominating the sector. The development prospects of the processed pineapple sector appear more complex given the fluctuations in purchase prices from producers. The less concentrated nature of the sector is more favourable to competition between the main Asian sources sharing this niche.

Trade Issues (disputes, negotiations, agreements)

Given the magnitude of the merchandise flows for both the fresh and processed sectors, there are few disputes. The most common ones involve the fresh pineapple, with its high perishability. The damage most often encountered is often attributable to the mode of transport. While reefer shipping is generally well managed, thanks to high-performance polythermal ships, transport in refrigerated shipping containers does give rise to some incidents. A malfunctioning refrigerated unit can lead to qualitative deterioration of the merchandise. In this case, the recipient commissions a joint inquiry to determine the cause of the damage and its economic impact. Sea shipments are generally covered by insurance paying out for devaluation of merchandise in case of a proven incident during transport. For processed merchandise, the same often applies. An inquiry is commissioned in order to determine the reason for the damage. This may be due to the transport, but also to a manufacturing flaw. In either case, the insurance can compensate for economic losses. In rare cases, if the survey cannot manage to determine the causes of the damage or if they remain incompatible, the case may be referred to a commercial court, under the terms of the contract of the various parties. Conversely, air-freight shipments are rarely insured given the short transport time (governed by the Warsaw Convention). Disputes arising for this type of merchandise are most often settled amicably between the supplier and customer, possibly after a survey. In case of disagreement during a dispute, the parties concerned may also appeal to an arbitration chamber to settle the dispute between them.

Opportunities

There exists great opportunities for expanding the cultivation of pineapple in India. Well established and organized Pineapple Farmers' Association, research and development support from Kerala Agricultural University and other Government research organizations are major support for the same. Indian farmers have knowledge and skill on pineapple cultivation and they are increasingly following commercial cultivation. The opportunity for niche organic market is increasing day by day and the traders in the market are generally vibrant and relatively organised and thus a good national and global market of pineapple is smoothly running. Farmers and promoters are interested in pursuing sustainable agriculture, which reduces costs and increases farmer's skills to manage food security concerns. Thus the production of pineapple within all the potential areas can be increased to hold the domestic as well as global market.

Challenges

Expanding area under cultivation and increasing productivity are very challenging. Identification of specific varieties for fresh fruit, canning and juice segments suitable for cultivation ensuring sustained economic return are urgently needed. Improved technology and production inputs for each segment are to be provided and institutional supports are necessary for the same. Technical and entrepreneurial support for post harvest management and value addition along with the storage and marketing is to be assured from Government sectors for the growers, processors as well as the FPOs. Crop specific training support to the stakeholders is also needed. Exploration of new fields viz, high-density planting, tissue culture, organic farming, etc. are in the present need in the field of pineapple cultivation. Staggering of pineapple production through chemical induction and round the year production, peak synchronizing with summer season when demand is highest should be promoted. Labour resource mobilization especially through NREGS can be done to promote agriculture and food security in rural areas for pineapple production as well as processing. Sustainable utilization of natural resources for long term pineapple production can mitigate the climatic hazards and market uncertainties causing farmers' suicides.

References

Agricultural Marketing Service of USDA www.ams.usda.gov
APEDA: https://apeda.gov.in
COMTRADE: http://comtrade.un.org
European Commission: http://ec.europa.eu
FAOSTAT: http://www.fao.org/faostat/.
Ministry of Agriculture, Govt. of India: https://www.indiastat.com/
United States Department of Agriculture (USDA): www.usda.gov

4

Status of Pineapple Production in North Eastern India

Abhijit Debnath[1], H. S. Datta[2], J. Uchoi[3], Harish G.D[4], R. Dangmei, R.K.[5] Chaurasia, D. Pasweth[6], C. A. Pariat[7], S.P. Bhutia[8], L. Sanga[9], S. Borang and E. Reang[10]

[1](Horticulture), KVK North Tripura
[2]College of Horticulture, Assam Agricultural University, Jorhat
[3]ICAR-National Bureau of Plant Genetic Resources, Regional Station Umiam, Meghalaya-793103
[4]Scientist, ICAR-National Bureau of Plant Genetic Resources, Regional Station, Umiam, Meghalaya-793103
[5]Agriculture Officer, Directorate of Horticulture and Soil Conservation, Sanjenthong, Imphal, Manipur-795001
[6]KVK, Zunheboto, Nagaland University
[7]KVK, Jaintia Hills, Meghalaya

Pineapple botanically called *Annanas comosus* L. is basically believed to be native to Brazil and thereafter it spread to worldwide over many centuries from its centre of origin. It is a subtropical fruit crops requiring humid climatic conditions with slightly acidic soils for its favourable growth and development. In India it reached around 1548AD. The juicy fruits are good source of Vit-A, B and C along with micronutrient Calcium. A single unit of pineapple fruits is reported to contain moisture-85%, sugar-13per cent, protein-0.6%,mineral mater-0.05%, fibre 0.3%, Ca-0.02%, phosphorus- 0.01per cent iron- 0.09.0 acids- 0.6%, Vit-A (60IU/100g), riboflavin (120mg/100g) and Vitamin-C (63mg/100g). All parts of pineapple are utilizable from delicacy of fruit for table purpose to making fabric *Pina* in some country like Philippine. Indeed the dry waste after juice extraction can also be used as good feed for cattle. India is such a country blessed with diverse agro climatic zones and eight classification of agro-climatic zone has been officially classified for suitability of different fruit crops based on their climatic requirement. One of those classifications is North Eastern subtropical Humid Zones which comprises of eight states namely Assam, Arunachal Pradesh, Manipur, Meghalaya, Nagaland, Sikkim and Tripura. Manipur is a land-locked state situated in the North Eastern Region of India. It is encircled by 9 hill ranges in all sides and has a small valley in the center (the capital Imphal is located in this valley). The State has 352 km long international border with Myanmar to the south-east and 502 km long border with the adjacent states of Nagaland on the North, Cachar district of Assam on the West and Chin Hills (Myanmar) and Mizoram on the South and the South-west and Surma Tract and upper Chindwin of Myanmar on the East. Manipur is next to Assam among NE states in respect of area of Pineapple cultivation. Owing to conducive soil and climate, Manipur is one of the major hubs for pineapple cultivation among the north east states where because of its popularity, it has surplus in area and production compare to other fruit crops in the state.

"Assam Pineapples Headed for Dubai", the news which created a great sensation in the print media in Assam when for the first time, a consignment of 15 metric tonnes of pineapples was exported from Lakhipur in Cachar district in refrigerated trucksto the international market in Dubai. Assam has fertile and organically rich soils, ample rainfall, water resources and great climatic diversity supporting diverse cropping possibilities such as Pineapple which is widely grown and very popular with their availability in the form of juice or slices round the year enable the consumers to enjoy them whenever they want. Pineapples are rich source of vitamin C, minerals and fiber but with low calories.

The tiny yet beautiful state of Tripura is situated in the North-eastern part of the country. The favourable climatic condition and physiographic land forms offers the excellent scope for cultivation of host of horticultural crops. The moderate acidic soils in the hill slopes, abundant sunshine and optimum temperature favour the production of finest quality pineapple in the state. The agricultural sector accounts for 33% of the gross state domestic product (GSDP) and provides employment to 42% of the total population.

Hon'ble President of India declared Queen Pineapple as a State fruit on 07.06.2018. Logo on Tripura Pineapple marketing was unveiled by Honb'le Agriculture Minister and Farmers Welfare, Govt of Tripura on 31st October, 2019. The Logo is developed by national Institute of Design, Bangalore and Amedabad, India. World's sweetest pineapple Tripura Queen Pineapple is a GI registered product by NERAMAC. The unique qualities of pineapple varieties especially Queen Variety (GI tagged) is an established fact. The pineapple is the traditional fruit crop of Tripura.

Pineapple is locally called as "Khihom" in Manipuri language. Like any other cash crops, pineapple has also found a place in the soils of Manipur and the state is also becoming the safe breeding ground for it. Without any exageration, pineapple production in the state has outdone any other fruits production. Over the years, its popularity too has been nagging the hearts of pineapple lovers of the state. Economically, the fruit has also become the backbone of a sizeable section of farmers who have been cultivating it as their major source of income. It has, interestingly, helped in keeping the ceaseless burning of many hearths of the state. Despite the lack of technical know-how to improve its production and preservation of the produces, numerous farmers' success stories have been scripted with a span of six decades since it was first cultivated in 1947.

Pineapple is an important fruit crop of North East (NE) region of India, and in Nagaland it occupies an area of 8697 ha with 115086.00 MT of production during the year 2018-19 (Statistical Handbook of Nagaland, 2019). Pineapple is being cultivated in the state and has been practiced since time immemorial but commercialization and adoption of modern technology has been of recent times only. The Nagaland Government having identified the potentiality of Pineapple production in the state and taken initiatives to increase its productivity. In Ao Pineapple is called by the name *Chiru*, Kuki calls it by "*Lengthei*, Angami by *Kihunyii*, Lothas by Ananas. Nagaland has conducive environment Pineapple production and has proved that the farmers can generate more income through Pineapple faming rather going for traditional method of farming (Jhum). The only concern is the how to make is more remunerative through development of Post harvest Technology, FPO formation and organic certification.

Meghalaya is one of the major Pineapple (*Ananas comosus* (L.) Merr.) producing state of India. It contributes 8% of the total pineapple produced in India. Meghalaya is situated in the north eastern region of India with a geographical area of 22,429 km^2. Pineapple is the most important fruit crop of the state. In most of the pineapple growing areas of the state, the age-old indigenous cultural practices are still in use. Although the yield is less these fruits are free from chemical residues. Traditional pineapple growing areas of Meghalaya have immense potential to be developed into certified organic pineapple producing hubs (Mathew *et al.*, 2011).

Cultivation of pineapple has been an age-old practice in Mizoram. It is an important fruit crop of Mizoram and being cultivated in a limited area within the state. It is locally called as Lakhuihthei; some of the successful pineapple growing areas in Mizoram are Buangpui, Serchhip (Serchhip district) and Muallungthu, Lengpui (Aizawl district).The Department of Horticulture, Mizoram has initiated area expansion scheme for successful cultivation of pineapple in this region.

Arunachal Pradesh which is popularly known as the land of rising sun in India has a long international border with Bhutan to the west (160km), China to the North & Northeast (1080kms) and Myanmar to the east (440km) which makes it strategically very significant for the country specially for international trading of horticulture produce in future. Among all the important crops grown by farmers in Arunachal Pradesh, a crop which succeeded to be in the priority list of the farmers is the Pineapple (*Ananas comosus*). Botanically, Pineapple is an herbaceous, perennial, self-sterile, monocotyledonous plant of about 90-100 cm in height, with spreading leaves, which give the plant a rosette appearance. The plant bears a single fruit terminally on a peduncle, protruding out from the centre of the rosette (Chadha *et al*., 1998). The state is steadily moving ahead owing to the hard work of the farmers and untiring efforts of Horticulture department by implementation of various centrally sponsored & state sponsored schemes for the welfare of the farming community of the state.

Agriculture is the mainstay of the rural population of Sikkim and 82% of people depend on it. Due to adverse geographical location and difficult terrain, the expansion of area under cultivation is limited. Farming is predominantly rice based but in Sikkim maize is the dominant crop. In terms of Fruit crop Sikkim mandarin is cultivated since time immemorial. The prevailing fruit crops mostly found are Sikkim mandarin, Guava, banana, papaya, kiwifruit, passion fruit and litchi. Pineapple is also one of the important horticultural fruit crops in Sikkim. However, it is not a focus crop of the state. In Sikkim Bhutia people calls it by the name "*Bar*", Lepchas calls it, Nepali by "*Bhui kotor*". It is the budding fruit crop of the State with the area expansion of 165 ha till date as shown in table no.2. Past few years the department of horticulture has extended some help to the grower in terms of providing planting material especially "queen" cultivar and other inputs.

Area, Production and Productivity

In Assam, pineapple is cultivated in an area of 16.30 thousand hectare with a production of 296.52 thousand MT with a productivity of 18.19 MT/ha (GoI,2018). The pineapples from Assam are in demand due to its organic by default tag and also considered the best in terms of sweetness, quality and good flavour. Pineapple is grown in almost all the districts in Assam with more concentration in Karbi Anglong, Dima Hasa district and Cachar district. The pineapple of Deithor in Karbi Anglong is famous for its size and sweetness and it is grown on the hilly terrain. The fruits are consumed fresh or made into value added products like jam, squash etc.

District wise area, production and productivity of pineapple in Assam (2016-17)

District	Area(ha)	Production(MT)	Average yield(kg/ha)
1.Cachar	1349	18613	13797
2.Hailakandi	1430	15154	10597
3.Karimganj	70	1069	15269
4.Goalpara	568	8828	15542
5.Dhubri	29	466	16106
6.Kokrajhar	590	9234	15650
7.Bongaigaon	281	5310	18895
8.Kamrup(R)	1605	23679	14753
9.Nalbari	75	1311	17480
10.Barpeta	65	1051	16171
11.Darrang	130	1300	10000
12.Sonitpur	808	12939	16013
13.Nagaon	259	4282	16534
14.Morigaon	43	770	17913
15.Baksa	320	6080	19000
16.Chirang	725	14368	19818
17.Kamrup(M)	289	7953	27518
18.Udalguri	665	11100	16692
19.Jorhat	119	1837	15441
20.Golaghat	243	3593	14789
21.Sivasagar	146	2306	15795
22.Lakhimpur	170	2628	15457
23.Dhemaji	175	2599	14854
24.Dibrugarh	101	1540	15243
25.Tinsukia	262	4729	18052
26.K.Anglong	2142	38840	18133
27.Dima Hasao	3608	67336	18663

District wise area, production and productivity of pineapple in Assam (2017-18)

District	Area(ha)	Production(MT)	Average yield(kg/ha)
1.Cachar	1440	20431	14188
2.Hailakandi	1350	14907	11042
3.Karimganj	66	1355	20530
4.Goalpara	582	9701	16668
5.Dhubri	29	467	16118
6.Kokrajhar	289	4875	16869
7.Bongaigaon	305	6240	20462
8.Kamrup(R)	1590	32592	20498
9.Nalbari	81	1417	17488
10.Barpeta	56	906	16179

District	Area(ha)	Production(MT)	Average yield(kg/ha)
11.Darrang	119	1787	15018
12.Sonitpur	781	12511	16019
13.Nagaon	247	3848	15574
14.Morigaon	25	448	17925
15.Baksa	372	7077	19025
16.Chirang	768	16012	20849
17.Kamrup(M)	252	5792	22984
18.Udalguri	872	15512	17789
19.Jorhat	151	3127	20700
20.Golaghat	256	4325	16896
21.Sivasagar	151	3040	20134
22.Lakhimpur	182	3609	19825
23.Dhemaji	185	3340	18054
24.Dibrugarh	98	2142	21855
25.Tinsukia	264	5432	20573
26.Karbi Anglong	2098	42263	20144
27.Dima Hasao	3695	73368	19856

District wise area, production and productivity of pineapple in Assam (2018-19)

District	Area(ha)	Production(MT)	Average yield(kg/ha)
1.Cachar	1320	19096	14467
2.Hailakandi	1280	14819	11577
3.Karimganj	58	1102	18994
4.Goalpara	571	9661	16919
5.Dhubri	30	531	17671
6.Kokrajhar	554	9532	17205
7.Bongaigaon	298	5908	19825
8.Kamrup(R)	1563	31791	20340
9.Nalbari	84	325	3872
10.Barpeta	50	872	17443
11.Darrang	110	1814	16495
12.Sonitpur	733	11930	16276
13.Nagaon	251	3909	15574
14.Morigaon	50	903	18051
15.Baksa	370	6959	18807
16.Chirang	1192	24435	20499
17.Kamrup(M)	247	5307	21485

District	Area(ha)	Production(MT)	Average yield(kg/ha)
18.Udalguri	661	11780	17822
19.Jorhat	98	1921	19595
20.Golaghat	230	3887	16896
21.Sivasagar	154	2966	19263
22.Lakhimpur	323	6191	19167
23.Dhemaji	175	3161	18064
24.Dibrugarh	97	1926	19851
25.Tinsukia	257	5287	20573
26.Karbi Anglong	2293	46191	20144
27.Dima Hasao	3856	76565	19856

District wise area, production and productivity of pineapple in Assam (2019-20)

District	Area(ha)	Production(MT)	Average yield(kg/ha)
1.Cachar	1355	19773	14592
2.Hailakandi	1320	15474	11723
3.Karimganj	62	1213	19561
4.Goalpara	576	9892	17173
5.Dhubri	28	500	17843
6.Kokrajhar	524	9184	17527
7.Bongaigaon	296	5977	20196
8.Kamrup(R)	1563	31932	20430
9.Nalbari	84	337	4011
10.Barpeta	49	904	18442
11.Darrang	110	1898	17257
12.Sonitpur	680	11205	16478
13.Nagaon	248	4179	16852
14.Morigaon	57	1097	19237
15.Baksa	379	7255	19143
16.Chirang	895	18501	20671
17.Kamrup(M)	245	5360	21878
18.Udalguri	658	11841	17995
19.Jorhat	133	2685	20186
20.Golaghat	229	4058	17722
21.Sivasagar	162	3218	19862
22.Lakhimpur	208	4099	19706
23.Dhemaji	183	3409	18630
24.Dibrugarh	94	1934	20578
25.Tinsukia	260	5252	20196
26.Karbi Anglong	2488	51250	20599
27.Dima Hasao	3720	75092	20186

(*Source*: Directorate of Horticulture & Food Processing, Department of Agriculture & Horticulture, Govt. of Assam)

The agro-climatic conditions of Tripura make it suitable for large scale pineapple cultivation. The crop accounts for 16% of the total area under fruit crop cultivation and 23% of the total fruit production in the state. During the period 2010-11 to 2014-15, the area under pineapple cultivation has become almost double in the state from 6800 ha to 12,000 ha. However, during the same period, the crop productivity decreased from 22.6 mt/ha to 14.3 mt/ ha and stagnated as against increasing year-on-year productivity in other pineapple producing States in India. The State average productivity of pineapple is in fact lower than the national average of 17 mt/ha and almost one-fifth of that in Karnataka, which has the highest yield in the country (53 mt/ha). The productivity is also very low compared to the leading pineapple producing States as Kerala (32.4 mt/ha), West Bengal (29.5 mt/ha) and Assam (17.6 mt/ ha). The decline in yield performance of the crop may be attributed to factors as technology gap, high dependence on monsoon, gradually declining soil fertility due to heavy rainfall and erosion and over-cultivation.

In Manipur, this fruits are cultivated in all the 16 districts. According to the 2017-2018 statistics (National Horticulture Board, 2018), The total area and production of pineapple in the state is 14160ha and 134.1(000MT) with 9.47MT/ha and occupies second largest area of its cultivation next to Assam state. Pineapple are grown in Nine district namely Bishnupur, Chandel, Churachandpur, Imphal East, Imphal West, Senapati, Tamenglong , Thoubal and Ukhrul Among the district Senapati (area-3724ha, production-39067MT) occupy the lion share in terms of both area and production followed by Churachandpur (area-2500, production-25000MT), Bishnupur (area-2060 ha, production-4488MT) and Thoubal (area-1665, production-20630MT) where there are more cultivation of this fruits crops. Over the last five years from 2014-15 to 2018-19, there is decrease in area of cultivation of the fruits crops from 14271 to 13994 ha. The reason is attributed due to high capital involvement in maintaining orchard and also other reason may be the uncertainty of low market price.

District Wise Area & Production of Pineapple during 2015-2019 in Tripura (Area in Ha, Prodn in MT & Productivty in MT/Ha)

Sl. No.	Year	South Tripura		Gomati		Sepahijala		West Tripura		Khowai		Dhalai		Unakoti		North Tripura		Total (State)		
		Area (Ha)	Prodn (MT)	Area (Ha)	Prodn (MT)	Area (Ha)	Prodn (MT)	Area (Ha)	Prodn (MT)	Area (Ha)	Prodn (MT)	Area (Ha)	Prodn (MT)	Area (Ha)	Prodn (MT)	Area (Ha)	Prodn (MT)	Area (Ha)	Prodn (MT)	Productivity (MT/Ha)
1.	2015-16	935	13084	895	12971	761	11036	802	11622	674	9435	2629	37728	655	9392	1419	21289	8769	126557	14.43
2.	2016-17	935	13099	895	12933	761	11171	801	11767	674	9786	2629	38173	734	10643	1419	21399	8848	128971	14.58
3.	2017-18	935	13093	895	112883	750	10995	750	10973	610	8857	2503	26324	884	12783	1401	21087	8728	126994	14.55
4.	2018-19	925	13512	926	13483	775	11284	863	12565	676	9843	3066	44641	908	13220	1486	21636	9628	140184	14.56

(*Source*: Directorate of Horticulture & Soil Conservation, Department of Agriculture & Farmers Welfare, Govt. of Tripura)

Area and production of pineapple (2014-15 to 2018-19) in Manipur

Districts	**2014-15**		**2015-16**		**2016-17**		**2017-18**		**2018-19**	
	A	**P**	**A**	**P**	**A**	**P**	**A**	**P**	**A**	**P**
Bishnupur	1725	3989	1902	4170.8	1980	4335	2034	4456	2060	4488
Chandel	755	9810	750	7410	597	5850	645	6335	367	3653
Churachandpur	2250	20250	2250	2250	2280	22500	2500	25000	2500	25000
Imphal East	1440	17950	1127	14042.4	1254	15629.8	1415	18327	1430	17738
Imphal West	694	8958	802	10086.1	801	10074	808	10284	813	10294
Senapati	3425	35985	1632	20324.9	3710	38213	3720	39045	3724	39067
Tamenglong	1028	2987	1002	3002.9	1006	3009	1013	3027	1072	3748
Thoubal	2624	30697	1632	20324.9	1644	20473	1659	20612	1665	20630
Ukhrul	330	6120	350	6747	359	6950	363	7020	363	7020

(*Sources*: Department of Horticulture and Soil Conservation, Govt.of Manipur, 2021)

As per the data available the productivity of Pineapple in Nagaland is 13.32 tons per ha which means the land is very fertile and adaptable and the fruits are of high of quality having TSS about 12-14, large sized with attractive color and flavor. The data presented in Table No.1 showed a decline of 1220 ha from 2014-2019 in the area under Pineapple production in the state. The decline in the area has not affected the productivity in the state. The productivity has been more or less similar in nature i.e, 14.86 to 13.23 from 2014-15 to 2018-19. The reason for decrease in the area is not known but a significant decrease in area reduction (537ha) was observed in Mokokchung district i.e. from 1127 ha in 2014-15 to 519 ha in 2018-19.

Area & production of major fruits during 2014-19 in Nagaland (A=area in hectare, P = Production in mt)

Year	Area & Pdtn	Kohima	Wokha	Mokok-chung	Zunhe-boto	Tuens-ang	Phek	Mon	Dima-pur	Kiphire	Long-leng	Peren	Nagal-and	Produc-tivity
2014-15	A	883.00	903.00	1127.00	463.00	463.00	493.00	436.00	2795.00	468.00	331.00	1551.00	9917.00	14.86
	P	12760.0	13374.0	15446.0	9011.0	8843.0	9280.0	9000.0	32395.0	9017.0	7732.0	20526.0	147384.0	
2015-16	A	257.0	908.0	1127.0	450.0	448.0	490.0	500.0	2750.0	468.0	350.0	1518.0	9337.0	13.78
	P	1271.00	13503.00	15210.00	8640.00	8775.00	9417.00	9000.00	32500.00	9360.00	870.00	20130.00	128676.0	
2016-17	A	260.00	908.00	1145.00	440.00	473.00	498.00	500.00	2850.00	468.00	365.00	1621.00	9528.00	13.92947
	P	1380.00	13508.0	15270.0	8448.00	8848.00	9493.00	9000.00	34500.0	9360.00	881.00	22032.0	132720.0	
2017-2018	A	240.0	637.5	990.0	360.0	473.0	495.0	500.0	2800.0	410.0	362.0	1628.0	8895.5	13.20443
	P	1273.0	10125.0	11535.0	6790.0	8405.0	9419.0	5750.0	34000.0	7380.0	812.0	21971.0	117460.0	
2018-2019	A	240.0	637.5	590.0	360.0	475.0	495.0	500.0	3000.0	410.0	362.0	1628.0	8697.0	13.23284
	P	1273.0	10125.0	8200.0	6790.0	8404.0	9419.0	5750.0	35000.0	7380.0	800.0	21981.0	115086.0	

(*Source*: Department of Horticulture: Government of Nagaland)

The total area under pineapple cultivation is around 12,202 hectare in the state Meghalaya. Production is around 1, 38,701 metric tonnes with an average productivity of 10.79 tonnes/ha. As per data available, Ribhoi district have the highest production and East Garo hills have the highest productivity during the last five years (2015-2020). The District wise area, production and productivity is shown in Table

District Wise Area, Production and Productivity of Pineapple (*Ananas comosus*) in Meghalaya

In Mizoram, three varieties of pineapple are grown such as Giant Kew, Queen and Mauritiusand it occupies an area of 5030 ha with 33150 MT of production; productivity of Pineapple in Mizoram is 6.59 tons per haduring the year 2018-19 (Area and Production of Horticulture crops, Dept. of Horticulture, Mizoram2019). The fruits are moderate in size which makes it very suitable for table and canning purpose, and is expected to have a good prospect in Mizoram future.

District wise area & production of pineapple during 2014-2019 in Mizoram (Area = ₹ 000Ha Production = ₹ 000MT)

Sl. No.	Year	Aizawl		Lunglei		Saiha		Champhai		Mamit		Kolasib		Serchhip		Lawngtlai		Total		
		Area (Ha)	Prodn (MT)	Area (Ha)	Prodn (MT)	Area (Ha)	Prodn (MT)	Area (Ha)	Prodn (MT)	Area (Ha)	Prodn (MT)	Area (Ha)	Prodn (MT)	Area (Ha)	Prodn (MT)	Area (Ha)	Prodn (MT)	Area (Ha)	Prodn (MT)	Productivity (MT/Ha)
1.	2014-15	1.48	10.98	0.37	2.73	0.39	2.90	0.31	2.30	0.34	2.45	0.48	0.35	0.49	6.77	0.22	1.62	4.091	30.14	7.37
2.	2015-16	1.57	11.34	0.47	3.07	0.49	3.25	0.41	2.65	0.43	2.79	0.57	0.69	0.59	7.11	0.32	1.97	4.867	32.87	6.75
3.	2016-17	1.57	11.36	0.48	3.08	0.50	3.27	0.41	2.64	0.43	2.77	0.57	0.69	0.59	7.14	0.32	1.98	4.867	32.89	6.76
4.	2017-18	1.58	11.34	0.50	33.11	0.51	3.25	0.44	2.68	0.46	2.80	0.58	0.71	0.59	7.11	0.31	1.97	4.970	32.98	6.64
5.	2018-19	1.59	11.40	0.51	3.13	0.52	3.27	0.46	2.70	0.46	2.81	0.58	0.72	0.60	7.14	0.31	1.98	5.03	33.15	6.59

(*Source*: Department of Horticulture: Government of Mizoram)

In Arunachal Pradesh, Horizontal area expansion of cultivated land under Horticulture crops is one of the thrust area of the department for enhancing the production and productivity of important Horticultural crops including pineapple in the state. As per *Arunachal Pradesh Horticulture Development* Vision 2030 document, a broad target has been set to triple the existing area under horticulture from current 1.1 lakh hectare to 3.0 lakh hectare by 2030. The district wise pineapple production and state production data is given below:

Districts	2015-16			2016-17			2017-18			2018-19			2019-20		
	Area (Ha)	Pro-duction (MT)	Produc-tivity (MT/Ha)	Area (Ha)	Pro-duction (MT)	Produc-tivity (MT/Ha)	Area (Ha)	Pro-duction (MT)	Produc-tivity (MT/Ha)	Area (Ha)	Pro-duction (MT)	Productiv-ity (MT/Ha)	Area (Ha)	Pro-duction (MT)	Productiv-ity (MT/Ha)
East Khasi Hills	936	7365	7.869	981	8386	8.548	984	8413	8.550	985	8423	8.550	986	8433	8.553
West Khasi Hills	245	1556	6.351	258	1770	6.860	258	1771	6.864	259	1778	6.864	260	1785	6.865
Ri Bhoi	3854	45706	11.859	4057	52941	13.049	4061	53000	13.051	4062	53018	13.051	4063	53035	13.053
West Jaintia Hills	61	488	8.0	62	534	8.6	63	543	8.6	63	543	8.6	64	552	8.62
East Jaintia Hills	25	224	8.96	26	244	9.38	28	262	9.35	28	262	9.35	28	262	9.35
West Garo Hills	2791	24567	8.802	2942	28104	9.553	2944	28125	9.553	2945	28137	9.553	2946	28149	9.555
East Garo Hills	869	16176	18.614	909	15486	17.036	912	15538	17.037	913	15556	17.037	914	15574	17.039
South Garo Hills	1182	8713	7.371	1231	9821	7.978	1232	9830	7.979	1233	9839	7.979	1234	9848	7.981
South West Garo Hills	583	5435	9.322	614	6182	10.068	616	6203	10.070	617	6214	10.070	618	6225	10.073
North Garo Hills	504	9622	19.091	529	10987	20.769	531	11029	20.770	532	11051	20.770	533	11073	20.775
South West Khasi Hills	527	3274	6.213	553	3742	3.742	554	3749	6.767	555	3757	6.767	556	3765	6.772
Meghalaya	11577	123126	10.22291	12162	138197	10.50918	12183	138463	10.78336	12192	138578	10.78336	12202	138701	10.78618

Area and production of pineapple (2019-20 to 2019-20) in Arunachal Pradesh. (Area in Ha, production in MT & productivity in MT/Ha)

Districts	2019-20		2020-21	
	A	P	A	P
Tirap	60	30	76	186
Lower Siang	121	683	143	647
Pakke Kessang	17.66	73.8	59.29	0
Namsai	60	482	60	482
Lapa Raada	12	120	27	122
Kamle	130	290	140	310
Shi-Yami	0	0	0	0
Lohit	100	500	100	600
Anjaw	0.45	3.30	2.45	3.30
Longding	15.5	19.94	15.50	22.30
Tawang	15	3.2	0	0
Bomdila	0	0	0	0
Seppa	15.30	184.74	20.30	154.74
P/pare	685.14	1990.25	655.18	1990.29
Zero	74.0	821.5	85	850
Siang	19	83.75	21	85
E/Siang	107	1505	117	1655
U/Siang	300	290	300	300
D/Valley	34	34	34.34	48.23
L.D/Valley	29.63	24.05	30.63	27.05
W/Siang	920	9200	920	9200
U/Subangsiri	243	142.6	243.0	142.6
Changlang	135.98	764.34	140.21	764.54
Kra Daadi	46	124	32	80
E/kameng	32	77.5	20.30	84.09
Total	3161.61	17447.17	3242.2	17754.14

(*Sources*: Directorate of Horticulture, Govt.of Arunachal Pradesh, 2021)

Data of pineapple production in Sikkim (2019-20)

	West District			South District	North District	East District
Variety	**Area (ha)**	**Production (tones)**	**Productivity (kg/ha)**	**No data available**		
Queen	165	2880.9	17460			

Table: Data of Pineapple in West District (2019-20)

Sl.No	Year	Area (ha)	Major cluster for pineapple in west sikkim
1.	2015-16	55	Timburbong, Mangsari, Mabong, Kamling and Birdang
2.	2016-17	65	
3.	2017-18	30	
4.	2018-19	15	
	Total	165	

N.B: pineapple variety Queen has been recently introduced in Sikkim for trial since 2015-16. The introduction of the variety Queen has been done only in West District of Sikkim covering few villages and Horticulture Birdang farm. Till date only 165 ha of area have been covered with the above data.

Concentrated growing pockets of North Eastern India

State	Districts
Assam	Karbi-Anglong, Nagaon, Cachar, N.C.Hills
Manipur	Senapati,Thoubal,Charchandpur, Bishnupur,East Imphal
Tripura	West Tripura,North Tripura,Dhalai Tripura
Nagaland	Kohima, Dimapur

(*Source*: APEDA)

Kew and Queen are the most popular varieties among the growers of the Assam state. Pineapple can greatly contribute to the economy of Assam and its unique taste gives immense value in the food industry. The congenial climate has favoured the cultivation of pineapple locally known as "Anaros" in the state.

Tripura is particularly famous for Queen Variety of pineapple, which is not only distinct from other pineapple varieties but also significantly different in quality of same variety grown in neighbouring districts. Recognizing the exclusivity of Tripura Queen Pineapple has been accorded Geographical Indication tag/ registration (2014-15).

The pineapple varieties 'Kew' and 'queen' are extensively grown in Manipur. The fruits of Kew variety are big size, broad and shallow eyes, light yellow flesh, almost fibreless and very juicy. Two type of pineapple varieties are grown in majority such as *Kew* and *Queen*. Queen is the dominant variety grown in the state. The fruits are available from the end of May and peak season is from July to August. Comparison with seasonality of pineapple of some other North Eastern states reveal that pineapple harvesting starts early in Manipur and continues after the season has ended in other states such as Tripura and Meghalaya. More than 90 percent of the area under pineapple cultivation in Manipur is under queen variety (NIAM report 2018). The unique characters of Kew variety is vigorous plant natures, having much lesser degree of spines and spines are confined only at the tip portion of the leaves. Leaves are characterized by brownish red along the entire length The size of the fruits are big, juicy, translucent, appears orange yellowish on ripening, cylindrical in shape and slight tapering toward the shoulder. The flattened nipples or the eye is the prominent characters of the fruits. The suckering nature is of shy type producing only 2-3slips and 1-2 suckers per plant per year. It is late variety and weight of the fruits can vary form 1.6-3kg according to the favourable soil, climatic conditions and also the agronomic practices. Flesh of the fruits is light yellow colour, very juicy and almost fibreless. This type of variety is very much suitable for canning industry and is available during August to October. Another variety which is famous and outstanding in the state is *Queen*, it is liked very much and consumed as at table purpose. Under this variety there can be sub variety but most of the queen variety in cultivation in the state is having morphological characteristics of leaves depicting spiny, small closed spines and serrated leaf margin with brownish colour. Fruits are small, singles, and its weight ranging from 1.25-2kg. The shape of the fruits is almost cylindrical, having prominent outgrowth of nipples or eyes unlike kew. The fruits when ripened have pleasant aromatic smell with shell having golden yellow colour and flesh deep yellow with pleasant texture, aroma and flavour. The fruits are harvested from May till June. They are not suitable for canning. In Nagaland, three varieties of pineapple is grown namely Giant, Kew and Queen. Giant Kew is the most common variety grown in the state Meghalaya followed by the Queen variety. It has been also followed in Arunachal Pradesh. Three varieties of pineapple are grown such as Giant Kew, Queen and Mauritius grown in Mizoram. In Sikkim, only variety Queen has been tried and area expansion is emphasised.

Cultivation Practices

Assam: Pineapple can be grown on any type of well drained acidic soil provided it is free-draining. The most typical characteristics of Assam soil is acidity and slightly acidic soil with pH range of 5.5 to 6.0 is considered optimum for pineapple cultivation. Warm and humid climate is preferable for pineapple cultivation and climatic condition of Assam is very much suitable for pineapple cultivation. It can be propagated by suckers, slip and crown. Planting material dipped in Mancozeb 75 WP @ 2g/l before planting. The time of planting is April-October. Planting is done in two-row beds. In each bed, plant at 60 cm from row to row and at 30 cm from plant to plant. Distance between rows of two adjacent beds should be 90 cm. Weeding is done at least three to four times a year. Black polythene (50 micron) is also used for controlling weeds. Covering the matured fruits with leaves adjacent to the fruits will reduce both sunburn and bird damage. There are two main seasons of flowering- March-April and June-July. Sporadic flowering may take place in other periods also.

Fruits usually ripen about 5 months after flowering. The growers do not use any kind of fertilizer and pesticides in these areas. Pineapple is grown as an intercrop in coconut and arecanut plantations and also as a pure crop in garden land. When grown as an intercrop, pineapple will give additional income to the growers. The pineapple growers are now adopting improved production technology of pineapple like high density planting, application of hormones for uniform flowering and other management practices. Harvesting is done when the fruits become yellow in colour. Over ripening on the mother plant or any injury to fruits at harvesting should be avoided. For distant market, the fruits may be harvested when they just change colour from green to yellow.

Tripura: the temperature ranges 7 to 27°C during winters to 20 to 36°C during summers. Average annual rainfall across the State varies from ranges from 1922 mm to 2855 mm lasting from month April/ May to August/ September. The north- eastern part of the state around Dharamnagar gets maximum rainfall. Soil type is mostly Red & Lateritic with a pH ranging 4.5 to 6.5.3 Owing to such climatic conditions, the State has highly conducive environment for cultivation of various horticultural crop as, Pineapple, litchi, mango, banana, jackfruit, citrus, cashewnut, betel leaf, arecanut, black pepper, coconut, tea, coffee, cucurbits etc.

Pineapple (*Ananas comosus* L.) is a natural fruit in the state with favorable sub-tropical agro-climatic conditions, fertile slightly acidic soil, and abundant rainfall. The undulating tilla land with varying degrees of slope is highly susceptible to erosion so practices to check soil erosion are vital. 'Queen' and 'Kew' ('Smooth Cayenne') are the main cultivars grown in different parts of Tripura. The area under pineapple cultivation is 5,180 ha and production is 1.06 million t with 'Queen' being the most popular for fresh consumption and occupying the largest area. Pineapple does not flower uniformly even after physiological maturity so 80% of pineapple fruit are harvested within only 2-3 months (May-June-July), resulting in a huge market glut in almost all markets of the state. During May to July the state usually receives high rainfall accompanied with high relative humidity and temperature, which results in rapid deterioration in the keeping quality and because of that the farmers get very low prices. In order to overcome this problem, year round production by using different grades of suckers and slips (300, 600 and 900 g), staggered planting from April to October at monthly intervals followed by chemical induction by application of ethephon at 25 ppm at 38-40 leaf stage result in off season flowering and fruiting. Moreover, spreading production into the off season plays a vital role in nutrition and livelihood opportunity for rural and tribal areas of the state for employment and income generation (Das *et al.*, 2011).

The main operation carried out by the farmers includes land preparation, sourcing and planting of planting material (for establishment of new orchards/ replacement of old ones), intercultural operations as weeding, application of growth regulators, removal of suckers, slips, crown pinching and crop harvesting. Post harvest the fruits are loaded onto wooden basket woven with bamboo strips and carried from orchard to nearby roadside aggregation points for sale to middlemen or loaded onto transport vehicles for transport to local haats/ rural/ primary market for sale. The fields are not ploughed but uniform rows are demarcated across the slopes in order to restrict the soil and nutrient erosion. The suckers are planted at uniform spacing by digging small pits. In practice, generally no organic compost is added to the soil except in cases where cow dung or other compost materials are available, they may be added it the pits. Considering the climatic conditions, the best time for plantation of crop is May to June. However, considering the lean availability of suckers in May-June and avoiding the rainy and winter months, the plantation is usually carried out from August to early October.

There are two systems of pineapple planting viz, single row system and double row system. Single row system is more common in the region. Generally about 25,000-30,000 plants are planted per hectare i.e. 4000 plants per *kani* (local unit) equivalent to 0.16 hectare. Double row system is recommended in which pineapple is planted across the slope at the spacing of 30 x 45 x 90 cm and can accommodate 43,500 plants per hectare i.e.6960 plants per kani, allowing maximum land usage and more profits due to increased population density and increased yield per hectare. Beside, as added advantage adoption of high-density planting checks weed infestation, protect fruits from sun-burn, increases per unit area production of propagules (suckers and slips) and support non-lodging of plants.

Manipur: Pineapple is the largest fruits that is produced, and rightfully could be honoured as the State Fruit. Manipur state has sandy and loamy soil or lateritic soils on hill tops soil with soil Ph of 5.5-6.0 and humid conditions during summer and is suited for pineapple cultivations. Pineapple is most suitably grown on the hilly slope of the state and in the hilly areas the soil is quite prone to soil erosion. The cultivation of pineapple requires intensive scientific cultural practices to give good yield. In Manipur, the land for showing is prepared by thorough plough just before the onset of the monsoon season in the month of November-February. The land are cleared from all vegetations and then ploughed. Basal foundation application of FYM is carried out before sowing the planting materials.

Planting material mostly used for growing are suckers. The pineapple growers of the state mostly procured the plantings material from their fellow farmers or either from their previously existing nearby orchards. For obtaining maximum production, the ideal recommended spacing for pineapple in double row system are 22.5cm(Between trench) x 60(between plants) cm x 75(between rows) cm for subtropical and mild humid and 25 x 60 x 90 cm for hot and humid which can accommodate 63000plant/ha and 53000plants/ha. Adopting technique has been reported to give higher yield without affecting the quality of and size of the fruits.(Chadha *et al* 1974). Adopting recommended spacing will help not only in giving good yield and quality but it will also aid in preventing soil erosion, prevent from sun burn, weed infestations and lodging of the fruits. However, the growers are generally adopting 1m (row to row) x 1m (plant to plant) spacing adopting much lower planting density of around 10000/ha-30000ha -than the recommended spacing, the reason being the drudgery of high cost of manpower, un-remunerative pricing and uncertainties involved in evacuation and trading. Adoption of this general spacing leads to vulnerability to soil erosion and subsequent washing ways of the nutrients with runoff water which makes the soil unproductive.

The hilly slopy nature of the terrain in the state may lead to soil erosion during monsoon years leading to runoff of available nutrients required for growth and development. Moreover per unit hectare of land, pineapple removes 123kgofN, P_2O_5 and 308kg K_2O for a crop yield of 40tonnes. Therefore for obtaining good quality fruits with higher yield the soil require proper nutrition management practices. Application of 250quinatl organic manure is recommended in the state and it play important role during the period when the soil is uncovered and thus exposed to attack by the abiotic factors. Application of organic manures help in maintains the soil physical structures. Under Manipur condition the recommended fertilizer dose is 738kgN, 123P_2O_5 and 738.1kg per hectare. Fertilizers are applied in split dose where 12g N; 2g P_20_5 and 12g K_2O per plant is recommended for increasing yield of pineapple (main crop) and 10g N per plant if ratoon crops are taken. The split doses of N, P_2O_5 and K_2O can be applied at an interval of 2 and 5months in the initial plant growth at the base of the plants; however when the plant initiates more growth, application in the soil becomes difficult at 8 and 11 months due to dense vegetative cover. At this stage the fertilizers solution are placed directly on the crown (Meitei *et al.* 2021).

To restrict competition for nutrients and moisture major Intercultural operations like weeding, mulching, removal of suckers and application of growth regulators are required to give good fruit yield and quality. Weeding is done at least three to four times in a year. Earthing up is an essential operation in pineapple cultivation aimed at good anchorage to the plants. Soon after harvest, earthing up is done leaving one to two suckers only. Weeds can be effectively controlled by application of Diuron (@ 2 kg/ha.) or a combination of Bromacil and Diuron @ 2 kg./ha each as pre-emergent spray and repeated with half of the dose, 5 months after first application. It was learnt that the farmers are mostly adopting hand weeding that too in limited extent. For effective mulching black polythene is good but, majority of the farmers cannot afford to spend more input on this, to compensate on mulching, they covered with the farm crop residues like paddy straw and weed as mulching materials. Paddy straw is effectively being used to cover the maturing fruits to avoid from sunburn. Most of the pineapple plantations are watered as rain fed and artificial system of irrigation is very rare. However, during hot summer months supplemental irrigation at 20-25 days interval for 4-6 irrigations would ensure good crops. After the main crops of pineapple are harvested, usually 1 sucker and not more than two suckers under any circumstances are allowed to retain to carry forward with the subsequent ratoon crops. The ratoon crops also are ensured of proper agronomic practices to give proper yield.

Nagaland: generally Pineapple is a humid tropical plant and grows well, both in the plains and also at elevations not exceeding 1500meters. However, it does not tolerate very high temperature and frost. It starts flowering from February to April and the fruits are ready from July to September. Sometimes, off-season fruits are harvested from September-December. The soils of Nagaland is slightly acidic soil with pH range of 5.5 to 6.0 and is considered to be the best for pineapple cultivation. The soil should be well drained and light in texture. It can grow in sandy, alluvial or lateritic soil. However, heavy clay soil should be avoided. The climate of Nagaland is very suitable for Pineapple farming. Heavy rainfall is best for pineapple growth and an optimum rainfall of 1500mm per year is good enough. Temperature range from 15.5 to 32.5^{0} C is good for production of quality fruits and can be successfully taken up till 1500 m above MSL. In Nagaland, the most common form of pineapple propagation is from suckers or slips. Suckers arising from the underground parts of the plant are commonly used. Slips arise from the fruiting stem and from the crown on top of the fruit. Suckers produce fruits in about 18 months, whereas slips and suckers propagated from disc cuttings take over two years.

In Nagaland, farmers cultivate Pineapple in Hilly terrain. The Jungles are cleaned and pits are dug out and based on availability of Planting materials and Knowledge level of farmers planting is done with a spacing of 30cm x 60cm x 90cm which is considered as the bestway to get more yield under rainfed conditions. HDP of Pineapple is also being practiced by some of the progressive farmers. In Nagaland, Pineapple cultivation is totally rainfed and its production depends upon the amount of rainfall. Generally, Nagaland receives enough rainfall for successful Pineapple production throughout the year. The pineapple farmers mostly utilize the Jhum field after cultivation in the first year. The fallow land is cleared and pits are dug out and kept for sun drying for two to three weeks after which the suckers are planted in each pit and left to grow. Because of hilly terrain and high rainfall Pineapple fields gets heavily infested with weeds and in one season at least two to three hand weeding is done which is very costly. Chemical weedicide /herbicides are often not used so that the pineapple remains organic. Earthing up is done in pineapple cultivation to provide good anchorage to plants.

Meghalaya: Pineapple is traditionally cultivated mostly in the hilly slope. The best season for planting pineapple is in the beginning of monsoon season. Plant spacing is 30 cm plant to plant within a row, 60 cm row to row and 90 cm trench to trench. High-density planting, besides increasing the yield, is associated with other advantages like less weed infection, protection of fruits from sunburn, helps to increase production of suckers and slips per unit area and non-lodging of plants. Close planting also saves on the cost of providing shade to the fruits, as it provides natural shade through upright orientation of the apical leaves and eventually results in uniformly coloured lustrous fruits.

The plantings system varies depending upon the topography of the land and rainfall. There are four planting systems in practice, namely, flatbed planting, furrow planting trench planting and contour planting. Pineapple can be grown on various types of soils including very poor soils. The flavour and quality of fruit grown on light soils is considered to be superior. The plant is particularly sensitive to soil being waterlogged. Therefore care should be taken to ensure proper drainage. It prefers soils with a pH range of 5.0-6.0. Pineapple grows in warm and humid climate. The optimum temperature is from 15° to 32°C for normal growth. High temperature over 35°C is unfavourable for the development of fruits, especially if the relative humidity is low. Exposure of the fruits to strong sunshine leads to sun scalding. It can be grown up to an elevation of 1,100 m above the sea level, if these places are free from frost, have a relatively high atmospheric humidity and an average rainfall of 760-1,000 mm.

Mizoram: In Mizoram, generally Pineapple is a humid tropical plant and grows well at elevations not exceeding 1500 meters. Mizoram has primarily sandy loam and clay loam soil rich in organic carbon and moderately rich in available potash. Due to high rainfall during May to September, acidic soil having pH ranging from 4.5 to 5.6 is considered to be well suited for pineapple cultivation. Although optimum annual rainfall for its commercial cultivation is 100-150 cm, it grows remarkably well under a wide range of rainfall in Mizoram. Crown is commonly used for propagation in Mizoram. In comparison with other propagating material like suckers and slips, crown takes longer time to come into bearing.

Sucker is also used as planting material; plants raised through suckers have vigorous growth and bears fruit early. Though it is not a common practice, slip are also used as propagating materials usually when suckers are not available. It takes about two years to bear fruit. Pineapple cultivation is totally rainfed under Mizoram condition and its production depends upon the amount of rainfall received. Farmers apply Urea 26 gm, SSP 25 gm and MOP 20 gm per plant per year. Urea has been applied in six split doses, the first split dose was applied two months after planting and the rest of the split doses applied at two months interval. The required amount of MOP also applied in two split doses. The first split applied along with the total amount of SSP at the time of planting while the second split applied at six months interval. Pineapple orchard gets heavily infested with weeds because of high rainfall and in one season at least two to three hand weeding is done which is very costly. Chemical weedicide/ herbicides are often not used so that the pineapple remains organic. Earthing up is done in pineapple cultivation to provide good anchorage to plants.

Auranachal Pradesh: The state is gifted with abundant natural resources and suitable agro-climatic zones for cultivation of various species of horticulture crops ranging from temperate, subtropical, tropical and evens the alpine. Arunachal Pradesh which is popularly known as the land of rising sun in India has a long international border with Bhutan to the west (160km), China to the North & Northeast (1080kms) and Myanmar to the east (440km) which makes it strategically very significant for the country specially for international trading of horticulture produce in future. Pineapple is one of the important focus fruit crops grown in the sub-tropical belt of Arunachal Pradesh and to some extent even in pockets of the temperate belt of Arunachal Pradesh (Saloni *et al.*, 2017).

Sikkim, Pineapple is grown mostly in the humid tropical region. Basically it can be grown throughout in the warmer regions. It is one of the important commercial fruit crops of India. July to September is the main growing season and it takes almost 18-24 months to mature for harvesting. Sandy loam soils with pH ranging from 5.0-6.0 are ideal for Pineapple cultivation. The soils of Sikkim are slightly acidic with the pH range of 5.2-6.5 and sandy alluvial to sandy loamy soil in the lower and middle belt of Sikkim. The lower attitudinal belt of Sikkim is found to be suitable for pineapple farming. It can be grown up to 1000m above mean sea level if the area is frost free. The rainfall required for pineapple cultivation ranges from 100-150cm. The optimum temperature required for successful cultivation is 22°-32°C. In Sikkim, propagation of Pineapple is mostly done using suckers and slips. Pineapple is cultivated in hilly terrain and according to the topography the layout is planned for cultivation. Spacing of 30cm X 60cm X 90cm is encouraged by the department and extension workers to yield more output. Timely irrigation and weeding is practiced for clean cultivation. The fruit takes about 18-24 months for harvest. The fruit is hand plucked and taken to store area to remove field heat and later transported to nearby local markets. Some of it is processed by the farmers to make wine, juice etc. and put up for sale in the home stays established in the area.

Scientific Advances to Improve Adoptability, Productivity and Economic Viability of Pineapple in North Eastern States

The Department of Agriculture, Govt. of Assam, initiated research in horticulture in 1957 with a scheme on pineapple at Horticultural Research Station, Kahikuchi. The centre initially had research thrust focused on the three mandated crops, *viz.*, Banana, Coconut and Pineapple.Pineapple is an important component of coconut based multistoried cropping and Agroforestry system research in the station under Assam Agricultural University. Central Plantation Crop Research Institute (CPCRI), Kahikuchi has developed technology of Arecanut Based High Density Multi Species Cropping System in which pineapple suckers are planted in the space between arecanut palms in the same row in beds of size 1x1m. *In-situ* moisture conservation practices in sloping lands by CPCRI revealed that catch pit with pineapple downstream was found to be the best conservation practice. Assam Agricultural University has conducted several significant research programmes on pineapple crop. Studies have been made for Standardization of organic nutrient schedule in pineapple cv. "Kew" and packaging technologies for long distance transport of pineapple has been developed to extend its shelf life. In a study conducted, it was reported that the best quality fruits were obtained in black polythene mulch of 50 micron thickness throughout the cropping period. A study on the effect of biotic polymer, microbial antagonist and UV radiation on shelf-life of Kew pineapple fruits revealed chitosan treatment was found to be the best in enhancing shelf life of the fruits during storage. Assam Agricultural University has also developed technology for packaging pineapple in which each perforated fibre board box accommodates 12 fruits of 'Kew' pineapple [50 cm (L) × 40 cm (H) × 35 cm (W) size] with damage resistant both on full body truck (up to 1000 km) and goods train (up to 2500 km) with no bruising loss and with decay loss of 0.9% and 2.78%, respectively. The nutritional and hormonal studies on pineapple were conducted by Guwahati University to improve Giant Kew variety of pineapple through proper scientific nutritional study.

In Tripura, Pineapple crop is disadvantaged due to irregular flowering. Therefore to increase productivity, flowering is induced in more number of plants through use of various practices such as a) Chemical Application of Etheral in Kew variety of pineapple @ 10-25ppm concentration with 2% Urea and 0.04 sodium carbonate. b) Planting at regular intervals and c) Planting suckers/slips/crown of different size. This practices resulting in all around the year availability of pineapple through off-season production. This also helps farmers to fetch premium price during lean production period. Staggering is primarily carried out from August to November, after harvesting of crop. Though farmers have now began to adopt this new improved technique, however, it needs to be promoted in big way not only to significantly improve the productivity, but also to make pineapple cultivation remunerative for the farmers who are now tending to shift from pineapple to rubber and arecanut cultivation in the region.

In 1980s decade significant research has led to the development of stagger system of planting at various dates of the calendar year at monthly interval from July to August followed by hormonal application of Ethrel(0.25g/L) and Calcium carbide(20g/L) after a year give the fruits throughout the year (Aich, 1981) in Manipur. Different asexual propagules such as sucker, slips and crown takes 15-18months, 16-18month and 18-20months comes to flowering, planted at different months can give production of the fruit crops round the year. For ease of cultural management, each planting material is planted at different blocks. Application of 50ppm Ethrel along with 2%Urea at 35leaf stage at the crown gives uniform flowering which is advanced by 94 days compare to normal flowering in Manipur conditions (Meitei *et al.*, 2021).

Though much information is not available regarding measures to improve productivity in Nagaland but many a farmers have started to adopt HDP of Pineapple and the department of Horticulture Nagaland has been providing assistance to the needy farmers. Study conducted showed that the cost of production is high in first year and from second year onwards Cost benefit ratio was more than 2.235 (Sema *et al.*, 2009). The research study in three locations over four years of trial revealed that average return from pineapple cultivation ha^{-1} was Rs. 53223.75 (Sema *et al.*, 2009).

Under Meghalaya conditions, Study conducted in Ri Bhoi District of Meghalaya show that an average total cost of establishing a pineapple orchard is Rs 17815 /ha. The largest single items of expenditure was incurred on digging and filling of pits which is at Rs 5549/ha (Rymbai *et al.,* 2019). A few farmers have practiced High Density Planting (HDP) in order to increase or improve productivity of pineapple in Mizoram. With the establishment of fruit juice concentrate plant in Chhingchhip (Serchhip district), Mizoram food and allied industries corporation (MIFCO), farmers cultivated pineapple with the hope of earning good income as feeders for the processing unit (Shweta *et al.*, 2017).

Under Arunachal Pradesh conditions, one of the major biological problem of pineapple production is the uneven flowering & fruiting and especially in commercial pineapple cultivation, uniform flowering and thereby, fruiting is very essential. However, the floral differentiation of pineapple plants has another singularity; it can be triggered artificially, by chemical substances that are also related to natural flowering (Cunha, 2005). As per Norman (1972), research by several workers has demonstrated that by the use of certain compounds, flowering is induced and consequently an early crop is obtained.

In Assam, The Department of Agriculture and Krishi Vigyan Kendras (KVK) of the state has adopted strategies to promote scientific cultivation of pineapple. Central Government schemes like Mission Organic Value Chain Development for North Eastern Region and 'Horticulture Technology Mission' implemented in Assam has resulted in increase in the area and production of pineapple. Priority should be given to showcase improved productiontechnology includinghigh density planting, mulching and Integrated Nutrient Management,staggered production, double and single row pineapple planter technology,pineapple processing, pineapple wine production which can trigger pineapple revolution. Organic farming technologies of pineapple focusing on non-chemical use for plant nutrition and plant protection and other organic farming requirements should be assessed and refined through the KVKs. Growers Association of pineapple should be formed in the districts in the line of Barak Valley Pineapple Growers Association for promoting production and value addition of pineapple. North Eastern Regional Agricultural Marketing Corporation Ltd (NERAMAC) procures pineapple from "HMAR Agro Organic Producer",a Farmer Producer Company (FPC) of Hmar Tribal community, located in Lakhipur, Silchar. NERAMAC plays an important role in marketing of pineapple pulp to Hindustan Unilever Ltd & frozen pineapple juice concentrate to M/s. Dabur Foods, Nepal. The Assam Hill Small Industries Development Corporation Ltd. under Govt of Assam has set up fruit processing unitand is producing natural fruit juice from the locally available raw-materials including pine apple in " Karbi Anglong and North Dima Hasao district and is marketed in the brand name of "KANCH".

Tripura pineapple is exported regularly every year at Middle East Countries like Dubai. Chief Minister of Tripura also initiated a special drive during COVID for enriching nutritional status of people named as "*Mukho mantri Corona Protisedhok Abhijaan*" scheme on 4th July 2020. Free pineapple juice distribution drive carried out in different places in Tripura for the common people under this scheme. Honb'le Chief Minister of Tripura 1st Flagged off Queen Pineapple of Tripura to Dubai on 3rd June 2018. MoU between THCL and Spice Jeet Merchandise, Gurgaon and Kremert Craft, Kolkata has benn singed on 18.05.2010 to export Tripura pineapple from state to foreign countries. Consignment of pineapple (Kew) was flagged of by Honb'le Chief Minister of Tripura from Kumarghat, Unakoti District, Tripura on 22nd July 2019 and the exporter was JSB Processed food export Pvt Ltd, New Delhi. Pineapple exported to Bangladesh on 11.06.19 from Sonamura, Tripura. Tripura pineapple of 20 MT (2nd load) and total of 14 thousand nos exported for Delhi on 25.05.2029. 20 MT exported to Delhi on 22.05.19 which is facilitated by Directorate of Horticulture, Govt of Tripura. Even during COVID- 19 situation also pineapple is exported from Tripura to Dubai. One Export agency named CMC, New Delhi has exported 7692 nos of pineapple from Darchoi, Unakoti Tripura during COVID also. Solar based cold room is installed at HRC, Nagicherra to facilitate pineapple export from Tripura. International conference and Buyer Seller Meet led at Guwahati where Tripura GI tagged pineapple was focused. Kew pineapple is exported to other states and other countries are approx 80 MT from Tripura on 06.07.2020. Tripura pineapple is also exported with the help of CSC platform. International Buyer Seller Meet was organized by Govt of Tripura for promotion and export of pineapple on 26.09.2019. Tripura Pineapple is also promoted at Dilli Haat on 14.07.18 to 18.07.18 by Govt initiatives. Pineapple festival was organized at Kolkata from 30.06.18 to 01.07.18 by Govt initiatives to promote Tripura pineapple. Several consignment and loads has been exported regularly from Tripura by different agencies time to time benefiting the pineapple farmers in Tripura. Tripura KVKs have also taken several initiatives through technology dissemination, group discussions, FLDs and OFTs.

Taking into account the enormous potential of horticulture and agricultural produce in the State and to curb the issue of lacking infrastructural facilities, a state of art multi fruit processing facility, Sikaria Mega Food Park, is being developed in the State with assistance of Ministry of food processing industries of Government of India. The facility is planned to be equipped with three tier storage and processing facilities at different stages of food processing namely

– Collection Centers(CCs), Primary Processing Centre (PPCs) and Center processing Centre(CPC). While the Common processing facility is being set-up at West Champamura in Agartala, the primary processing centres are proposed to be set-up across the State at Sonamura (Sipahijala district), Hrishyamukh (South Tripura), Chandipur (North Tripura District), Manu (Dhalai district), Bhodhjungnagar (West Tripura).

In Manipur State, Locally known as 'lengthei', pineapple has been cultivated in all the 9 districts of Manipur. As per survery conducted by the Manipur Pineapple Festival Committee in August 2008, about 2400 acres of land has been under pineapple cultivation in Manipur. Churachandpur District, with over 700 acres of land in cultivation is recorded to have the most extensive cultivation in Manipur. About 350-400 farmers are actively engaged in pineapple cultivation in Churachandpur District alone. State Government has taken several initiatives like MIDH Schemes for area expansion and export of organically grown pineapple export to Western and Arab countries though farmers groups , FPOs etc.

The state government of Nagaland has developed infrastructure facilities for processing Pineapples. The state is having 3 nos. of plants for Canned Pineapples at Dimapur having capacity of 5MT per day with 960 MT/ season and another factory for Pineapple Fruit Pulp at Dimapur which functions from the month of July-September and November to December with a capacity of 10 MT.

Through the intervention of Department of Horticulture, Govt of Meghalaya farmers have started to adopt HDP of Pineapple and the department under MIDH Schemes in certain district is giving subsidies to the needy farmers for the cost of planting materials, drip irrigation and cost of INM/IPM in pineapple. The KVK and Horticulture Department is training the farmers on standard package of practices for pineapple and also encourage the farmers to follow double row planting technique and mulching technique (Paddy Straw, Plastic).

In Mizoram, Several important and useful programmes that could further improve the production of horticulture crops have been taken up by Horticulture Department. Some have already shown noticeable results, while other programmes are still being implemented; pineapple growers are provided the assistance through MIDH scheme such as planting materials, irrigation facilities, rain water harvesting structure, plant protection measures etc. which will have profound impact towards improvement of their farm produced.

In Arunachal Pradesh, due to good production and quality of pineapple and visualizing the future potentiality, the govt. of Arunachal Pradesh established a processing unit in the district which produces pineapple juice concentrate that are sold to other big companies in the country for making squash and other products. Pineapple cultivation in the state has been accelerated with the launch of TMNE. Growers Prefer 'Giant Kew' for its size and quality and also due to easier intercultural operation, especially weeding, because of its non –serrated leaves (Saloni *et al.*, 2017).

Sikkim State Horticulture Department has taken up Pineapple cultivation and till date has expanded its area up to 165 ha. People have been given awareness on the benefit of Pineapple cultivation. Sikkim Government has even supplied good quality planting materials "pineapple variety queen" to the interested growers. Shade house for all Horticultural produce has been established in different wards. To overcome the problems relating to marketing, State has established processing units and formation of FPOs have also been encouraged.

Constraints of Pineapple Production in North Eastern States

Pineapple sector in Assam is facing several challenges like difficult terrain, non availability of proper transport system, inadequate infrastructure including lack of modern cold storage, marketing support and processing units. Non availability of tissue culture plants, erratic flowering, climate change uncertainties are some of the issues that needed to be addressed. Production issues such as the unavailability of farm labour, low labour productivity under the manual operation are some production constraints which can be overcome by farm mechanization.

The pineapple sector is faced with several constraints in Tripura. Most important one is being low adoption of improved technology and inadequate infrastructure facilities, particularly for post-harvest operations (including processing) and marketing of produce. Moreover, as the small and marginal farmers constitute about 96% of the farming community, therefore, most of the agricultural activities in the State are confined to subsistence production. As fragmentation of land holding still continues as a part of social phenomenon, the average size of holding has been declined from 1.25 hectares in 1976- 1977 to 0.49 hectares in 2010-111 (Census 2011). Other factors as less intensive use of agri-inputs negligible seed/ varietal replacement, calamities as recurring floods etc are also threatening to sustainability of livelihood from agriculture sector in the State. Lack of post-harvest processing facilities results in high wastages and hardships for farmers during the peak production season. Being land locked, the State has disadvantage of restricted movement of goods, which affects the marketing of agricultural produce, particularly the perishable produce. Due to hilly and forested terrain, the scope for area expansion under agricultural crops is limited. Further, there is increasing pressure on land due to population rise and its requirement for promotion trade and other industrial purposes. Commercial scale processing of pineapple is found to be absent in the State, therefore, the produce is mostly sold directly to local traders and buyers, who visit the production clusters during production season. Same time of harvesting period along production clusters during May to August lead to excess supply than seasonal requirement, resulting in huge market glut in almost all regional, State and national level markets reducing the market price and thus, decreased farmer returns.

In absence of appropriate post harvest storage infrastructure and processing, the farmers are compelled to sell and dispose the produce at very low prices. It was also learnt during the discussion that not all plants in a plot flower during the year; moreover there is lack of uniform flowering even after physiological maturity, resulting in lower productivity than achievable potential. Farmers use minimum chemical fertilizers in the cultivation mostly because of enhanced growth of weeds with the usage of fertilizers in low density cropping. This may affect the productivity of the land. The emerging boom in the demand for organic food products provides significant opportunities for Tripura. Pineapple is primarily a rain fed crop in Tripura. Irrigation during dry months (November to March) may help in off season production of the fruit. Farmers suffer huge losses in the peak season because of imbalance in demand-supply scenario and lack of farm level infrastructure for proper storage. Many times there are about 10% loss in between harvesting and loading in trucks. Every year distress selling is done by the farmers. The small size of landholding effectively limits the cultivation and marketing. Also farmers lack awareness on value-addition aspect of the produce. The downstream industry being weak and almost absent, there is little scope for higher price realization among small farmers particularly during market glut.

Manipur occupies in sixth Position in terms of its area of cultivation After West Bengal, Assam, Tripura, Karnataka and Nagaland (Horticulture statistics at a glance, 2018). Though it is cultivated in large area in the state and considered good cash crops and economic backbone of the state, like any other fruits crops one of the major bottleneck that is faced by farmers is the market. The farmers may get the net profit (Rymbai *et al.*, 2012; Hatai *et al.*, 2013; Kichu *et al.*, 2014), but however they do not receive their due for what they deserved. Like other North Eastern states, In Manipur there is also lack of good road for transportation of the produce and most of the stake holders may not be willing to bear the transportation cost leading to low sell of the fruits. The fruits being perishable in nature over the process of transportation may suffer from many physical brushes and damages which are one of the major challenges faced leading to low prices in the market.

There is an urgent need for standardization of technology in Nagaland to bring down cost of production of fresh pineapple and its adoption by growers, assured market for the produce and to produce pineapple throughout the year may go a long way in promoting pineapple industry in India. The processing industry for pineapple is limited to a specific location and processing cost of pineapple are high due to high cost of fruit, sugar, containers and overheads and non-availability of fruits throughout the year. Marketing of fresh pineapple is a big problem due to its highly perishable in nature. Mature pineapple fruits cannot be stored for more than 4-5 days after harvesting and any injury to fruits during harvesting and transportation results in poor quality of pineapples and low market demand.

Bulk of the pineapple produced is consumed in fresh form in Meghalaya. Marketing of fresh pineapple also poses problems due to its highly perishable nature. Mature pineapple fruits cannot be stored for more than 4-5 days after harvesting. Therefore, it is necessary to take ample care to avoid any injury to fruits while transporting to major consumption centres.

Marketing of fresh pineapple is a big problem in Mizoram due to its highly perishable in nature; any injury or bruise during harvesting and transportation results in poor quality of pineapples and low market demand. In addition to this, the processing industry for pineapple is very limited in the state.

Despite vast potential in Arunachal Pradesh, the horticulture sector has failed to establish its credibility for improving productivity, generating employment, improving economic condition of the small and marginal farmers in rural areas due to major constraints like inadequate nursery base, poor quality of seeds and planting materials and dependence on other States for procurement of basic plantation materials, small and uneconomical average farm size, higher order of perishability of horticulture produces leading to high degree of losses, lack of modern and efficient infrastructure facilities, poor technological support, poor post-harvest management, absence of nursery Act & Regulation, absence of marketing and storage facilities, absence of packing technologies and inadequate storage facilities, vulnerability of crops to pest and diseases, non-emergence of watershed approach and inadequate irrigation facilities (Draft Annual Plan 2011-12, Department of Planning, Govt. of Arunachal Pradesh.pp-45). Lastly, there is lack of proper fund support for research and development activities in the state.

In Sikkim, due to the lack of awareness about the importance and nutritional value of Pineapple fruit crop, lack of quality planting materials and standardized management technology Pineapple is not considered as an important fruit crop in Sikkim.

Opportunities of Pineapple Production in the North Eastern States

Assam offers ample of opportunities for expansion of areas and processing of pineapple. Initiatives have been taken by Horticulture Technology Mission for the development of pineapple sector in Assam along with well-organized Pineapple Growers Society in some districts. Some districts like Karbi Anglong, Cachar and Dima Hasao are the traditional belt of pineapple where the farmers adopt organic practices. State Horticulture department are providing extension support to the pineapple growers. A well-coordinated research and development support are provided to the growers from Assam Agricultural University through Horticultural Research Station, Kahikuchi and Krishi Vigyan Kendras located in the districts. Farmers have knowledge on pineapple cultivation and the planting material is locally available. The tag of Organic by default helps to explore new markets apart from assured market. In Tripura, the undulating tilla land and lower foothills (which are susceptible to soil erosion) are particularly suitable for pineapple as cultivation on slope helps avoid water stagnation, which is otherwise not suitable for pineapple cultivation. Also growing pineapple on these slopes helps check soil erosion. Considering the voluminous pineapple production in the state, the crop is assessed to have the maximum potential for processing, compared to other fruit crops. Owning to high climatic suitability, the farmers in the selected production in Tripura are observed not to be using any chemical inputs (fertilizer, pesticide) for pineapple cultivation. Production is mostly carried out organically. This may also be attributed to factor, that mostly the tribal population in the region had been earlier practicing jhum cultivation, where much emphasis was not given on chemical input application.

Beside all challenges being addressed, there is a large scope and opportunity in Manipur. There should be more establishment of cold storage such as modified atmospheric storage which can be also reliable for other fruits, set up of cold storage and cold chain technology will help in capturing the market in the offseason. Set up of large scale industries to process the produce into different value added products not confining to just rural self help group will greatly boost in marketing. To compete and capture the export oriented market, The fruits need to have some specific requirement for its appearance and quality, for export of fruits pineapple According to APEDA (Agricultural and Processed

Food Products Export Development Authority) it is graded into different code as A,B,C,D and Baby and as Extra classes, class I and class II based on the weight of the fruits ranging with (<700-2750g) and without crown (580-<2280) and other quality parameters excepting the queen variety, the weight of the Kew fruits meet it specific demand. Pineapple growers in collaboration with scientific experts with already well set up of vast network of Research institute and farm science centre will also help to cater the need of developing the market specific oriented produce by adopting good agricultural practices such as adopting high density planting, using different types of planting material and planting at different times of year, Management of nutrients. Inter-culture, weed control etc, integrated management of pests and diseases, inducing of flowering are making fruits available throughout the year. This technology for year round production of fruits with good agricultural practices can makes it more vibrant and viable to provide income generation even during the off season. Manipur lies to the nearest destination to all the South East Asian countries such as Myanmar, Thailand, Laos, Cambodia, Indonesia, Philippines and Vietnam. Set up of Agricultural export zone for promotion of pineapple export at North Bengal will also make it more reliable as Manipur is nearer to this destination. Transportation through road and location of International connectivity through air has led to open scope for marketing of the produce.

Under Nagaland situation, to overcome all these problems emphasis needs to be given for regulation of markets for pineapple and integration of production, marketing and processing activities would help in decreasing marketing cost and thereby encouraging cultivators for self marketing. Formation of FPO, support from the government in hilly areas, development of infra-structural facilities (transport and communications), improvement in packing, storage and handling facilities, subsidization of inputs are the various aspects which need attention. In Nagaland almost the crop production system is organic by default and if organic certification process of Pineapple growing areas can be taken up it will boost the demand of certified organic Pineapple resulting in more demand and income to the farmers.

Standardization of technology to bring down cost of production of fresh pineapple and its adoption by growers, assured market for the produce and to produce pineapple throughout the year may go a long way in promoting pineapple cultivation in Meghalaya. In Meghalaya, steps like regulation of markets for pineapple and integration of production, marketing and processing activities would go a long way in decreasing marketing cost and thereby encouraging cultivators for self marketing. Development of infra-structural facilities (transport and communications), primary markets, and improvement in packing, storage and handling facilities, and subsidization of inputs are the various aspects which need attention.

Documentation of the activities encouraged during the adoption of farm activities and further to increase the farm efficiency in Mizoram. Emphasis also needs to be given for regulation of markets for pineapple and integration of production, marketing and processing activities that would help in encouraging pineapple growers.

Despite many constraints, the state Arunachal Pradesh has huge potential in horticulture owing to its vast available land resource and existence of diversity in agro-climatic zones which makes it a sleeping giant for growth of horticulture in the state. According to the *Arunachal Pradesh Horticulture Development Vision-2030* document, about 18,00,000 hectares is identified suitable for horticulture crop cultivation at various altitude and agro-ecological zones (about 15% in tropical, 65% in subtropical where pineapple comes and 25% in temperate zones). However, only 6% of total area i.e. 1,11,000 hectares only has been covered so far.

Strategies for Improving Production of Pineapple in North Eastern India

Proper planning and strategies should be adopted for increasing production and quality of pineapple in Assam. A few points have been chalked out for development of the pineapple sector in the state.

i. Identification of potential belts & formation of Pineapple Growers societies and their training/awareness/ exposure.

ii. Development of commercial plantation with commercial varieties as monocrop/ intercrop

iii. Initiation of Pineapple Mission or establishment of Pineapple Promotion Council. Formation of Pineapple Fruit Processing Companies should be promoted.

iv. Adoption of improved package of practices and modern technologies like plastic mulching, fertigation through drip etc.

v. Agronomic manipulation for yield and extension of harvesting period to escape glut and supplement processing sector.

vi. Adoption of Integrated Pest and Nutrient Management technologies.

vii. Reduction in postharvest losses with appropriate technologies.

viii. Facilitate marketing / market information/ market tie-up / for domestic and export market.

ix. Extending hassle financial support to the growers

ix. Quality improvement and Value addition with Establishment of commodity processing unit

Tripura, the region produces about 50 percent of pineapple produced in the country. The advantageous position of NER in terms of fertile and organically rich soils, abundant rainfall, water resources and great agro-climatic diversity supports the cultivation of best quality pineapple in the region. Pineapple produced in the region is qualitatively different and is said to be among the best in the world as they are sweet (high TSS) with less fibre. Though large quantities of pineapple are produced in NER, exports from the region are minimal. Only small quantities of pineapple are exported even to the neighbouring nations of Bangladesh and Myanmar. Tripura has exclusive advantage and potential for development of integrated value added chain for pineapples produced in the State as the Kew and Queen variety produced in the State are both suitable for processing/ canning purpose. The changing consumption patterns and preferences in the country and world are creating a huge demand for processed products. Moreover, 90-95 percent of the productions is being organic, opportunity existing for catering to niche organic product market. The recently accorded Geographic Indication registration of Tripura Queen Pineapple furthers add to the distinctiveness and assurance of quality of the product's origin and may be leveraged for boosting export of pineapples from the State, besides fetching premium price in international market.

Despite the advantage, the overall level of crop productivity in the State is found to be lower, compared to national and NER productivity levels. The pineapple growers need to be more extensively encouraged for adopting modern cultivation practices as high-density planting, hormone application for uniformity in flowering, staggering through chemical induction for round the year pineapple production and other management practices. Moreover, if the premium for the GI and organic produce has to be realized by the farmer, the State has to ensure better participation and integration of smallholder pineapple growers along the value chain. The existing marketing and trading system for pineapple in the State also need to be made more vibrant and organized. Processing of pineapple into other value added products is negligible.

Pineapple is the major leading fruit crops grown in the state of Manipur. The conducive soil and climatic conditions make suitable fruit production of this crop. Intrusion of viable and feasible technology and its adoption by the growers will lead to more production and remuneration provided there is good road infrastructure, set up of cold storage technology and large scale industries for processing. As per the data available the area under Pineapple cultivation has decrease in the state which may be due to introduction of other cash crops. However, farmers have shifted from Jhum cultivation to Pineapple production which has helped in improving the socio-economic condition of the farmers.

In Nagaland, Dimapur district is having a large area under Pineapple cultivation and all other district too has a considerable area under Pineapple cultivation. Formation of Pineapple FPO and organic certification will help them to get premium price of their commodities. Establishment of processing industries in every district is need of the hour to attract the farmers in increasing Pineapple production within the state leading to improvement in socio-economic conditions of the farming community.

Though pineapple is an excellent material to be preserved in different forms; the production used for processing is very less in Meghalaya. This is in contrast to the principal producing countries, where over 95% of the pineapple is absorbed by the processing industry. The pineapple processing business is very lucrative since it helps in value addition as well as it put aside the problem face by farmers during storage and transportation of fresh fruits. However this should be accompanied by setting up processing units at every major pineapple producing district of the state.

In Mizoram, to standardize the package of practices (production technology) and establishment of processing industry is basic necessity for pineapple cultivation in Mizoram in order to improve and increase the productivity of pineapple.

Area, production and productivity of pineapple in Sikkim is growing considerably at lower rate compared to the other north eastern states during the last few years. Recently we can see that the area under pineapple has been increased at faster rates while production is still low. It is obvious that the area production and productivity contrast within the NE States is mainly due to lack of awareness about the fruit crop, management technology and lack of provision of good quality planting material. Various strategies to increase the area and production of Pineapple have been adopted by the Sikkim Government

References

Aich, K. 1981. Ph.D. Thesis submitted to Bidhan Chandra Krishi Viswavidyalaya, Kalyani.

Akali Sema, K. K. Jha and Amodh Sharma. 2009. Cultivation of Pineapple for upliftment of rural economy in Nagaland. March 2009; Acta Horticulturae 822:99-106; DOI: 10.17660/ActaHortic.2009.822.11

Anonymous. 2005. Pineapple Cultivation in Tripura, 2005, ICAR

Anonymous. 2010 Draft Annual Plan 2011-12, Department of Planning, Govt. of Arunachal Pradesh.pp-45.

Anonymous. 2015. Area and Production of Horticulture Crops (2015), Department of Horticulture, Government of Mizoram.

Anonymous. 2016a. Area and Production of Horticulture Crops (2016), Department of Horticulture, Government of Mizoram.

Anonymous. 2016b. Statistical Handbook of Nagaland, 2016, Directorate of Economics & Statistics Government of Nagaland, Kohima

Anonymous. 2017a. Economic Review of Tripura 2015-16 (http://ecostat.tripura.gov.in/economic_review_15_16.pdf)

Anonymous. 2017b. Area and Production of Horticulture Crops (2017), Department of Horticulture, Government of Mizoram.

Anonymous. 2017c. Statistical Handbook of Nagaland, 2016, Directorate of Economics & Statistics Government of Nagaland, Kohima

Anonymous. 2018a. Horticulture Statistics at a Glance (2018). A report: www. nhb.gov.in.

Anonymous. 2018b. Area and Production of Horticulture Crops (2018), Department of Horticulture, Government of Mizoram.

Anonymous, 2018c. Pineapple value chain analysis and market assessment for imphal east district, Manipur(2018). A report submitted to Mission for Integrated Development of Horticulture MoA&FW, Govt. of India by. CCS National Institute of Agricultural Marketing (A Govt. of India organization under Ministry of Agriculture and Farmers welfare.

Anonymous. 2018d. Pineapple value chain analysis and market assessment for Unakoti & Dhalai District Tripura. In: Report submitted to Mission for Integrated Development of Horticulture Moa&FW, Govt. Of India, 2018.

Anonymous. 2018e Statistical Handbook of Nagaland, 2016, Directorate of Economics & Statistics Government of Nagaland, Kohima

Anonymous. 2019a. Area and Production of Horticulture Crops (2019), Department of Horticulture, Government of Mizoram.

Anonymous. 2019b. Package of Practices of Horticultural Crops. Prepared jointly by Assam Agricultural University, Jorhat & Department of Agriculture.p:1-2.

Anonymous. 2019c. Statistical Handbook of Nagaland, 2016, Directorate of Economics & Statistics Government of Nagaland, Kohima

Anonymous. 2021. State Horticulture Department, Govt. of Sikkim.

Anonymous. 2022. Arunachal Pradesh Horticulture Development Vision-2030.

Borthakur, D. N. 1993. Plant genetic resources of North East. Indain Journal of Hill Farming, 6(1): 1-18.

Chadha, K. L., Reddy, B. M. C and Shikhamany, S. D. 1998. Pineapple. Indian Council of Agricultural Research: pp-1, 4-5, 8-9.

Chadha, K. L., Shikhamany, S. D. and Melanta, K. R. 1974. Pineapple Cultivation, Indian Journal of Horticulture, 19:3-5.

Cunha, G. A. P. 2005. Applied aspects of pineapple flowering. Bragantia Campinas., v.64, n.4:499-516.

Das, P. and Hazarika. B. 1999. Effect of black polythene and rice husk mulch on chemical composition of pineapple, Journal of Applied Horticulture, 1(2):125-127

Das, S. C. Prakash, J., Suresh, C.P., Das, A. and Bhattacharjee, T. 2011. Pineapple cultivation in hilly Tripura with year around production: improving livelihood opportunities in rural areas of Tripura. Acta Horticulturae, 902 (902):291-298

Dayohimi Rymbai, Ram Singh, S.M. Feroze, P.M. N. Rani, R. Bardaloi and Monika A. (2019). Pineapple cultivation in Meghalaya and Livelihood: An Economic Analysis; Horticulture for economic prosperity and nutritional security in 21st Century Westville, New Delhi, pp. 287-299.

FAOSTAT, Food and Agriculture Organization (http://www.fao.org/faostat/en/#data/QC/visualize accessed on 24.08.2017

Hatai L.D, Naik D. 2013. Marketing strategy and supply chain analysis of pineapple: A case study in West Garo Hill of Meghalaya. National Seminar on Agri-business potentials in North Eastern States Souvenir, pp. 24-25. November 8-9.

http://apeda.in/agriexchange/Market%20Profile/one/PINEAPPLE.aspx.

http://pineappleindia.com

http://www.aau.ac.in/research/directorate/directorate-of-research(agri)/hrs-kahikuchi/about/10/122

http://www.nhb.gov.in. National Horticulture Board

https://diragri.assam.gov.in/sites/default/files/swf_utility_folder/departments/diragri_medhassu_in_oid_4/portlet/level_2/11%28B%29.14.pdf

https://diragri.assam.gov.in/sites/default/files/swf_utility_folder/departments/diragri_medhassu_in_oid_4/menu/information_and_services/HORTI%20VISION%202020.pdf

https://tmnehs.gov.in/writereaddata/Technical_Details_MM-I.pdf

https://www.business-standard.com/article/current-affairs/university-generated-technologies-reach-rural-entrepreneurs-in-assam-115041400643_1.html

https://www.neramac.com/commercial-activity/

https://www.pratidintime.com/assam-pineapples-headed-for-dubai/

https://www.sentinelassam.com/news/governor-all-praise-for-pineapple-growers-of-hmarkhawlien/

Joy, P. P. 2013. Pineapple sector in Kerela:Status, opportunities, challenges and stakeholders. Pineapple Research Station,Vazhakulam,Kerala.

Kichu, Y. and Sharma, A. 2014. Economics and constraint of pineapple cultivation in Dimapur district of Nagaland. Agri-Business Potentials in India. Experiences from the Hill States, pp, 217-223.

Mathew, B., Pereira, L. S. and Suresh, C. P. 2011. Pineapple production in meghalaya (india) - indigenous cultural practices and status. Acta Horticulturae, 902: 275-280DOI: 10.17660/ActaHortic.2011.902.30

Meitei, W.I., Jayalaxmi, H. and Sophia L. 2021. Economic Production of Kew Pineapple In Manipur. Scientific paper inc. E-pao. http://www.e-pao.net.

Mili, J.P. 2014. Shelf life enhancement of Kew Pineapple and Khasi Mandarin using biotic polymer, microbial antagonist and UV radiation.

Mishra, A. K., Bundela, D. S., Satapathy, K. K. 2004. ENVIS Bulletin: Himalayan Ecology, 12(1): 2

Norman, J. C. 1972. The influence of some compounds on the growth and flowering of Ananas comosus (L.) Mer. cultivar Sugarloaf. Ghana Journal of Agricultural Sciences, 5:213-219.

Alam, P., Tariq M. Usmani, M. D. 2020. Growth of Pineapple Cultivation: A Spatio-Temporal Analysis in India, International Journal of Trend in Scientific Research and Development, 4(2): February 2020

Patowary, U. 1975. The nutritional and hormonal studies on pineapple. PhD Thesis submitted to Gauhati University

Rymbai, D, Singh R, Feroze S. M., Rani P. M. N. and Bardoloi, R. 2013. Utilization pattern and marketed surplus of pineapple in Ri-Bhoi district of Meghalaya. Indian Journal of Agricultural Marketing, 27 (2): 35-39.

Shweta Saloni, Sindhu, Chauhan, K. and Tiwari, S. 2017. Pineapple production and processing in north eastern India, Journal of Pharmacognosy and Phytochemistry, SP1: 665-672

5

Status of Pineapple Production in Eastern India

Agniva Halder

Precision Farming Development Centre, Department of Agriculture and Food Engineering, Indian Institute of Technology Kharagpur, West Bengal

The states in the Eastern India where pineapple is grown include West Bengal, Bihar and Odisha. In between this states the ideal condition for pineapple is found in most parts of north Bengal and some north eastern districts of Bihar. West Bengal ranks first in production of pineapple in India (Figure 2).

West Bengal and Bihar

West Bengal ranks ninth in fruit production in India with a total production of 3850.5 thousand ton from 261.02 thousand hectare area (Table 1). Major fruit crops are mango, banana, pineapple, guava, litchi, jackfruit, mandarin, papaya and sapota. A large number of other fruit crops such as Jamun, bael, custard apple, ber, lime, lemon, plum, karonda, aonla, etc. are also grown in various parts of the state (Mitra, 2013). Pineapple is one of the important fruit crop of West Bengal covering an area of 11,410 ha with an average productivity of 30.24 t ha^{-1}(Table 2). The state of West Bengal, having five distinct agro-climatic zones, is ideally suited for growing several tropical and subtropical fruits. These agro-climatic zones are:

I. Hill zone II. Terai zone III. Alluvial zone IV. Laterite zone V. Coastal zone

In between these five agro-climatic zones commercially pineapple is grown mostly in Terai zone which is situated at the foot hills. High rainfall, light acidic soils and prolonged monsoon are characteristics of this zone. So the districts are Darjeeling, North Dinajpur, Jalpaiguri and some parts of Alipurduar. Besides terai zone other places where pineapple is grown in small scale or for experimental purpose are described below.

- In Baruipur and neighbouring areas of South 24 Parganas district there was a trend of pineapple cultivation with some local cultivars like Haricharanvita and Baruipur (Pandey, 2005). But last 10 years there is no record of pineapple cultivation, the farmers shifted to mainly karonda and other fruit crop orchards.
- In North 24 Parganas district some pineapple plantations are found under the mango and jackfruit orchards in blocks like Habra 1, Bongaon and Gaighata.The varieties are mainly local which has not so good taste and very low production.
- At Duma area of Gaighata block of this district some farmers are cultivating pineapple under ATMA scheme and the variety is Queen.
- In Birbhum district cultivation of Kew and Mauritius pineapple cultivars showed excellent result at Horticulture Farm of Institute of Agriculture, Visva-Bharati, Sriniketan. Pineapple cultivars Kew and Queen gives satisfactory result at the farm of RRSS for Red Laterite Zone-BCKV, Shekompur of this district.
- In Bankura district a new trialon pineapple cultivation with Giant Kew variety has been started at Ranibandh block under Saal (*Shorea robusta*) forest.
- In Krishi Vigyan Kendra Kalyan,Jahajpur , Purulia, they have plantation of variety Kew, it is also showing good result but fruit colour is not so attractive.
- In Nadia district the farmers of some blocks like Krishnagar (1&2), Hanshkhali etc. had been planted pineapple plantation of some local cultivars under *Khaya senegalensis* (local name-Lambu) orchard.

Bihar ranks eighth in fruit production in India with a total production of 5117.2 thousand ton from 296.69 thousand hectare (Table 1). Major fruit crops are Mango, Banana, Guava and Litchi. Other fruit crops such as Makhana, Pineapple etc. are also grown in small scale in various parts of the state. Pineapple is a minor fruit crop of Bihar it is cultivated in an area of 4260 ha with an average productivity of 27.02 t ha^{-1}(Table 2). Based on soil characterization, rainfall, temperature and terrain, four main agro-climatic zones in Bihar have been identified.

These are: Zone-INorth Alluvial Plain, Zone-II North East Alluvial Plain, Zone-III A South East Alluvial Plain and Zone-III B South West Alluvial Plain, each with its own unique prospects. Pineapple is cultivated only two districts of Kishanganj and Purnia from North East Alluvial Plain of Bihar. The major pineapple belt of Bihar and West Bengal coincide together so the cultivation practices, cultivars and problems are same in this two states.

Cultivation Practices Followed

The main variety is Kew (70%) other is Queen and some local types (Jaldhup, Haricharanvita).

The basis propagating material used by the farmers is suckers and slips from their own plants of previous season. They collect this suckers and slip after harvesting of crop in the month of July August. Then they dried this in a shady place for 4to 6 week. Planting of this plants are done in pre winter or pre monsoon season. According to the farmers the best planting material is suckers/ground suckers of weight around 450g. These suckers give yield in 23-25 month after planting. They called suckers as A-grade planting material and slips as B-grade planting material for selling it. As slips (weighted around 350g) gives yield after 15 to 20 month. For this belt, it was reported that nitrogen at 12-16 g, phosphorus at 2-4 g, and potassium at 10-12 g per plant should be optimum for a plant for a plant density 51,000per ha. However, the highest yield with N, P and K ar 600, 400 and 600 g per hectare, respectively was reported in plant density of 64,000 and 72,727 per hectare (Bal, 2014). Besides chemical fertilizers farmers also use adequate amount of organic nutrient sources FYM, oil cakes, vermin compost etc. For flowering induction the farmers use growth regulators like auxin substances as NAA and planofix (Verma, 2017).

Constrains

It is found that progress of pineapple cultivation in West Bengal particularly in North Bengal zone is not encouraging. The major constraints which need to be seriously thought of, like Low opportunity cost, Low income potentiality as the crop can generate only one time income in two years, high price fluctuation due to the perishable nature of the crop, labourer migration problem, rather high capital-intensive nature (maintenance cost is around 40,000 per acre/ year), replacement of pineapple area with allocation (of the replaced area) under tea which is fast growing as a competitive crop in the area, pest and disease problems, non-uniformity in fruit size and weight etc. are briefly described below

- The main problem for this pineapple growing belt is marketing at proper time as the local market has not much demand for the crop. So the farmers have to export most of their production in Kolkata, Gaya UP, Delhi, Mumbai etc. Especially in the COVID-19 condition export in other states is mostly impossible.

- Other notable problem is labourer crisis as this crop is labourer extensive. Due to these two causes the farmers are not able to harvest the fruits timely. So the fruits and plants are rotten in the field. It is very harmful as they collect planting materials from previous season's crop.
- The uncertainty of rain is now become routine thing. So, in case of low rainfall year the yield reduced as adequate irrigation facility is not available many of the farmers. But in the year of high rain fall water stagnation is a great problem for the plants.
- A very bad practice followed by the pineapple farmers of this belt is excessive use of potash and nitrogenous fertilizers. They never follow the dose recommended by the state horticulture department. So soil become sterile, especially in some areas in Kishanganj block of Bihar the soil condition is extremely bad.

Agri-Export Zones for Pineapple of West Bengal and Bihar

The northern part of West Bengal and two districts of Bihar has been declared as the Agri-export Zone on Pineapple for Eastern India by the Govt. of India since 2001.

The catchment areas of market of pineapple in West Bengal and Bihar are mentioned below:

West Bengal	Darjeeling	Takdah, Pedory, Thode, Sivok, Mirik, Matigara, Baghdogra, Ragtong, Sukna, Batasi.
	Jalpaiguri	Damdin, Ramshai, Goyerkata, Neora Nadi, Mainaguri.
	Dakshin Dinajpur	Kushmandi, Bansihari, Gangarampur, Balurghat, Hilli, Harsura, Patiram, Badalpur.
	Uttar Dinajpur	Daspara, Ramganj, Dhantola, Panjipara, Kanki, Karandighi
	Coochbehar	Natabari Hat, Jamaldaha Hat, Haldibari, Sitalkuchi,Tufanganj, Bhainkuchi, Dinhata.
Bihar	Purnia	Amour, Qasba, Banmankhi, Barhara, Kothi, Dhamdaha, Bhawanipur, Rupauli.
	Kishenganj	Dighalbank, Bahadurganj, Kochadhamin, Pothia, Thakurganj.
	Saharsa	Mariches, Nauhata, Mahishi, Chapram, Salkhua, Sonbarsa Raj, Sour Bazaar.
	Madhepura	Kumarkhad, Singheshwar, Sankararpur, Murliganj, Alamnagar, Chausa, Puraini

Odisha

Odisha ranks eleventh in fruit production in India with a total production of 2402.3 thousand ton from 340.48 thousand hectare (Table 1). The agro climate of Odisha state is very favourable for different long and short duration fruits. Under long duration fruits Coconut, Cashewnut, Mango, Sapota, Jackfruit, Orange, Lime, Litchi are most important crops. Similarly, fruits like Banana, Papaya and Pineapple are the prominent short duration fruit crops. Actually like Bihar pineapple is a minor fruit crop of Odisha it is cultivated in an area of 970 ha with an average productivity of 11.88 t ha^{-1}(Table 2). Based on soil characterization, rainfall, temperature and terrain, ten main agro-climatic zones in Odisha have been identified. Pineapple is cultivated in only three districts Gajapati, Kandhamal (Zone-NorthEasternGhat) and Khurda (Zone-East & South Eastern Coastal Plain). Farmers of Kandhamal district are cultivating mainlylocal cultivars which have very small fruits but very testy. Production in this areas are mainly done by using organic inputs. Pineapple productivity for this state is comparatively very low (almost 1/3rd) than WB and Bihar. The production is mainly consumed by local market or other cities in the state. Many of the farmers of the state have pineapple plantation for extraction of pineapple leaf fiber.

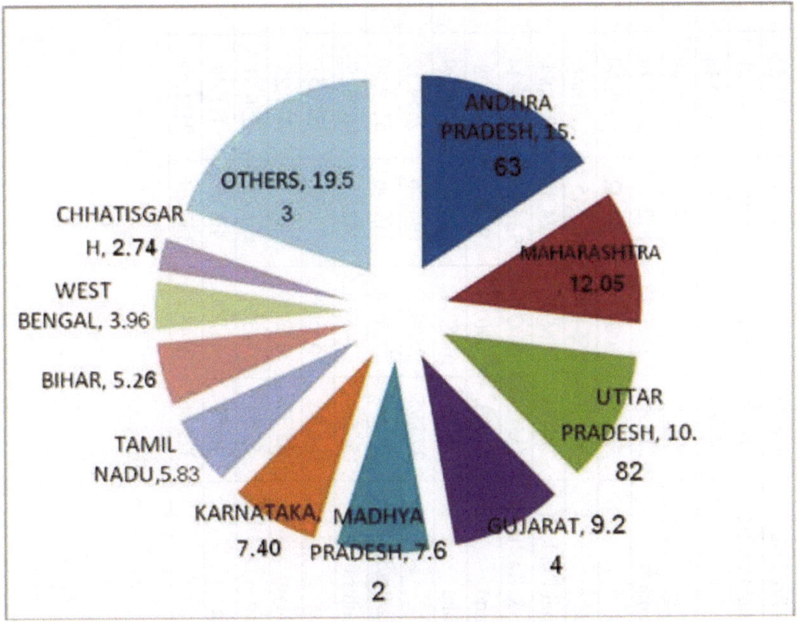

Fig. 1: Production Share of Leading Fruit Producing States: Graph for 2017-18
Source: Horticultural Statistics at a Glance 2018

Table 1: State-wise area and production of fruit crops for the last five years

A: Area in '000 Ha P: Production in '000 MT

State/UT's	2013-14		2014-15		2015-16		2016-17		2017-18	
	A	P	A	P	A	P	A	P	A	P
Andhra Pradesh	640.05	10510.56	545.86	9121.62	577.03	10088.82	609.38	13612.92	650.52	15215.85
Arunachal Pradesh	89.09	321.26	90.00	331.40	66.21	306.27	48.71	124.38	48.13	125.70
Assam	144.68	2007.80	145.21	2030.14	145.71	2077.77	142.89	2024.84	147.26	2123.62
Bihar	302.07	4013.58	301.00	3990.00	306.18	4230.63	307.79	4234.62	296.65	5117.12
Chhattisgarh	212.89	1930.18	192.09	2071.14	209.87	2294.73	220.86	2480.40	229.67	2666.20
Gujarat	370.76	8001.96	384.42	8300.60	392.52	8477.17	411.23	8937.42	422.41	8996.02
Haryana	50.59	554.90	60.45	703.68	60.92	737.82	61.60	770.97	64.02	793.35
Himachal Pradesh	220.71	866.34	224.35	751.94	226.80	928.83	229.20	611.88	230.85	565.31
Jammu & Kashmir	355.21	2073.94	336.45	1779.44	286.23	2115.72	338.82	2241.06	327.43	2355.19
Jharkhand	94.03	890.04	94.14	898.08	96.54	961.19	100.79	1047.97	104.30	1081.69
Karnataka	396.00	6652.42	400.17	6799.89	415.37	7023.69	418.34	7218.38	431.64	7133.94
Kerala	376.95	2889.50	194.66	2554.12	243.15	2532.94	251.35	2509.13	309.45	2045.75
Madhya Pradesh	203.79	5696.00	219.96	6119.00	276.82	5783.06	351.36	6935.60	354.06	7416.91
Maharashtra	1565.00	13457.92	742.28	11089.53	736.69	9749.80	705.12	10630.08	753.18	11728.66
Manipur	54.05	515.69	55.62	521.57	51.12	467.76	50.58	478.77	47.61	455.59
Meghalaya	35.30	348.00	36.33	377.25	36.59	395.40	37.37	426.86	32.81	316.51
Mizoram	57.55	343.90	60.27	350.91	55.01	330.28	62.56	339.05	63.19	340.51
Nagaland	40.56	411.00	40.56	411.00	37.05	374.13	39.19	388.49	39.50	380.52
Odisha	325.86	2148.27	327.29	2156.49	340.80	2386.94	340.97	2432.27	340.48	2402.30

State/UT's	2013-14		2014-15		2015-16		2016-17		2017-18	
	A	**P**	**A**	**P**	**A**	**P**	**A**	**P**	**A**	**P**
Punjab	76.59	1541.24	77.75	1644.64	85.46	1790.94	86.66	1818.19	90.63	1908.85
Rajasthan	37.40	581.78	39.44	735.60	43.27	681.57	57.50	995.60	57.22	761.95
Sikkim	16.02	24.05	0.02	0.03	17.53	23.48	18.55	25.56	19.36	54.90
Tamilnadu	328.55	7369.86	285.69	5963.93	274.85	6635.10	334.74	6181.77	291.37	5680.52
Telangana	364.48	4440.98	361.78	5287.70	331.49	4319.87	166.18	1200.30	167.13	1939.39
Tripura	68.38	786.35	71.77	819.12	75.74	854.05	57.84	559.92	53.75	547.52
Uttar Pradesh	378.97	6887.45	372.34	7558.99	468.89	10296.14	474.87	10302.76	476.64	10539.78
Uttarakhand	171.62	678.49	204.96	785.97	175.33	659.10	177.32	662.85	178.65	669.94
West Bengal	223.50	2909.71	228.25	3313.70	249.22	3516.71	253.02	3585.30	261.02	3850.56
Others	15.69	123.98	16.58	134.23	18.28	143.16	18.61	140.75	17.32	143.36
Total	**7216.31**	**88977.13**	**6109.67**	**86601.68**	**6300.67**	**90183.04**	**6373.39**	**92918.04**	**6506.23**	**97357.51**

Source: Horticultural Statistics at a Glance 2018

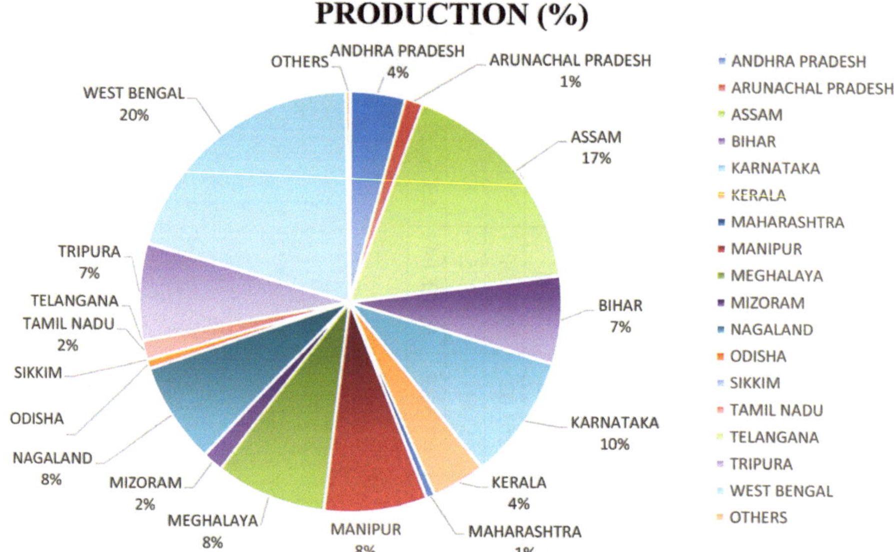

Fig. 2: Production share of pineapple producing states :graph for 2017-18

Table 2: State-wise area and production of pineapple for major producing states of eastern India A: Area in '000 Ha, P: Production in '000 MT

Sl. No.	States	2015-16		2016-17		2017-18	
		A	P	A	P	A	P
1	West Bengal	11.00	330.07	11.20	336.11	11.41	345.15
2	Bihar	4.30	116.30	4.31	116.58	4.26	115.13
3	Odisha	0.95	11.68	0.94	11.55	0.97	11.53

Source: Horticultural Statistics at a Glance 2018

Table 3: District wise area and production of pineapple for major producing states of eastern India A: Area in '000 HaP: Production in '000 MT

States	Districts	2015-16		2016-17	
		A	P	A	P
1. West Bengal	Darjeeling	4.53	141.50	4.56	141.56
	Dinajpur Uttar	3.27	96.79	3.41	102.23
2. Bihar	Kishanganj	2.10	59.60	2.11	59.88
	Purnia	1.70	43.70	1.70	43.70

Source: Horticultural Statistics at a Glance 2018

References

Bal, J. S. 2014. In: Fruit growing. Kalyani Publishers, pp 32-140

Mitra, S. K. and Pathak, P.K. 2008. Tropical and subtropical fruit production in West Bengal, India. Acta Horticulturae, 975: 309-313.

Pandey, S. N. 2005. In: Pineapple leaf fibres. New Delhi (India) ICAR, pp. 5.

Saxena, M. 2018. Horticultural Statistics at a Glance 2018. , Govt. of India Ministry of Agriculture & Farmers Welfare Department of Agriculture, Cooperation & farmers Welfare Horticulture Statistics Division, Government of India Controller of Publication, New Delhi, India, pp. 1-458.

Verma, O.P., Roychowdhury, S., Rautaray, S. K., Raychaudhuri, M., Antony, E., Ambast, S. K. and Brahmanand, P. S. 2019. Fitting Pineapple (Ananas comosus L.) with Rainfed Rice in the Cropping Sequence in Eastern India. National Academy Science Letters, pp.1-4.

6

Status of Pineapple Production in Southern India

Kosgi Mounika[1], Peddaveeri Pravalika Reddy[2] and Kanne Sushmitha[3]

[1]Department of Horticulture, School of Agricultural Sciences and Rural Development, Medziphema, Nagaland University, Nagaland, India
[2]Department of Plantation, Spices, Medicinal and Aromatic crops, Bidhan Chandra KrishiViswavidyalaya, Mohanpur, West Bengal, India
[3]Department of Horticulture and Post-Harvest Technology, Institute of Agriculture, Visva-Bharati, Sriniketan, West Bengal, India

---

Pineapple [*Ananas comosus* (L.) Merr.] is one of the most importanttropical fruit belonging to the family Bromeliaceae and is native to Brazil. It is cultivated in High rainfall and humid coastal regions of peninsular India and Hilly areas of North eastern states. In India, pineapple is commercially cultivated in the states like West Bengal, Assam, Karnataka, Bihar, Tripura, Kerala, Nagaland, Goa, Arunachal Pradesh, Manipur and Meghalaya on large scale whereas on small scale in the states like Gujarat, Andhra Pradesh, Maharashtra, Tamil Nadu, Orissa and Uttar Pradesh etc. It is popularly known as 'Ananas' in India and has lot of health benefits due to the presence of a proteolytic enzyme called bromelin, which is used to tenderize meat (Coppensd' Eeckenbrugge, 2011). It acts as antiamoebic, arbortifacient and also used in correction of stomach disorders (Leal and Coppensd' Eeckenbrugge, 1996). It is consumed both fresh and processed but mostly as fresh (70%) (Loeillet, 1997). Portuguese are the first people to introduce pineapple to India. The greenhouse cultivation of pineapple was first started in 1500s by Europeans, which led to the varieties like Smooth Cayenne and Queen and transported to all tropical and subtropical regions of the world (Collins, 1951). Due to its ease in transport of its propagules and high tolerance to drought, pineapple cultivation has spread worldwide. But due to its short shelf life, in temperate regions it is consumed more as processed food. The first commercial product made out of pineapple is jams and sweets (Thevet, 1957).

Total production of pineapple in the world in 2018 was recorded as 27924.29MT (FAO, 2018). Among other countries India stands in sixth position with 1706.00MT production with global share 6.11%. Costa Rica is the leading producer of pineapple in the world (3418.16 MT) with 12.24% share) followed by Philippines, Brazil, Thailand, Indonesia and India (APEDA, 2018).

Table 1: Production and percent share of pineapple in the world in the year 2018. (Production: '000MT)

S.No	Country	Production	Share (%)
1.	Costa Rica	3418.16	12.24
2.	Philippines	2730.99	9.78
3.	Brazil	2650.48	9.49
4.	Thailand	2113.38	7.57
5.	Indonesia	1805.51	6.47
6.	India	1706.00	6.11
7.	Nigeria	1664.51	5.96
8.	China P Rp	1573.47	5.63
9.	Mexico	999.59	3.58
10.	Colombia	900.39	3.22

Source: APEDA, 2018.

National Scenario

In India pineapple is grown in almost all states except for few. Total area and production under pineapple cultivationis recorded as 87.345 thousand Ha and 1458.121 thousand MT (HAPIS, 2018-19). West Bengal is the leading producer of pineapple in India with 353.215thousand MT production followed by Assam (313.833 thousand MT) (HAPIS, 2018-19).

Table 2: All India area and production of pineapple in 2018-19. (Area: '000Ha, Production: '000MT, Productivity: MT/Ha)

S.No	State/UT'S	Area	Production
1.	Andhra Pradesh	4.551	72.816
2.	Arunachal Pradesh	2.659	16.883
3.	Assam	16.905	313.833
4.	Bihar	4.173	111.828
5.	Goa	0.313	5.238
6.	Kerala	0.094	0.718
7.	Maharashtra	0.400	11.255
8.	Manipur	13.994	131.638
9.	Meghalaya	12.192	138.578

10.	Mizoram	0.005	0.027
11.	Nagaland	8.698	115.086
12.	Odisha	0.336	4.356
13.	Sikkim	0.172	0.862
14.	Tamil Nadu	1.637	41.564
15.	Telangana	0.001	0.040
16.	Tripura	9.628	140.184
17.	West Bengal	11.588	353.215

Source: Hapis Database, 2018-19.

Scenario of Southern States

In South India, pineapple is mostly grown in Andhra Pradesh, Karnataka, Kerala and Tamilnadu and in very negligible amounts in Telangana.

Table 3: State-wise area and production of pineapple in southern states of India. (Area in'000Ha (A), Production (P) in'000MT, Productivity (PR) in MT/Ha)

S. No	States	2015-16			2016-17			2017-18		
		A	P	PR	A	P	PR	A	P	PR
1.	Andhra Pradesh	3.79	55.38	14.60	4.29	64.41	15.0	4.52	71.33	15.78
2.	Karnataka	2.48	155.41	62.74	2.69	164.26	61.06	2.62	163.73	62.42
3.	Tamil Nadu	0.77	26.22	34.00	0.80	26.26	32.95	0.83	25.74	31.12
4.	Telangana	-	-	-	-	-	-	0.00	0.05	50.00
5.	Kerala	10.97	305.67	27.86	17.29	310.32	17.95	8.22	69.72	8.49

Source: Horticulture at a glance, 2018.

Kerala

In Kerala, pineapple is cultivated in an area of around 8.22 thousand ha with 69.72 thousand MT of production. Areahas been decreased drastically from 10.97 thousand Ha to 8.22 thousand Ha, production from 305.67 thousand MT to 69.72 thousand MT and productivity from 27.86 MT/Ha to 8.49 MT/Ha from 2015-2018 (NHB, 2018).It is one of the major South Indian state in pineapple cultivation and is extremely well suited for growing under natural conditions without fertilizers and pesticides. Its cultivation has been taken up fifty years ago in Kerala and the main reason for its adaptation is the congenial humid climate. It is mostly grown as intercrop in rubber plantations which provides additional income and also controls the soil erosion .The finest quality 'Mauritius Pineapple' comes from Kerala. Not only in India but also in other countries, the produce ofKerala has much demand as a fresh fruit because it is considered as the best in quality, sweetness and has good flavour.Almost every district in Kerala grows pineapple but the majority of production comes from Ernakulam district particularly around Vazhakulam. Ernakulam, Kottayam and Idukki districts of Kerala has major area in pineapple cultivation. Keeping in view of drastic increase in pineapple cultivation, Kerala government has implemented many schemes and also supported research and development by establishing the pineapple research station (PRS) one at Vazhakkulam (Ernakulam district) and other at Thrissur under Kerala Agricultural University(KAU). 'Amritha', a hybrid between Kew and Ripley Queen has been released by KAU. Due to the growing demand of export, pineapple growers are slowly shifting to the internationally accepted variety MD-2 which has long shelf life compared to Mauritius variety (10-12 days). MD-2 is best suitable for processing and has shelf life of 30 days.

Vazhakulam Pineapple: In Vazhakulam area pineapple is planted year round except in heavy monsoon days and it is known as one of the biggest pineapple market in India from where fruits are transported to rest of the Indian states. Vazhakulam pineapples belongs to Mauritius variety under Queen group and are locally known as 'Kannara'. It weighs around 1.2-1.5Kg, with deeply placed eyes, conical in shape, unique aroma, sweet in taste and has the capability to withstand long distance transport with 14-16°Brix and acidity of 0.50-0.70%. The other varieties like Kew and Queen are also grown but to a least extent. Kew and Queen are mostly used for processing purpose whereas Mauritius is consumed fresh. Vazhakulam Pineapple got geographical index tag (GI tag) which boosted the export in Kerala nationally and internationally.

Table 4. District wise area and production under pineapple cultivation in Kerala during 2018-19. (Area in '000Ha, Production in'0000MT)

S.No	District	Area	Production
1.	Thiruvananthapuram	0.0942	0.71751
2.	Kollam	0.0952	0.6428
3.	Pathanamthitta	0.1925	0.9594
4.	Alappuzha	0.0577	0.4205
5.	Kottayam	1.5868	15.0285
6.	Idukki	1.3023	13.8087
7.	Ernakulam	5.3756	58.5715
8.	Thrissur	0.0659	0.4731
9.	Palakkad	0.0506	0.2859
10.	Malappuram	0.0397	0.2157
11.	Kozhikode	0.1139	0.7188
12.	Wayanad	0.0238	0.1116
13.	Kannur	0.1003	0.5068
14.	Kasargad	0.0536	0.3716

*Source:*Agriculture statistics, 2018.

From the above table it is evident that the highest area under pineapple cultivationis under Ernakulam district i.e., 5.37 thousand Hawith total production of 58.571 thousand MTfollowed by Kottayam and Idukki.

Tamil Nadu

In Tamil Nadu pineapple is mainly grown in Kolli hills of Nammakal district around 0.748 thousand Ha and in Kanyakumari district (0.106 thousand Ha). The major variety grown is Mauritius due to its taste and local preference and it is mostly cultivated as monocrop. Total area under pineapple cultivation in 2018-19 in Tamil Nadu was recorded as 1.637 thousand Ha and production as 41.564 thousand MT.

Table 5: District wise Area and Production of Tamil Nadu for 2018-19 (Final Estimates)

(Area in'000Ha, Production in; 000MT, Productivity in MT/Ha.)

S.No	District	Area	Production	Productivity
1.	Dindigul	0.001	0.032	32.00
2.	Kanyakumari	0.106	3.685	34.76
3.	Namakkal	0.748	26.019	34.78
4.	Salem	0.002	0.059	29.50
5.	Theni	0.012	0.409	34.08
6.	Thiruchirappalli	0.001	0.022	22.00
7.	Vellore	0.001	0.036	36.00

Source: Hapis Database, 2018-19.

Karnataka

Total area under pineapple cultivation in Karnataka in 2015-16 was recorded as 2.48 thousand Ha which has been increased to 2.69 thousand Ha in 2016-17 and decreased slightly in 2017-18 to 2.62 thousand Ha. Similar data has been recorded in case of production which was like 155.41 thousand MT in 2015-16, 164.26 thousand MT in 2016-17 and 163.73thousand MT in 2017-18 (NHB, 2018). In Karnataka pineapple is grown in districts like Shimoga, Uttara Kannada, Dakshina Kannada, Udipi and Chikmagalur. Major area and production falls under Shimoga district followed by Uttar Kannad districts. Harvesting period of pineapple in Karnataka is done in the months of September-October. The important variety grown in UttaraKannad district is Kew.

Table 6. District wise data of area and production in Karnataka during 2017-18. (Area in '000Ha, Production in'0000MT)

S.No	District	2017-18	
		Area	**Production**
1	Shimoga	1.455	86.750
2	Uttar Kannada	0.421	31.650
3	Chikmagalur	0.023	1.406
4	DakshinKannad	0.380	21.814
5	Hassan	0.020	0.980
6	Kodagu	0.023	1.380
7	Mandya	0.003	0.125
8	Mysore	0.001	-
9	Udipi	0.297	19.623

Source: Hapis Database, 2018-19.

Andhra Pradesh

Pineapple cultivation is constricted to only few districts in Andhra Pradesh, among them Srikakulam is leading in both area (3.205 thousand Ha) and production (51.280 thousand MT). Total area and production under pineapple cultivation in Andhra Pradesh in 2018-19 is reported as 4.519 thousand Ha and 71.327 thousand MT. Simhachalam is the local variety grown in Srikakulam and Araku valley of Vishakhapatnamdistricts. This variety is particularly used inspite of its small size, for its fast growing and easy multiplication of suckers compared to other varieties. Trails have been made to grow other varieties like Kew but failed due to the delay in fruit setting.

Table 7: District wise data of area and production in Andhra Pradesh during 2017-18. (Area in '000Ha, Production in'0000MT)

S.No	District	Area	Production
1.	Srikakulam	3.205	51.280
2.	Vishakhapatnam	1.143	18.288
3.	East Godavari	0.014	0.224
4.	West Godavari	0.187	2.992
5.	Chitoor	0.002	0.032

Source: Hapis Database, 2018-19.

Telangana

Very negligible amount of land is used to grow pineapple in Telangana and it is mostly confined to ashwaraopet area in Khammam District.According to 2018-19 data total area under pineapple cultivation is recorded as 0.001 thousand Ha and 0.040 thousand MT production. Kew is the most preferable variety grown in Khammam district.(NAPIS,2018-19).

Opportunities

- Pineapple cultivation has the potential to improve the livelihood of the rural people through employment opportunities.
- It can provide additional income as an intercrop in rubber and coconut plantations.
- The agro-climatic conditions prevailing in various parts of the country appear to be favourable for bringing more area under pineapple cultivation.
- A good number of improved varieties are now available with higher yield and long shelf life.
- Supply chain of pineapple from production to marketing is developing.
- Pineapple has huge opportunities for export in international market.
- Presence of research centres in Kerala facilitated in release of new varieties and gave scope of improvement work in pineapple.

Limitations

- Lack of knowledge among farmers regarding modern farm management techniques.
- Poor and lack of post-harvest management facilities and machinery.
- Fluctuation in product price due to the lack of availability of fruit year round.
- Poor storage facilities.
- Lack of availability of quality planting material.
- Nil or less knowledge or awareness among farmers regarding the varieties suitable for fresh or processing purpose.
- High cost of cultivation and lack of financial assistance by government in certain states.
- Huge labour requirement.

References

Anonymous. 2018a. Area and Production of Different Fruits in India, National Horticulture Board, India. www.nhb.gov.in

Anonymous. 2018b. Horticulture Statistics at a glance, National Horticulture Board, India, www.nhb.gov.in

Anonymous. 2018c. Agriculture statistics, Department of Economics and Statistics, Government of Kerala.

Anonymous. 2018d. APEDA Agri Exchange database. Agricultural and Processed Food Products Export Development Authority, 2018. http://agriexchange.apeda.gov.in/India

Anonymous. 2019. Area and Production of Different Fruits in India, Horticulture area and production information system, India. Aps.das.gov.in.

Collins, J. L. 1951. Notes on the origin, history, and genetic nature of the Cayenne pineapple. Pacific Science, 5(1): 3–17.

Coppensd'Eeckenbrugge, G., Sanewski, G.M., Smith, M.K., Duval, M.F. and Leal, F. 2011. AnanasIn: Kole, C. (Ed.), Wild Crop Relatives: Genomic and Breeding Resources: Tropical and Subtropical Fruits. Springer Science & Business Media Science, pp. 21-41.

FAO STAT. 2018. Food and Agriculture Organization. FAO STAT database collections, agricultural data, food and agriculture organization of the United Nations. http://www.faostat.fao.org/

Leal, F. and Coppensd'Eeckenbrugge, G. 1996. Pineapple. In: Janick, J. and Moore, J.N. (Eds.), Fruit Breeding. John Wiley & Sons, New York, pp. 515–606.

Loeillet, D. 1997. Panorama du marchémondial de l'ananas: l'importance de l' Europe (The world pineapple market: the importance of Europe). Acta Horticulturae, 425: 37–48.

Thévet, A. 1957. Les Singularités de la France Antarctique. Maspero, Paris.

7

History, Taxonomy and Genetic Resources of Pineapple

Polu Parameshwar[1], V.S.S.V. Prasanna[2], Susan Subba[3] Diya Ghosal[4], Nilesh Bhowmick[5]

[1]Teaching Associate, College of Horticulture, Malyal, Sri Konda Laxman Telengana State Horticultural University, Telengana
[2]Principal Research Engineer, Linkoping University, Sweden
[3]Research Scholar, Department of Pomology & Postharvest Technology Faculty of Horticulture, Uttar Banga Krishi Viswavidyalaya, Pundibari Cooch Behar, West Bengal
[4]Research Scholar, Department of Pomology & Postharvest Technology Faculty of Horticulture, Uttar Banga Krishi Viswavidyalaya, Pundibari Cooch Behar, West Bengal
[5]Department of Pomology & Postharvest Technology, Faculty of Horticulture, Uttar Banga Krishi Viswavidyalaya, Pundibari, Cooch Behar West Bengal, India

History

Pineapple was discovered a little later in time than expected. It had been a staple of the American diet for years before Christopher Columbus, an Italian explorer, discovered it on November 4, 1493. (www.library.ucf.edu; www.levins.com). That's when the Guadeloupe Island natives served him and his colleagues this delectable fruit, pineapple. The varieties cultivated at the time were all seedless, and many typical wild cultivars were missing (Collins, 1948), demonstrating the area's early vegetation. He brought the fruit to Europe. The European travellers were so fascinated by this fruit that they frequently mentioned it in their accounts (Morrison, 1963; Collins, 1960).

It was known as *'nanas'* or *'ananas'* by the people of South America (*'ananas'* meant *'excellent fruit'* in Guarani). People in the area literally cultivated it according to a strict protocol, and they were especially keen to select superior varieties in order to increase fruit yield and quality. They gained a thorough understanding of crop agronomy and production. They also put into practise some of the processing standards that are still in use today around the world (holidayandtraveleurope.blogspot.in). Pineapple wine was made, fibre was extracted, and pharmacological needs such as emmenagogue, abortifacient, antiamoebic, vermifuge, stomach disorders, and poisoning arrow heads were met (Leal and Coppens d'Eeckenbrugge, 1996). In 1980, Antoni and Leal proposed that the pineapple originated in an area further north, between 10° N and S latitudes and 55–75° W longitude, which included parts of north western and eastern Brazil, Columbia, Guyana, and Venezuela. During colonisation, Spanish and Portuguese navigators were instrumental in the distribution and spread of pineapple to all corners of the globe (www.daf.qld.gov.au 2015).

Although some records suggest that pineapples were cultivated near Sydney as early as 1824, the first pineapples (rough-leafed) are believed to have been introduced to Australia from India in 1838 by a German missionary (www.daf.qld.gov.au.,2015). In 1841, an English horticultural journal first mentioned the Cayenne pineapple variety. Long before it reached French Guiana, it is believed to have been cultivated by the Maipure Indians in the upper Orinoco River valley. Its origin, as well as the manner and time of its occurrence, are unknown. Although it arrived in Jamaica via Florida in 1870, the Cayenne variety arrived in 1870.

During the nineteenth century, Australia was largely responsible for the pineapple canneries in Hawaii. France was the centre of European propagation. During the first half of the twentieth century, Hawaii was a significant source of distribution. It was brought into South Africa from Ceylon. Although the first Cayenne peppers arrived in Hawaii from Florida in 1885 and Jamaica in 1886, their growth was relatively rapid in Australia, and they were able to supply planting materials to expand the pineapple industry in the Hawaiian Islands from 1890 to 1895. The Cayenne variety collected in the Hawaiian Islands during the decade from 1885 to 1895. Plants were introduced from 11 tropical countries (Florida, England, Jamaica, Bahamas, Trinidad, Puerto Rico, Mexico, Australia, Singapore, Samoa, and Algeria), four of which were known to contain the Cayenne variety (Collins, 1934).

Some historical artifacts: The European gardeners put in more effort to design a hot house for pineapple development (www.levins.com). In 1530, King Ferdinand of Spain was recorded as consuming pineapple. Pineapple was first discovered in South Africa in 1655. (www.hort.purdue.edu). King Charles II of England posed for an official portrait by Hendrick Danckurts in 1675, in which he is shown receiving a pineapple from his royal gardener, John Rose. Pineapple cultivation was unsuccessful in England until 1712. Greenhouse society flourished in England by the 1700s (www.hort.purdue.edu). Near Leyden, Le Cour established the first pineapple greenhouse. His 'Pineapple horticulture treatise' which included the practise of 'forcing' the plants to flower. This may be the first time an artificial flower has been induced in a pineapple. Captain Cook cultivated pineapple in the South Pacific and adjacent islands in 1777. More Spanish explorers arrived in Hawaii later, planting pineapples and other fruits among other things. In the early 1800s, Francisco de Paula Marin, a Spanish adventurer who arrived in Hawaii in 1794 and became King Kamehameha I the Great's trusted friend and adviser, experimented with growing pineapples. He was well-known for his one-of-a-kind gardens, which he created using seeds from Spain (www.dolefruithawaii.com).As early as 1816, the "Wild Kailua" pineapple was discovered growing in the Kona region (www.dolefruithawaii.com). Lutheran missionaries in Brisbane, Australia, brought plants from India in 1838. (www.hort.purdue.edu). Nundah was the first place in the world to introduce commercial agriculture in 1840. (a part of Brsibane). Perrotet was the first to identify smooth cayenne' from French Guiana in 1819. (Perrotet, 1825). The cultivars 'Smooth cayenne' and 'Queen' were exported from Europe to tropical and sub-tropical regions. They are now shipped from the West Indies to European countries. The first reference in the literature to the Cayenne variety appears to be the short notice carried in the Gardeners' Chronicle (England) of March 6, 1841, under the column heading of "Foreign Correspondence". James Drummond Dole who pioneered the industry and became popularly known as the "Pineapple King" established a company in Hawaii in 1851, the company built its reputation on its dedication to "quality, and quality, and quality" (www.dolefruithawaii.com). James Dole, created the nucleus of what would eventually become the largest pineapple industry in the world (Larsen and Marks, 2010). In 1901, James Dole founded the Hawaiian Pineapple Company (HPC), which is credited with producing the first commercial pack of 1,893 cases of canned pineapple in 1903 (Auchter, 1951). By1900, half a million cases had been shipped (www.hort.purdue.edu). In 1959, the Malaysian Pineapple Industry Board was created. There has been steady improvement since then. Until 1966, when the first major cannery was established, the pineapple was a minor crop in Thailand. In 1996, the trademarked Del Monte Gold 'MD-2' pineapples were formally launched to the US and European markets. The 'MD-2' pineapple was designated the Outstanding Fruit Cultivar of 2010 by the American Society for Horticultural Science (Anonymous, 2010).

Taxonomy

The pineapple belongs to family Bromeliaceae, the genus *Ananus* and the closely related genus *Pseudananas* has no such crown. Both the genera share the basic ch.no X=25. common throughout the Bromeliaceae but *Ananus* is typically diploid (2n=2X=50) while *Pseudananas* is tetraploid (2n=4X=100). Some triplod genotype with 2n=3x=75(e.g. Cayenne BR59,Spanish GU75-2 and DOS indios BR47)have also been reported to exist.

Among the wild species of Ananas, *A.bracteases, A.frizmulleri* and *A. parguazensis* seem closely related both *A. bracteases ,A.frizmulleri* produce fleshy edible (through seedy) fruits.*A.bracteases*, in fact, has been under cultivation in parana-paraguay basin (Collins,1960).

Wild Species

***A. comosus var. bracteatus* (Lindl.)** is a very vigorous and spiny plant, producing a mediumsized fruit with long bracts. It is cultivated as a living fence. Its fruit was also collected for juice and it is still found as a sub-spontaneous plant in ancient settlements of southern region of South America. A variegated variant has become a common ornamental of tropical gardens (Carlier *et al*, 2006). It was cultivated as a living hedge and harvested for fibre and fruit juice, or for traditional medicine, in southern Brazil and Paraguay (Bertoni, 1919). Indeed, its dense, long and wide leaves are strongly armed by large antrorse spines, forming impenetrable barriers. It is very robust and still thrives in abandoned plantations, but it seems unable to colonize new habitats. The syncarp is of intermediate size (0.5–1 kg), borne by a strong scape. The inflorescence has a spectacular appearance.

A. fritzmuelleri shares an additional trait with A. macrodontes, as it exhibits retrorse spines at the leaf base. According to Camargo (1943) and Smith and Downs (1979), it was also used in making fences. It is a very rare form whose diversity has not been documented, only one clone being conserved in Brazil, by EMBRAPA (The Brazilian Agricultural Research Corporation) and the botanical garden of Rio de Janeiro. Nuclear and chloroplast DNA data confirm its closer proximity with *A. macrodontes*. The chromosome number is 2n = 2x = 50 (Camargo, 1943).

A. comosus var. parguazensis (Camargo & L.B. Smith) is another wild form with wider leaves, constricted at their base, antrorse and retrorse spines and a globose fruit. It is very similar to *A. comosus var. ananassoides*, from which it differs by wider leaves, slightly constricted at their base, and larger spines. Its distribution mostly corresponds to the basins of the Orinoco and upper Rio Negro, with a few observations in eastern Colombia and in the north eastern

Amazon (Coppens d'Eeckenbrugge *et al.*, 1997). It grows in lowland forests, under canopies of variable densities, from clearings or river banks to dense forest. It seems restricted to shadier environments, because of lower water use efficiency (Leal and Medina, 1995).

Genetic Resources

Previously, Hume and Miller (1904) classified varieties in Florida into three groups based on general character similarity: the Queen group, which had six varieties, the Red Spanish group, which had seven varieties, and the Cayenne group, which had three varieties. According to Knight (1980) and Samson (1986), all pineapples are divided into five categories based on their unique characteristics.

Different Groups in Pineapple

Smooth Cayenne

Smooth Cayenne pineapple originated before 1820 and was the type indigenous Venezuelans cultivated. The Spanish call it Cayena Lisa. Since then, this type became a leading pineapple in several countries.In most areas, it constitutes a mixture of clones due to new introductions from mutations, lack of rouging and other various sources. Local selections are mostly known by their areas of origin, such as 'Sarawak' in Malaysia, 'Champaka' in India and widely grown in Hawaii. The group is susceptible to mealy bug wilt and nematodes. 'Giant Kew' grown in west Bengal, Goa and Meghalaya states of India belongs to this group. 'Kew' cultivated in the north eastern states and in the southern region of India also is a cayenne member (Paull and Duarte, 2011). Fresh fruit cylindrical shape, shallow eyes, yellow flesh, sweet, mild acidic taste, low fibre, high yield Smooth leaves. These cultivars are grown throughout the world. Some are more widespread in various regions based on disease resistance, flavor or harvest scheduling.

Hilo: It is a sub-variety of Smooth Cayenne, selected in Hawaii by Collins(1960), and its cultivation limited to Hawali only it differs from Smooth Cayenne Its plants are usually smaller, and slips are not produced Shoots are more and develop early .Fruit shape is cylindrical, size is slightly smaller with larger fruitlets. Flesh colour is more yellow having higher percentage of translucence. Fruit deteriorates more rapidly after reaching full ripeness.

Kew: It is a late-maturing variety and is the leading commercial variety in India. Kew is valued particularly for its canning quality. The plants are vigorous and leaves are long with straight margins. The upper surface is dark green with a superficial brownish-red mottling and the lower surface is silvery-grey or ashy-grey in colour Leaves often have a short sector of small spines at the tip and also at the base, near attachment to the stem, where they are irregularly arranged. Fruit weighs 15-25 kg, and is oblong in shape, slightly tapering towards the crown. Eyes are broad and shallow, making fruits more suitable for canning. The fruit is yellow when fully ripe and flesh is light yellow, almost fibreless, and very juicy with 0.6-1.2% acid, and its total soluble solids content varies from 12 to 16°brix. Normally fruit will have one crown but occasionally more are present Slips arising on peduncle are 0 to 10 and the number of suckers produced per plant vary from 0 to 2. This shy-suckering habit is a disadvantage in its multiplication.

Giant Kew: This variety grown in certain regions of India is synonymous to Kew excepting in size of the plant and the fruit, which are larger than Kew as the name signifies.

Charlotte Rothschild: In Kerala and Goa, a variety Charlotte Rothschild is under cultivation, which is similar to Kew in fruit characteristics and taste.

Red Spanish

Red Spanish pineapple is commonly grown as a commercial cultivar in the West Indies, Mexico and Venezuela as well as in Florida. Breeders crossed Red Spanish and Smooth Cayenne pineapples to create modern cultivars that are grouped in the Red Spanish class. The plants are generally small to medium, spiny-leaved, vigorous and resistant to mealy bug wilt, but susceptible to gummosis caused by the larvae of the Batrachedra moth. It is acceptable for the fresh fruit market but not favoured for canning, due to deep eyes and poor flesh colour. 'Red Spanish' or 'Espanola roja' is the major cultivar in the Caribbean region. 'Singapore Spanish', or 'Singapore Canning' and 'Nanas Merah', are the principal canning pineapple in West Malaysia because of their adaptability to peat soil. The flesh has a bright yellow colour. Other Malaysian cultivars are 'Masmerah', a spineless type with large fruit, and 'Nanas Jabor', a Cayenne–Spanish hybrid that is susceptible to fruit marbling and cork spot. 'Cabezona', a natural triploid, is an exception, having large plants and fruit weighing 4.5–6.5 kg. It is grown primarily in the Tabasco State of Mexico and a small area of Puerto Rico where local consumers prefer the larger fruit. The Puerto Rico clone PR 1-67 is suspected to be hybrids between 'Red Spanish' and 'Smooth Cayenne', as these were the only clones grown in adjacent fields. The fruit has light yellow flesh with adequate sugar and resistance to gummosis, is fairly tolerant to mealy bug wilt, and has good slip production and good shipping qualities (Paull and Duarte, 2011).

Singapore Spanish: It is grown in Malaysia for canning industry. This is also known as Singapore Canning and Nenasmerah. The leaves are around

50 in number. They are 100 cm long and are slightly narrower than those of Cayenne. This variety has smooth leaves with a few spines near the tip. The fruit is cylindrical in shape, weighing about 1.6-2.3 kg. The fruitlets range between 70 and 110 and have slightly protruding surface with deep eyes. The ripe fruit is reddish orange and flesh is golden-yellow, fibrous and good flavoured, contributing to the quality of the canned product. Crown is 10-30 cm long, frequently multiple and fasciated.

Masmerah: Grown mainly in Malaysia, this is a selection from Singapore Spanish. The plant resembles a typical Singapore Spanish, excepting that it is more vigorous. The leaves are generally spineless. The longest leaf is about 120 cm long and 6 cm wide at the mid-point. Leaves tend to be more erect than those of Singapore Spanish. Inflorescence bears numerous purple florets which are subtended by red bracts. Fruits are borne on a 30-40 cm long peduncle which is slightly thicker than that Singapore Spanish. The fruit is cylindrical, weighs 15-30 kg, and bars1-12 slips. As Masmerah is closely related to Singapore Spanish, the skin of the fruit is also thick. The flesh is translucent and intensely golden. The TSS content is about 10°brix and acidity is around 15%. The crown is large, 50 cm or longer.

Queen: Queen is the most preferred table purpose variety in India. However, small fruit size, deep set eyes in large numbers and weak fruit stalk make this less preferred in processing industries (Jai Prakash, 2009). This group generally produces smaller plants and fruit with spiny, shorter leaves than the 'Cayenne' group. 'Queen' is grown in South Africa, Australia and India for the fresh fruit market. 'Z-Queen' or 'James Queen' is reported to be a mutant of 'Natal Queen' and is a natural tetraploid. Mauritius of the Queen group is cultivated broadly in Kerala, the southern state of India. It is used as a fresh fruit, for processing and in export (Paull and Duarte, 2011).The Queen or "common rough" pineapple is more compact in size and more tolerant of cold. Cultivars in the Queen class are still grown worldwide. Fresh fruit plants and fruits small, grown in South Africa, Australia, and India. Fruits are conical, sweeter, less acid, low fibre. Spiny, shorter leaves compared to Cayenne group

Mauritius: It is grown in some parts of Meghalaya and Kerala in India. Fruits are of medium size and are of 2 types, deep yellow and red. Fruits of yellow variety are oblong, fibrous, and medium sweet compared to red type. Mauritius is exclusively grown for table purpose. Leaves are yellowish green, spiny throughout the margin. Crown also is spiny in both the types.

Ripley Queen: It is a selection from Queen and is grown in Australia. It has pale-green foliage heavily tinged with red, and fruit is more conical in shape than Queen, with a distinctly flattened top. Fruits have pale-coloured skin and a richer flavour.

Alexandra: This variety is grown in Australia. It is a local selection made in Queensland from the Ripley Queen and is similar to parent variety, but is more vigorous and produces somewhat larger fruits.

MacGregor: It is a local selection at Queensland in Australia from variety Queen. It is sturdier, broad leaved, more vigorous and produces larger and better shaped fruits.

Abacaxi: Abacaxi types are numerous in Brazil, the Bahamas and South Florida.Not suitable for export as fresh or canned. Fruit conical, sweet, tender, Juicy, Spiny leaves, grown in Latin America and in the Caribbean region .This group is grown mostly in Latin America and in the Caribbean region. It was also called as the Pernambuco group (Py *et al.,* 1987). The fruit is not considered suitable for canning or for fresh fruit export, but the juicy, sweet flavour of the fruit is favoured in the local markets. 'Perola', 'Pernambuco', 'Eleuthera' and 'Abacaxi' are the principal clones in Brazil, along the eastern Espirito Santo in the south through Bahia and Pernambucoto Paraibo (Paull and Duarte, 2011) Abacaxi pineapples are often regarded as the most fragrant and delicious of all. They bruise easily and don't make good commercial plants for canning, but are ideal for fresh markets.

Corbezona: This is grown in Puerto Rico for trade of fresh fruits and for shipping to New York .It is seedless, where as Cayenne and Red Spanish produce some seeds when grown together in Puerto Rico.

Maipure: This group is cultivated in Central and South America as fresh fruit for the local markets. Their clones may be of interest to breeders in the western hemisphere as they constitute a gene pool of adapted forms almost unused in breeding programmes (Paull and Duarte, 2011). Fresh fruit cylindrical, sweeter than cayenne, fibrous, tender, juicy, Smooth leaves, cultivated in Central and South America.

Clones: 'Maipure', 'Perolera', 'Lerija', 'Monte Liro', 'Rondon'.

Indigenous Types of India

Jaldhup and Lakhat: These are two local types, both being named after the places of their maximum production. Both fall in Queen group of fruits, being small erthan Queen. Lakhat is markedly sour in taste, whereas Jaldhup has its sweetness well-blended with acidity. Fruits of Jaldhup have a characteristic alcoholic flavour of their own and can be easily distinguished from other fruits of the Queen group.

Simhachalam: In Vishakhapatnam district of Andhra Pradesh, a local variety Simhachalam is grown.

Baruipur Local: It is largely grown in Baruipur, Sonarpur and Joynagar areas of south Bengal. The plants are moderately vigorous. Leaf-margins are heavily serrated. It has heavy slip and suckering habit. Plants are very hardy and can stand adverse climates. Fruits are of small size (weighing between 1 and 2 kg), conical in shape and taper towards the crown. The crown is single and big. Rind colour of the ripe fruit is reddish-yellow. Eyes are prominent, irregular and deep set. Flesh is yellow, fibrous, little stringy and sour to taste. It is neither suitable for canning nor for table purpose.

Haricharanvita: It is grown in some pockets of Siliguri sub-division of Darjeeling district and Sadar sub-division of Cooch Behar district. Its plants are vigorous, leaves are long and slender, and leaf margins are heavily serrated. It has heavy slip and suckering habit. Plants are very hardy and can stand adverse climates. Fruit weight ranges from 0.75 to 2.0 kg. Fruit is conical in shape and tapers towards the crown. The rind of the ripe-fruit is greenish yellow; skin very thick, and eyes are prominent, irregular and deep set. The flesh is yellow, fibrous, and sweet to taste but a little stringy. It has 1 or 2 crowns per fruit which are of big size. This variety is not suitable for commercial production (Chadha *et al.*, 1998)

Hybrids

Amritha: It is a selection from a cross of Kew× Ripley Queen and developed from Kerala Agriculture University, Vazhakulam, Trichur.The leaves are short stiff, spiny along the margins, and thickly covered with a whitish bloom on both surfaces. Fruits are similar to queen and suitable for table purpose. It ripens 10 days late compare to Queen. Fruit weight 0.8-1.1kg. Peduncle is short, fruit lets or eyes are small, prominent, deep set. The total soluble solids content varies from 16 to 18°Brix and acidity between 0.5 and 0.6 percent.

MD-2: The world's pineapple fresh fruit export industry went through a remarkable transformation after Del Monte Corporation introduced 'MD-2'pineapple to consumers in the United States and Europe, officially in 1996 (Frank, 2003). Until the mid1990s, the world's fresh fruit export industry was relatively small and was based mainly on cultivar Smooth Cayenne. Prior to the introduction of 'MD-2', the focus of the pineapple export industry was on canned 'Smooth Cayenne' pineapple. With the major producers focused on 'Smooth Cayenne' pineapple, the unanswered question is what circumstances made it possible for Del Monte Corporation to become the first multinational pineapple company in the world to break the industry fixation on 'Smooth Cayenne' and risk growing a new and untested hybrid. Williams and Fleisch (1993) reported thatPineapple Research Institute (PRI) hybrids had been developed that, when compared with 'Smooth Cayenne' had better resistance to *Phytophthora* rots, pineapple wilt, nematodes, pink disease and internal brown spot. Higher levels of vitamins C and A, reduced acid content in winter ripened fruit, better harvest peaking, higher yield, better cannery recovery from a ton of fruit, faster plant growth a range of distinctive flavours.

Kew | Queen | Mauritius

Baruipur Local | Amritha | MD-21

CO-2: Developed as an inter specific cross between Pineapple Research Institute hybrid clone 58-1184 and Pineapple Research Institute (PRI) hybrid clone 59-443, and has been named hybrid CO-2. This hybrid clone generally resembles the parent varieties, but is distinguished there from in that this plant produces a sweeter, more vitamin C rich fruit, which is low in parasitic fruit diseases, and tolerant or resistant to the physiological disorder called internal browning. This plant has more uniform yellow color, good fruit appearance after refrigerated storage, and produces acceptable crop yields of saleable fresh pineapple (Bartholomew *et al.*, 2010).

References

Anonymous. 2010. 'MD-2' Pineapple named the American Society for Horticulture Science 2010 Outstanding Fruit Cultivar. Pineapple News, No. 17:2.15

As'adah, M., Rahayu, T., Hayati, A. 2016. Metode Pemberian Kolkisin Terhadap Research.

Auchter, E. C. 1951. People, research, and social significance of the pineapple industry of Hawaii. Pineapple Res. Inst. Haw., Honolulu, HI.

Bartholomew, D. P., Coppens de' Eeckenbrugge, G. and Chen, C. C, 2010. Fruit and nut register list No. 45. HortScience, 45: 716–756.

Bertoni, M. S. 1919. Contribution à l'étudebotanique des plantescultivées. I. Essaid 'unemonographie du genre Ananas Anal CientParag (Serie II), 4: 250-322.

Camargo, F. 1943. Vida e utilidade das Bromelia´ceas. Bol Tec InstitutoAgrono´mico do Norte, Belem, Para´, Brazil.

Carlier, J. D., Nacheva, D., Coppens, d' Eeckenbrugge, G. and Leitão, J. M. 2006. Genetic mapping of DNA markers in pineapple. Acta Horticulturae, 702: 79-86.

Chadha, K. L., Reddy, B. M. C., Shikamany, S. D. 1998. Pineapple .Book-Jan-1998

Collins, J. L. 1934. Introduction of pineapple plants into Hawaii and some brief accounts of pioneer pineapple growing. Pineapple Quarterly, 4: 119-120

Collins, J. L. 1948. Pineapples in ancient America. Science Monthly, 65: 372–377.

Collins, J. L. 1960. The pineapple, botany, utilization, cultivation. Leonard Hill Ltd, London, 294

Coppenx D'Eeckenberugge, G., Bernascooni, B., Messiaen, B. and Duval, M. F. 1997. Using incompatability alleles as genetic markers to identify pineapple varieties. Acta Horticulturae, 425:161-169.

Frank, R. 2003. Going for 'The Gold' Turns Pineapple World Upside Down. Wall Street Journal, Oct. 7, p. 1.

Hume, H. H. and Miller. H. K. 1904. Pine apple culture II: Varieties. Bulletin of Fla. Agricultural Exponential Station, 70:37-62.

Jai Prakash , Bhattacharyya, S. And Chattopadhyay, K, Roy, S., Das, S. P. and Singh, N. P. 2009. PQM-1: A newly developed superior clone of pineapple for north eastern India as evident through phenotype, fruit quality and DNA polymorphism. Scientia Horticulturae, 120 :288–291.

Knight, R. Jr. 1980. Origin and world importance of tropical and subtropical fruits.AVI Publishing inc., Westport, Connecticut, USA pp.1-120.

Larsen, J. L. and Marks, T. A. 2010. Hawaiian pineapple entrepreneurs 1894–2010. Creative Co., McMinnville.

Leal, F. and Coppens, d'Eeckenbrugge, G. 1996. Pineapple. In: Janick J and Moore J N (eds) Fruit Breeding. John Wiley & Sons, New York, pp. 515–606.

Leal, F. and Medina, E. 1995. Some wild pineapples in Venezuela. Journal of Bromeliad Society, 45:152–158.

Morrison, S. E. 1963. Journals and Other Documents of the Life and Voyages of Christopher Columbus. Heritage Press, New York

Paull, R. E. and Duarte, O. 2011. Tropical Fruits, CAB International, 2nd Ed., (1). pp. 327-365.

Perrotet, S. 1825. Catalogue raisonné des plantesintroduitesdans les colonies francaises de Mascareigneet de Cayuenne, et de cellesrapports'eesvivantes des mersd'Asie et de la Guyane, au Jardin des Plantes de Paris. Memoires Society Linneas, 3 (3): 89–151

Py, C., Lacoeuilhe, J. J. and Teisson, D. C. 1987. The Pineapple: Cultivation and Uses. G.P. Maisonneuve and Larose, Paris, pp. 127-137.

Samson, J. A. 1986. Tropical fruits,3rd Edn., Longman Londan, pp.190-215.

Smith, L. B. and Downs, R. J. 1979. Bromelioideae (Bromeliaceae) Flora Neotropica. New York Botanical Garden, New York, USA.

Williams, D. D. F. and Fleisch, H. 1993. Historical review of pineapple breeding in Hawaii. Acta Horticulturae, 34:67-76.

8

Reproductive Biology of Pineapple

[1]Santhoshkumar G M, [2]Bidyarani Devi Senjam, [3]Polu Parameshwar, [4]Novin Chamling, [5]Dinabandhu Samanta and [6]Nilesh Bhowmick

[1]J C Diwakar Reddy Horticultural College, Tadipatri, Andhra Pradesh
[2]ICAR- National Research Centre for Orchids, Pakyong, Sikkim
[3]College of Horticulture, Malyal, Sri Konda Laxman Telengana State Horticultural University, Telengana
[4]NICRA, Darjeeling KVK, Uttar Banga Krishi Viswavidyalaya, Kalimpong
[5]Department of Pomology & Postharvest Technology, Faculty of Horticulture, Uttar Banga Krishi Viswavidyalaya, Pundibari, Cooch Behar West Bengal
[6]Department of Pomology & Postharvest Technology, Faculty of Horticulture, Uttar Banga Krishi Viswavidyalaya, Pundibari, Cooch Behar West Bengal, India

Pineapple is a monocarpic perennial herb. Each stem flowers only once and dies after fruiting, similar to bananas, before a side shoot takes over. After the pineapple plant reaches a certain size, natural inflorescence initiation (flowering) occurs in response to decreasing day length and temperature (Bartholomew, 2003). Flowering of Smooth Cayenne pineapples grown under an 8-hour day length was encouraged by a night temperature of 20°C, whereas flowering was delayed by temperatures of 15 and 20°C, and totally inhibited by 30°C.Flower initiation usually starts in November and lasts through the winter. Flowering in nature is highly irregular, and some plants can fail to produce fruit. Thus, growth regulators such as ethylene (25-50 ppm) and NAA (25 ppm) are used in industrial practise to force plants to bloom. The numerous radish-purple sessile flowers are spirally arranged on the axis, each with a pointed bract. To explain the emergence of developing inflorescence, this stage is also known as the red bud stage. The spike-like inflorescence contains 100 to 200 florets. The petals are white at the base and bluish-purple above the calyx at anthesis. They are ligulae form in shape, about 1.6 cm long and 5 mm wide, and are about 1.6 cm long and 5 mm wide (Collins, 1960).The outer flower petals are very close together, allowing only small insects to reach the flower; they have 14 to 20 locules, resulting in a capacity of 6000 to 7000 seeds per 150 floret fruit. After pollination, fruits with 2000 to 3000 seeds are not uncommon (Collins, 1960).

Molecular Mechanisms of Flowering

Flowering is a unique, integrated and very complex process under multi-factorial controls. The eco-physiological to biophysical aspects of flowering have been studied extensively over several decades (Glover, 2007). Flowering occurs as a result of reprogramming the identity of the shoot apical meristem cells (Blázquez *et al.*, 2006). Rather than initiating stems, leaves and axillary buds, a reproductive (or floral) meristem gives rise to an inflorescence, i.e. flowers and their subtending bracts (Öpik and Rolfe, 2005). The reprogrammed identity of the floral meristem is due to signals from flowering pathways that converge to a restricted group of genes named "floral pathway integrators". Physiological studies revealed that the expression of such genes occur in response to both endogenous factors (autonomous and gibberellin pathways) and environmental cues (photoperiod and vernalization) (Amasino and Michaels, 2010; Wigge, 2011). In addition, less well characterized factors such the carbohydrate metabolism (Moghaddan and Van den Ende, 2013), temperature (Blázquez *et al.*, 2003), light quality (Valverde *et al.*, 2004), stress (Wada and Takeno, 2010) and hormonal changes (Davis, 2009) may also play a role in the flowering "decision." Four major floral pathways have been characterized in the intensively studied Arabidopsis thaliana (Kim *et al.*, 2009), which has been used as a model system because of its short generation time, ease of manipulation and a small genome for a complex, multicellular, eukaryote organism.

The autonomous pathway for floral inhibition and induction is central to the entire process of floral transition. Flowering is repressed in Arabidopsis through the expression of the Flowering Locus C (FC) gene and FC relatives in the shoot apical meristem (Simpson, 2004). FC represses the expression of the floral pathway integrator genes Flowering Locus D (FD), Flowering Locus T (FT), and Suppressor of Constans 1 (SOC1) (Kim *et al.*, 2009). The vernalization pathway is found in those plant species where flowering is promoted by prolonged exposure to cold temperatures. This pathway acts by silencing the repressor FC in the shoot apical meristem (Amasino and Michael, 2010; Kim *et al.*, 2009), which leads to the activation of the floral pathway integrators FD, FT and SOC1. In the photoperiod pathway, inductive photoperiods promote the circadian expression of a CONSTANS (CO) protein acting between the circadian clock and the control of flowering time (Suárez-López *et al.*, 2001). CO protein directly activates the expression of the FT gene, which is normally in the leaves. During the last decade several FT gene homologs were identified in different plant species and the protein produced by this gene family (FT protein) was shown to be a floral-promoting signal that can be transported from the leaves to the apical meristem (Xu *et al.*, 2012). In the meristem, FT and FD proteins interact and initiate flowering by activating SOC1, which integrates the photoperiod, temperature, hormone (gibberellin pathway), and autonomous (agi-related) signals into a unitary signal at the shoot apical meristem (Lee and Lee, 2010). SOC1 regulates LEAFY (LFY), which controls the expression of a wide variety of genes required for organ development (Amasino and Michaels, 2010; Blázquez *et al.*, 2006; Kim *et al.*, 2009; Siriwardana and Lamb, 2012).

Plant growth regulators and other substances interact with various genes to promote or inhibit flowering. In species where gibberellins (GAs) promote flowering, they do so by directly activating SOC1 and such floral meristem-identity genes as LFY (Mutasa-Göttgens and Hedden, 2009; Wigge, 2011). Nitric oxide (NO) promotes the vegetative phase by suppressing genes related to the photoperiod pathway and enhancing FC expression through the autonomous pathway (He *et al.*, 2004; Seligman *et al.*, 2008). Ethylene biosynthesis in response to environmental stress delayed flowering in Arabidopsis by repressing GA synthesis (Achard *et al.*, 2007). However, in the Bromeliaceae family, ethylene biosynthesis, as well as treatment with ethylene, induces flowering (Rodriguez, 1932; Kerns, 1968; De Proft *et al.*, 1984), but no studies of the mechanism of action were found. The silencing of ACC synthase, an enzyme responsible for ethylene biosynthesis, delays floral initiation in *Ananas comosus* (Trusov and Botella, 2006).

Other hormones such as abscisic acid, jasmonic acid, salicylic acid and the cell division regulators auxins, cytokinins and brassinosteroids were also shown to be players in the floral transition. Over the past 15 years, tremendous advances in the knowledge of flowering pathways revealed that several critical events need to occur under certain environmental and endogenous conditions before plant reproduction takes place. Whereas the picture in Arabidopsis thaliana is close to complete, much less is known about the genetics of flowering in other plant families. For instance, there are some striking differences between the circuitry and components of vernalization in Arabidopsis and cereals while the photoperiod pathway is much more conserved within plant families (Higgins *et al.*, 2010; Kim *et al.*, 2009) Also, comparative genomics of flowering time pathways between monocots (e.g. *Oryza* sp., *Brachypodium* sp.) and dicots (e.g.

Arabidopsis sp.) suggest that it will be possible to integrate knowledge of gene function into a widely accepted model (Higgins *et al.,* 2010). Nevertheless, studies on the manipulation of flowering related genes to improve crops are appearing worldwide. For instance, photoperiod-insensitive cultivars are being produced to improve cereal production (Tsuji *et al.*, 2011) whilst others attempt to reduce natural flowering induced by environmental cues to schedule and synchronize fruiting by silencing a specific ACC synthase gene (Trusov and Botella, 2006). Recently, Flachowsky and Hanke (2011) produced an early flowering transgenic apple tree, for breeding purposes, using a FUL-like gene from silver birch to reduce the long vegetative phase. Improved knowledge of flowering pathways can have an important effect on agronomical practices in the future, by introducing crops with synchronized flowering mechanisms adapted to different environments in order to achieve maximum efficiency in crop planning and labor costs and to maintain a scheduled crop to be marketed according to the consumer's demands.

The aims were to determine (a) whether differences in pineapple fruit quality among individual fruits are associated with differences in vigour of the individual plants within the crop at the time of artificial flower induction; and (b) whether the side shoots produced by the plant during the generative phase account for the fruit quality heterogeneity. Two pineapple cultivars were considered: cv. Sugarloaf and cv. Smooth Cayenne. Plant vigour at the time of artificial flower induction was measured by three variates: the number of functional leaves, the D-leaf length and their cross product. Fruit quality attributes measured at harvest time included external attributes (weight and height of fruit, infructescence and crown) and internal quality attributes [total soluble solids (TSS), pH, translucent flesh]. Results showed that the heterogeneity in fruit weight was a consequence of the heterogeneity in vigour of the plants at the moment of flower induction; that effect was mainly on the infructescence weight and less or not on the crown weight. The associations between plant vigour varieties at flower induction and the internal quality attributes of the fruit were poor and/or not consistent across experiments. The weight of the slips (side shoots) explained part of the heterogeneity in fruit weight, infructescence weight and fruit height in cv. Sugarloaf. Possibilities for reducing the variation in fruit quality by precise cultural practices are discussed.

Heterozygosity

Most domesticated cultivars of the pineapple (*Ananas comosus*) are considered highly heterozygous. Some pineapple breeders have associated slow progress with this heterozygosity (Williams and Fleisch, 1993; Chan *et al.*, 2006).

Heterozygosity is the breeder's source of diversity and it is essential for crop improvement. However, a high level of heterozygosity will also mean it is necessary to develop large populations of seedlings and apply selection pressure for a large number of traits. Heterozygosity is naturally maintained in pineapple through a strong incompatibility.

Controlled Pollination

Pineapple flowers bloom early in the morning (6-9 a.m.) and fade by evening. Anthesis occurs in a sequential order, beginning with the lower flower at the inflorescence's base. Every day, five to ten flowers open. Flowering progresses up the inflorescence axis over the course of 2 to 3 weeks, eventually completing the process. Almost all varieties contain viable pollen, but they cannot be incompatible with one another. As a result, after self-fertilization, they do not produce seed. Cross pollination between cultivars, or between cultivars and wild forms, leads to natural fertilisation and abundant seed development. In South America, hummingbirds are the primary pollinators. However, since they aren't found in other parts of the world where pineapple is cultivated, pineapple fruits are normally parthenocarpic and seedless.

Incompatibility and Seed Setting

The only self-incompatible species in the genus is *Ananas comosus. Pseudananas*, like other wild Ananas species, are partially self-fertile. Brewbaka and Gorrez (1967) discovered that the pineapple has a gametophytic S-allele form of incompatibility that causes polyploidy without breaking down the incompatibility framework (Pickersgill, 1976). Several clones from the Red Spanish and Singapore Spanish varieties were found to be incompatible by Coppens D'Eeckenbrugge *et al.* (1997). In addition, incompatibility is common in pernambuco clones, confirming the group's heterogeneity. Despite its morphological resemblance to the yellow-fleshed Pernambuco clones, the incompatible Pernambuco genotypes were cross compatible with Queen, suggesting that they were genetically distinct from the other members of the same clone (Coppens D'Eeckenbrugge *et al.,* 1997). Smooth Cayenne has a good self-incompatibility response as well. The fruit of the pineapple develops parthenocarpically. It's unclear if pollination is needed to start the development process. Pollen tube growth stops before reaching the ovary, preventing self pollination. Fruits with seeds are rarely picked, but it's unclear if they're the product of insect cross-pollination (hummingbirds, honeybees, and pineapple beetles) or an environmentally mediated physiology shift that breaks the usual barrier to self-pollination. Fruits can produce seeds as a result of natural cross-pollination between clones. Self-compatible *Ananas comosus* mutants also grow fruits with up to 3000 very hard seeds.

References

Achard. P., Baghour, M., Chapple, A., Hedden, P., Van Der Straeten. D., Genschik, P., Moritz, T. and Harberd, N. P. 2007. The plant stress hormone ethylene controls floral transition via DELLA-dependent regulation of floral meristem-identity gene. Proceedings of National Academy of Science USA, 104: 6484-6489.

Amasino, R. and Michael, S. 2010. The timing of flowering. Plant Physiology, 164: 516-520.

As'adah, M., Rahayu, T., Hayati, A. 2016. Metode Pemberian Kolkisin Terhadap Res

Blázquez, M. A., Ahn, J. H. and Weigel, D. 2003. A thermosensory pathway controlling flowering time in Arabidopsis thaliana. Nature Genetics, 33: 168-171.

Blázquez, M. A., Ferrándiz, C., Madueño, F. and Parcy, F. 2006. How floral meristem are built. Plant Molecular Biology, 60: 855-870.

Brewbaker, J. L. and Gorrez, D. D. 1967. Genetics of self-incompatibility in the monocotgenera. Ananas (pineapple) and Gasteria. American Journal of Botany, 54: 611-616.

Chan, Y. K. 2006. Hybridization and selection in pineapple improvement: the experience in Malaysia.: V International Pineapple Symposium, Acta Horticulturae, 702.

Collins, J. L. 1960. The pineapple, botany, utilization, cultivation. Leonard Hill Ltd, London, pp. 294.

Coppenx D'Eeckenberugge, G., Bernascooni, B., Messiaen, B. and Duval, M. F. 1997. Using incompatability alleles as genetic markers to identify pineapple varieties. Acta Horticulturae, 425:161-169.

Davis, S. 2009. Integrating hormones into the floral-transition pathway of Arabidopsis thaliana. Plant, Cell and Environment, 32: 1021-1210.

De Proft, M., Jacobs, L. and De Greef, J. 1984. Endogenous ethylene production and flowering of Bromeliaceae. Ad. Agricultural Biotechnology, 9: 165-166.

Flachowsky, H. and Hanke, M. 2011. Genetic control of flower development in apple and the utilization of transgenic early flowering apple plants in breeding. I Workshop on Floral Biology and S-Incompatibility in Fruit Species.

He, Y., Tang Hao Y., Stevens, R. D., Cook, W. C., Ahn, S. M., Jing, L., Yang, Z., Chen, L., Guo, F., Fiorani, F., Jackson, R. B., Crawford, N. M. and Pei, Z. M. 2004. Nitric oxide represses the Arabidopsis floral transition. Science, 305: 1968-1971.

Kim, D. H., Doyle, M. R., Sung, S. and Amasino, R. M. 2009. Vernalization: winter and the timing of flowering in plants. Annalytical Revews on Cell and Developmental Biology, 25: 277–299.

Lee, J. and Lee, I. 2010. Regulation and function of SOC1, a flowering pathway integrator. Journal of Experimental Botany, 61: 2247- 2254.

Moghaddam, M. and Van den Ende, E. 2013. Sugars, the clock and transition to flowering. Frontiers in Plant Sciences, 4: 1-6.

Mutasa-Göttgens, E. And Hedden, P. 2009. Gibberellin as a factor in floral regulatory networks. Journal of Experimental Botany, 60: 1979– 1989.

Pickersgill, B. 1976. Pineapple (In) Evolution of crop plants (Simmonds 0 Longman,) London, pp. 14–18.

Rodriquez, A. G. 1932. Influence of smoke and ethylene on the fruiting of pineapple (Ananas sativus) Schult. Journal of Department of Agriculture, 16: 5-18.

Seligman, K., Saviani, E., Oliveira, H., Pinto-Maglio, C. and Salgado, I. 2008. Floral transition and nitric oxide emission during flower development in Arabidopsis thaliana is affected in nitrate reductase-deficient plants. Plant Cell Physiology, 49: 1112-1121.

Siriwardana, N. and Lamb, R. 2012. The poetry of reproduction: the role of LEAFY in Arabidopsis thaliana flower formation. International Journal of Developmental Biology, 56: 207-221.

Suárez-López, P., Wheatley, K., Robson, F., Onouchi, H., Valverde, F. and Coupland, G. 2001. CONSTANS mediates between the circadian clock and the control of flowering in Arabidopsis. Nature, 26: 1116-1120.

Trusov, Y. and Botella, J. R. 2006. Silencing of the ACC synthase gene AcACS2 causes delayed flowering in pineapple [Ananas comosus (L.) Merr.]. Journal of Experimental Botany, 57: 3953–3960.

Trusov, Y. and Botella, J. R. 2006 Delayed flowering in pineapples [Ananas comosus (L.) Merr.] caused by co-suppression of the acacs2 gene. Acta Horticulturae, 702: 29–36.

Tsuji, H., Taoka, K., Shimamoto, K. 2011. Regulation of flowering in rice: two florigen genes, a complex gene network, and natural variation. Current Opinion on Plant Biology, 14: 45-52.

Valverde, F., Mouradov, A., Soppe, W., Ravenscroft, D., Samach, A. and Coupland, G. 2004. Photoreceptor regulation of CONSTANS protein in photoperiodic flowering. Science, 303: 1003-1006.

Wada, K. and Takeno, K. 2010. Stress-induced flowering. Plant Signal Behaviour, 5: 944-947.

Wigge, P. 2011. FT, a mobile developmental signal in plants. Current Biology. 21: 374-378.

Williams, D. D. F. and Fleisch, H. 1993. Historical review of pineapple breeding in Hawaii. Acta Horticulturae, 334:67-76.

Xu, F., Rong, X., Huang, X. and Cheng, S. 2012. Recent advances of Flowering Locus T gene in higher plants. International Journal of Molecular Sciences, 13: 3773-3781.

9

Recent Advances in Pineapple Improvement

[1]Nilesh Bhowmick, [2]Santhoshkumar G.M. and [3]Dinabandhu Samanta

[1]Department of Pomology & Postharvest Technology, Faculty of Horticulture, Uttar Banga Krishi Viswavidyalaya, Pundibari, Cooch Behar West Bengal, India
[2]Department of Fruit Science, J C Diwakar Reddy Horticultural College Tadipatri, Andhra Pradesh
[3]Department of Pomology & Postharvest Technology, Faculty of Horticulture, Uttar Banga Krishi Viswavidyalaya, Pundibari Cooch Behar, West Bengal

The pineapple, botanically known as *Ananas comosus* Merr, is the most famous edible member of the Bromeliaceae family. It is native to southern Brazil and is known by a variety of names, including "pina" in Spanish, "abacaxi" in Portuguese, "ananas" in Dutch and French, "nanas" in Southern Asia and the East Indies, "po-lo-mah" in China, "sweet pine" in Jamaica, and merely "pine" in Guatemala (Morton, 1987). After banana and mango, the pineapple is the third most important tropical fruit of the world (Chen *et al.*, 2019; Nath *et al.*, 2019). The pineapple became a major cultural icon of luxury after its introduction to Europe in 17th century. Pineapple has been grown commercially in greenhouses and on many tropical plantations since the 1820s. Furthermore, it is the world's third-largest tropical fruit producer. Hawaii was a major pineapple producer in the twentieth century, with Costa Rica, Brazil, and the Philippines accounting for nearly a third of global pineapple production. Pineapples grow as a small shrub, with the un-pollinated plant's individual flowers fusing to form multiple fruits. The plant is usually propagated from a side shoot or an offset produced at the top of the fruit and it matures in about a year.

Crop Improvement

Pineapple is usually propagated through asexual method by using slips, crowns or suckers. Sexual reproduction in pineapple is barely used due to its self-sterile nature and failure in formation of seeds. Seed set can occur in case of self-fertilization but the seedlings are with low vigour due to inbreeding depression (Robert *et al.,* 2003). Pineapple is heterozygous in nature, on this wise hybridization is possible in this crop. Gametophytic self incompatibility is reported in this crop where pollen tube growth is repressed in the uppermost part of the style. Diploid chromosome number of the crop is 2n=50. The genome size of the crop is estimated to be 526Mb and total genes were anticipated to be around 25,862 in number (Zhang *et al.*, 2014). Conventional breeding in pineapple is not at its advantage to improve the fruit as it takes nearly 2 years from one generation to successful one. Many cultivars in pineapple are developed through spontaneous mutations or naturally evolved without involvement of conventional breeding strategies. Though many interspecific and intergeneric crosses were carried out in the crop, the results in the improvement of the crop were limited due to its heterozygous nature and strong segregation and recombination which restricted the scope of hybrid breeding.

In vitro propagation technique is one of potential methods for improvement of the crop which was first developed by Wang during 1977. Explants such as shoot base, leaf base, leaf segments, excised leaf buds, embryogenic tissues were used on MS medium in inculcating traits such as herbicide resistance, wilt disease resistance, mealy bug resistance and in controlling the flowering. Shoot tips are obtained *in vitro* can be used for mutation experiments by using suitable mutagens which can also give superior traits like spineless leaves, disease resistance and improved fruit quality. Also, soma clonal variations that arise during in vitro culturing of the plants can lead to development of elite cultivars. Guillermo *et al.* (2011) revealed that somaclonal variants have desired attributes such as decreased transpiration rate, increased leaf photosynthetic parenchyma compared with donor plant. During past few years specific horticultural traits in the pineapple were incorporated using transgenic approach without altering other agronomic important traits (Smith *et al.*, 2002). Transformation of the pineapple can be achieved by different methods such as co culture with *Agrobacterium tumefaciens* as an indirect method of gene transfer or direct method such as by micro projectile bombardment technique.

Breeding Objectives

1. High yield with early maturity and early ripening
2. Adapted to local environmental condition
3. Hardy vigorous plant with large and vigorous root system
4. Spineless leaves for easy of handling
5. Short and strong fruit stalk to keep to keep fruit erect and minimize sun-sorch
6. Absence of collar of slips
7. Absence of multiple crowns and faciation
8. Fruits for canning as nearly cylindrical as possible
9. Flesh firm with few fiber, crisp, juicy of good flavor and aroma and deep yellow colour
10. Early formation of one is two shoot and presence of one or two slips at least 2 cm below the fruit base
11. Small ratio of leaf to fruit
12. Short leaf length and narrow angle of leaf emergence
13. Resistant to pest and diseases (Mealy bug, Nematode and Heart rot and Root rot (Kumar, 2006)

Inheritance Pattern

There are three types of leaves in pineapple viz. :(i) leaves with spiny tips, (ii) spiny leaves with spines along the entire leaf margin, and (iii) the 'piping' type, which is a fully spineless leaf-form. A portion of the lower epidermis, with its trichomes, stomata, and corrugated surface, is hayed over the edge of the leaf and extends onto the upper leaf surface in the piping form. This results in a thin stripe or piping on the upper surface of the leaf that is silvery-white on the lower surface. The phenotypic expressions of a single pair of alleles (S = spiny tips and s = spiny leaves), with spiny tips being dominant, are the spiny tip and spiny leaf characters. Spiny tips are produced by both heterozygous Ss and homozygous SS genotypes. Spiny plants are produced by recessive ss genotypes in progeny (Collins, 1960). Another non-linked pair of alleles controls the piping and non-piping characters, with the gene P (piping) being epistatic to S and s. In the presence of S or s genes, the homozygous PP genotype produces more pronounced piping than done the Pp has no apparent phenotype, so such genotypes, whether they exist, cannot be differentiated phenotypically from spiny tips or spiny plants (Collins, 1960). Some cultivars can also display a condition called ‘closed flower’. While this is uncommon it can occur to varying degrees and appears to be modified by the environment. An application of 10 ml of 400ppm gibberellic acid (GA) just before the first petals open can overcome closed flower (Williams and Yamane, 1970). Germination success and vigour of seedlings can be expected to be reduced in seedlings with more closely related parents. Usually some screening of the seedlings approximately 6 weeks after germination is necessary to remove weaker plants, of which up to 40% of seedlings can be discarded at this time.

Selection Criteria

1. Spineless leaves
2. High slip production
3. Good sucker production
4. Upright sucker types
5. Disease tolerance
6. High yellow pigment in the winter
7. Square shoulder rather that pyramidal shape to increase slice production in cannery
8. Single crown
9. Small core (core is an extension of peduncle)
10. Medium sized eye
11. Uniformity of ripening in each fruit
12. High soluble solids or sugar content and good sugar acid ratio
13. Yellow flesh for fresh fruit market

Breeding Methods

Clonal Selection

The majority of pineapple improvement has resulted via simple selection of mutant clones inside established cultivars and hybridization between cultivars followed by selection of strongly heterozygous progeny. Selection in the Singapur Spanish population in Malaysia has led to a new cultivar, “Masmerah” is more vigorous, has more leaves, which stand more erect and bears heavier fruit than the parent cultivar (Wee, 1974). Several such selections have been made in different pineapple growing areas. Local selections are mostly known by their areas of origin, such as 'Sarawak' in Malaysia. Champaka is a selection of 'Smooth Cayeme' originating in India and widely grown in Hawaii. The 'Z-Queen' or 'James Queen' is reported to be a mutant or 'Netal Queen' and is a natural tetraploid

Many commercial pineapple cultivars are improved through selecting genetic variants that occur occasionally in experimental plots or production fields. Such variants are perpetuated as clones whose genetic make-up is identical and 'fixed' by vegetative propagation until new variation arises. Clonal planting form the basis of modern pineapple culture, and many different clones are used throughout the world. All 'Smooth Cayenne' clones in Hawaii, including 'F200' and the 'Champaka' clones were developed through clonal selection (Py et al., 1987). In Australia, clonal selection began in 1950 with the selection of 100 plants from commercial fields. Four clones viz. 'C8', 'C10', 'C13' and 'C30' were eventually released to the industry in 1975 and, in conjunction with the Hawaiian clone 'Champaka F180', form the basis of the present-day industry. Several clones of the cultivar 'Queen' have also been established in Australia, including 'Ripley Queen', 'Alexander' and 'McGregor', through a process of clonal selection by private growers. In Malaysia, an improved cultivar, 'Masmerah', with higher vigour and yield, was obtained by clonal selection from a commercial field of 'Singapore Spanish' (Wee, 1974). Clonal selection has also been used in the Philippines to develop a new clone of 'Smooth Cayenne' less prone to the green-ripe disorder.

Clonal selection is tedious and rigorous work. It has been the object of constant efforts by modern pine growers. As a result, 'Smooth Cayenne' potential has been maintained at its best. The importance of clonal selection to avoid the accumulation of negative quantitative and qualitative variation is illustrated by the comparison of different strains of the 'Smooth Cayenne' (Benega et al., 1995). But clonal selection is not a creative method. Old 'Smooth Cayenne' clones in French Guyana are not very different from the most modern clones. Most mutations are deleterious and no significant positive new traits have appeared, with the possible exception of wilt resistance selected by our Mexican colleagues (Torres et al., 1989). Clonal selection has generally been neglected in other varieties, with the exceptions of 'Queen' in Australia and South Africa and 'Espanola Roja' in Cuba and Puerto Rico. In most places, heterogeneity is the rule. Many cultivated clones of 'Espanola Roja' and 'Singapore Spanish' have reverted to the spiny condition and even such important varieties as 'Singapore Spanish', 'Perola' and 'Perolera', although well adapted to the conditions of important local markets and/or zones of production, suffer from strong defects as a proliferation of crowns and slips. If we want to maintain a minimal genetic diversity in this crop, we should select performing clones in varieties of regional importance. What is needed for efficient clonal selection is a better knowledge on the selection parameters (importance of somaclonal variation, efficiency of clonal selection) and fast multiplication methods. On the first of these themes, published data are very poor despite the constant efforts of growers and breeders. The second theme was the subject of the session devoted to biotechnology. Micro-multiplication and acclimatization techniques are commonly and satisfactorily used in pineapple now. Most of the work presented at the symposium was oriented to optimize efficiency and reduce production costs (Almeida et al., 1995; Daquinta et al., 1995; Dominquez et al., 1995; Firoozabady et al., 1995; Ventura et al., 1995), and to improve and complement acclimatization through inoculation with endo-mycorrhizal fungi or Azotobacter (Gonzalez et al, 1995). Somaclonal variation is generally limited. With standard in vitro techniques, its importance is probably more related to the multiplication rate per se than to the particular physiology of the plants in the growth chambers. However there remain some risks that the progress in homogeneity, quality and quantity obtained through the selection of specific subclones may be partially lost in the micropropagation process.

PQM-1: Queen is the most preferred table purpose variety in India. However, small fruit size, deep set eyes in large numbers and weak fruit stalk make this less preferred in processing industries. A completely different plant type with distinct fruit character was identified from a clonal population of Queen which seems to be a natural mutant. The identified mutant was maintained and multiplied through vegetative propagation at Lembucherra farm of ICAR Research Complex, Tripura centre. This new clone (PQM-1) is distinct from 'Queen' and other cultivars. Distinctness of this new clone was confirmed on the basis of standard morphological and biochemical descriptors as well as through DNA polymorphism. PQM-1 was found uniform at DNA level and complement the information of phenotypic and quality parameters by showing stable differences with Queen in ISSR banding pattern. Natural mutation in DNA level might be the probable cause of its origin due to its late maturity, fruit quality and usefulness in both canning and fresh consumption. This certainly would emerge as new alternative to pineapple growers of this region. The cylindrical shape of the PQM-1 was also found suitable for table purpose and in canning industry. Yellowish golden flesh colour of the fruits of PQM-1 might be preferred over 'Queen' by the canning industry to prepare slices.TSS is the most important quality trait of pineapple. Total soluble solids of PQM-1 were comparatively higher (17.5°Brix) than Mauritius and Kew but less than

Queen. However, PQM-1 might have been preferred than Queen for its late maturity and its usefulness in both canning and fresh consumption. Almost, similar trend was noticed in average sugar percentage where PQM-1 (12.96) exceeded Kew but could not supersede Queen. The average acidity was higher than Kew and slightly lesser than Queen and Mauritius (Jai Prakash , 2009)

Yeppoon Gold: Yeppoon Gold is a clone of 'Smooth Cayenne'. It can produce marketable fruit during winter, and its fruits are sweeter and less fibrous than those of other cultivars. Because of these features, it is anticipated that the demand for its propagating material will increase over the next five years.

Selfing

Fifty percent of alleles at each locus will be homozygous with each cycle of selfing (Allard, 1960). Selfing of 'Smooth Cayenne' was investigated by the PRI in the late 1930s to early 1940s to identify desirable recessive alleles. Seedy clones of 'Smooth Cayenne' were first identified and then taken through 2 cycles of selfing by which time the resultant plants were too weak to develop an F_3 population (Collins, 1960). Despite considerable effort, no desirable types were ever produced. More recently, selfing has also been proposed as a strategy to develop more homozygous cultivars for use as parents (Dujardin, 1991). However, more than one cycle of selfing has the potential for losing some favourable alleles. One cycle of selfing is likely to have some advantage in a simple recurrent selection program when used in combination with out-crossing to other foundation cultivars. To date, selfing has only been used on a minor scale to develop parental lines with homozygous alleles for Fusarium spp. resistance and piping leaves (Cabral and Rheinhardt, 2004). To date although selfing has only demonstrated limited success, it deserves further investigation because of its promising potential benefits. The immediate bottle-neck is the genetic incompatibility that exists in most domesticated pineapple. Attempts in Australia to produce seed by selfing of late generation cultivars resulted in few selfed seed that were usually weak and even with in vitro techniques, few germinated (Sanewski, 2009). Of the few selfed lines that fruited, none were good enough to be useful as parents. To obtain useful parents requires the production of a reasonable number of seedlings to allow screening for recessive undesirable alleles. Thus, methods to overcome self-incompatibility are essential. The application of 1% NAA (Bhomik, 1980) and irradiation of pollen (Marr, 1964; Subramanian *et al.*, 1981) have produced limited success and might be worth revisiting. The cut-style technique of pollination has also been unsuccessful. Anecdotal evidence suggests self-incompatibility in pineapple might be at least partially overcome by high temperature (Kerns *et al.,* 1968; Williams, 1970) as it is in some other species including alfalfa, tulip, beet, Easter cactus (Boyle *et al.*, 1994), rye and chrysanthemum. This has yet to be tested experimentally in pineapple.

Hybridization

Most of the improvement of the pineapple has taken place by simple selection ofmutant clones within existing cultivars and by hybridization between cultivars followed by selection of the highly, heterozygous progeny. The development of a new hybrid through hybridization is a slow process. AF_1 plant takes four years to produce fruit and, in the case of an inter-specific cross, four backcross generations will probably be necessary to produce a plant suitable for commercial use. Thus a new cultivar may be developed 25 years after the initial cross and then needs some more time to be multiplied and distributed (Pickersgill, 1976). In spite of being a long term venture, hybridization in pineapple has resulted in several promising hybrids (Cabot,1987; Fitchet, 1989; Loison-Cobot, 1990b, Chang, 1995).

- A cross between Red Spanisbánd Cayenne has led to the development of two new hybrids PR-1-67 and PR-1-56 in PuertoRico (Ramirez, 1970). This hybrid shows better plant vigour and resistance to wilt disease. However, self-fertile somatic mutants obtained from Cv. Cayenne show a loss of vigour on selfing and heterosis on crossing.
- A hybrid "Tainung13 has been obtained from a cross between Smooth Cayenne x Tainung 8(Chang, 1995). Its average plant height is 127.5 cm, leaf length 104 cm, leaf width 5.1 cm and leaf margin spineless (base and tip have small spines). It bears cone shaped fruits, average weight 1.2 kg, flesh-firm, green yellow and slightly fibrous. It also has some degree of wilt tolerance in the field.
- Gopimony and Balakrishnan (1984) have tried to combine the large fruit size of Kew and good dessert quality of the Ripley Queen. Among the 16 hybrids obtained one (Hybrid-8) appeared to bear the desired fruit qualities.
- Sanewski and Giles (1997) have developed a hybrid resistant to blackheart. This intergroup hybrid (bearing No.53-116) was without blackheart after storage at 10°C for 14 days followed by20°C for 8 days.
- In Taiwen two dwarf and high yielding hybrids namely Typhone1 and Typhone 3 have been released. The fruit shape is cylindrical in Typhone 3.

Evaluation of 19 hybrids from the cross Cayenne x Perolera has led to selection of 3 promising clones. Clone 13 (C-13) has cylindrical fruits, a nice orange colour, small crown, normal fruit firmness and medium ascorbic acid content. C-15 has large fruits, brilliant orange colour and a good ascorbic acid content, while C-5 has a small crown, normal fruit firmness and good ascorbic acid content (Yapi and Coppens, 1997).

Characteristics of Few Potentials Hybrids of Pineapple

Amritha

In India, hybridization work was carried out at the Kerala Agriculture University Pineapple Research Centre in Vazhakulam, Trichur, which resulted in the production of a high-yielding hybrid, Amritha (Kew × Ripley Queen). The leaves are short stiff, spiny along the margins, and thickly covered with a whitish bloom on both surfaces. Fruits are similar to queen and suitable for table purpose. It ripens 10 days late compare to Queen. Fruit weight 0.8-1.1kg. Peduncle is short, fruit lets or eyes are small, prominent, deep set. The total soluble solids content varies from 16 to 18OBrix and acidity between 0.5 and 0.6 percent.

MD-2: The world's pineapple fresh fruit export industry went through a remarkable transformation after Del Monte Corporation introduced 'MD-2'pineapple to consumers in the United States and Europe, officially in 1996 (Frank, 2003). Until the mid1990s, the world's fresh fruit export industry was relatively small and was based mainly on cultivar Smooth Cayenne. Prior to the introduction of 'MD-2', the focus of the pineapple export industry was on canned 'Smooth Cayenne' pineapple. With the major producers focused on 'Smooth Cayenne' pineapple, the unanswered question is what circumstances made it possible for Del Monte Corporation to become the first multinational pineapple company in the world to break the industry fixation on 'Smooth Cayenne' and risk growing a new and untested hybrid. Williams and Fleisch (1993) reported that Pineapple Research Institute (PRI) hybrids had been developed that, when compared with 'Smooth Cayenne' had better resistance to Phytophthora rots, pineapple wilt, nematodes, pink disease and internal brown spot. Higher levels of vitamins C and A ,reduced acid content in winter ripened fruit ,better harvest peaking ,higher yield ,better cannery recovery from a ton of fruit ,faster plant growth a range of distinctive flavours.

Imperial: The pineapple disease known as fusariosis is the most important constraint of the pineapple industry in Brazil. Its control is based upon integration of several cultural practices associated with chemical control, and growing resistant cultivars. In order to do that, Embrapa Cassava & Tropical Fruits is carrying out a pineapple breeding program aimed at obtaining fusariosis resistant cultivars. Up to the present that breeding program has generated and evaluated 53,397 hybrids. Among them, the hybrid PE × SC-56, that showed good quality fruits, besides being resistant to fusariosis, was released as a new pineapple cultivar in 2003, named 'Imperial', and was recommended for planting mainly in regions where that disease is one of the main limiting factors for pineapple production. The 'Imperial' pineapple for commercial planting in areas where fusariosis is the main production limiting factor. The 'Imperial' pineapple plant is of intermediate height with spineless dark green leaves. The fruit is small and cylindrical, with a yellow peel at ripening. The pulp is yellow, with high sugar content, moderate acidity, a high level of ascorbic acid and a very good flavour. 'Imperial' also shows resistance to internal browning. The following characteristics may be considered as unfavorable: slow plant growth, peduncle with small diameter, small fruit (1.6 kg), rough fruit surface due to uneven fruitlets (influenced by climatic conditions), slips occasionally attached to the fruit base. Growing 'Imperial' pineapple does not require fungicide applications to control fusariosis, thus resulting in reduction of production costs, lower environmental impact and safer fruit production (Cabrala and Matos, 2009).

Vitória: Among the problems that limit pineapple production in Brazil, fusariosis due to the fungus *Fusarium subglutinans* f.sp. *ananas* (Syn.: *F. guttiforme*), is the most serious one causing yield losses estimated from 30% to 40% of marketable fruits, and 20% of planting material (slips and suckers). After preliminary evaluations under controlled conditions, some promising hybrids were introduced to the Experimental Farms of Incaper, where a Recurrent Clonal Selection of the hybrid PRI × SC-08 was carried out during the past 10 years, originating the Incaper access EC-099, that gave rise to the cv. Vitória, released to growers at the Sooretama Experimental Farm of Incaper, Sooretama-ES, in November 2006 (Ventura *et al.,* 2009). The 'Vitória' pineapple is a fusariosis resistant hybrid, generated at Embrapa Cassava & Tropical Fruits, by crossing between 'Primavera' and 'Smooth Cayenne', and transferred to experimental stations of the Instituto Capixaba de Pesquisa, Assistência Técnica e Extensão Rural (INCAPER), where it was evaluated for other field characteristics. 'Vitória' has spineless light green leaves, fruits of intermediate size (1.5 kg), cylindrical shape, yellow skin when ripe and high sugar content (15.8°Brix). 'Vitória' was recommended for commercial cultivation. Twenty two promising genotypes have already been selected and are currently under evaluation showing potential to be recommended as new ornamental cultivars (Cabrala *et al.,* 2002).

The qualitative and quantitative characters of pineapple hybrids from crossing between Cayenne x Queen and their reciprocal crossing had also been done. This research was conducted at the Indonesian Tropical Fruits Research Institute. Genetic materials used were 75 hybrid plants from Cayenne x Queen crossing and 40 hybrid plants from Queen x Cayenne crossing. The qualitative and quantitative characters were observed on each of sampled plants. The results showed that the distribution of spines on the leaves and the shape of eyes were affected by female parents. The percentage of hybrid plants possessing spiny leaves was 14.3% lower on the Cayenne x Queen crossing than on the Queen x Cayenne crossing. Similarly, the percentage of plants bearing fruit ≥1019.5-g was 14.2% lower on the Cayenne x Queen crossing. In contrast, plants bearing fruit with broad eyes and sugar content ≥ 16° Brix were 23.8% and 6.7%, respectively, higher on the Cayenne x Queen Hybrids than on the Queen x Cayenne hybrids. The hybrid plants possessing the combined characters of spineless or spiny at the leaf tip, fruit ≥ 1000 g, sugar content (TSS ≥ 16°Brix), and broad eyes are CQ-4, CQ-20, CQ-22, CQ-26, CQ-41 and QC-26. Except for the projected eyes, these characters are also possessed by the hybrids CQ-16, CQ-46, and QC-8. These results can potentially provide valuable information on the choice of female and male parents for the programs of pineapple variety improvement.

Dole-14: A new pineapple variety developed from a cross made by Dole between the commercial variety MG-03 and P-1972, a patent is extended or adjusted under 35 patented clone, and has been designated Dole-14. This new variety differs from its progenitors in having higher carotenoid content, medium acidity, low fibre content, improved aroma, distinct fruit flavour and flesh colours, and a conical to cylinder shaped fruit with a long conical crown. It also shows high tolerance to NDF and less susceptibility to shell cracking. The plant is characterized by long spineless leaves with piping that are pale to greyish green in colour. When unripe, fruit shell is pale green turning to uniform yellow colour when ripe, and the flesh develops a deep golden colour at maturity.

A25-34 and A54-47: The performance of five pineapple F1 hybrids selected for early bearing was evaluated with three standard cultivars. The results showed that two of the hybrids, viz. A25-34 and A54-47, were early maturing and fruit were harvested 2 weeks ahead of the standard cultivar Gandul. The primary difference between early and late genotypes arose from the period for fruit development after inflorescence emergence ('red-heart'). This ranged from 94–97 days for early genotypes to 112–113 days for late genotypes. There was not much difference in the time between 'forcing' (application of flower hormone) to emergence of inflorescence. All genotypes flowered 30–34 days after hormone application. There was a linear relationship between fruit weight and plant weight for all, but one of the genotypes – A04-16. This genotype was unique because it had the highest mean fruit weight but the smallest plant resulting in a very high fruit-to-plant weight ratio of 0.60. There is good scope for reduction of crop cycle time with this genotype because it was not necessary to extend the growing period to get large plants for production of economic-sized fruit. Two other genotypes, viz. A25-34 and A12-45 that showed a 'flat' gradient in the regression of fruit on plant weights, also appeared to have potential for early harvesting. The increase in fruit size after a certain time was marginal compared with the increase in plant size and therefore, the growing period in the field may be reduced (Chan and Lee, 1999).

Interspecific and intergeneric Hybridization

Both interspecific and intergeneric hybridization have been carried out at different research stations. When Ananas comosus is crossed with *Pseudo ananas* only 5-10 per cent of seeds are fertile and produce generic hybrids. These hands are very vigorous, harcy, and self-fertile. They produce an abundance of hybrid seed (Collins, 1960). *Pseudo ananas* shows little variation when selfed, indicating that it is relatively homozygous. The Ananas species have a diploid number 50 chromosomes while Pseudo ananas has 100 chromosomes. The F1 hybrids (*Ananas* x *Pseudo ananas*) have predominantly 100 chromosomes, with a very few triploid plants having 75 chromosomes. These triploids are smaller than the tetraploids but resemble, the Pseudo ananas more than they do the than as plants. They are highly self sterile, but produce a few seeds from cross pollination. When *Pseudo ananas* is crossed with spiny species (or varieties) of *Ananas* all the progeny are spiny. When it is crossed with Cayenne having spiny- tip leaves, the hybrid plants show an irregular combination of spiny and smooth leaves. Also, classification of leaf type during the juvenile stage does not correspond with the characters in mature plants. The cumulative action of multiple geris at the tetraploid is expected to be involved in such phenotypic expressions (Collins,1960)

Interspecific hybridization is usually accompanied by some degree of sterility. This is reflected by decreased viability of the seeds produced by the hybrid plants. *Ananas comosus*, with its numerous varieties produce seedless fruits because of self-incompatibility of the germ cells. Most of these varieties, however are cross-fertile and produce F1 seeds and fertile F1, plants when crossed with other species of *Ananas*.

Pineapple cultivars are heterozygous and hybridization between them often leads to development of fertile seed that presents a wide spectrum of genotypes. This segregating F1 population provides an excellent source of gene recombinants for selection and cloning of new, superior individuals(Chan, 2006).Most of the improvement of the pineapple has taken place by simple selection of mutant clones within existing cultivars and by hybridization between cultivars followed by selection of the highly, heterozygous progeny. A cross between Red Spanish and Cayenne has led to the development of two new hybrids PR-1-67 and PR-1-56 in PuertoRico (Ramirez, 1970).This hybrid shows better plant vigour and resistance to wilt disease. However, self-fertile somatic mutants obtained from cv. 'Cayenne' show a loss of vigour on selfing and heterosis on crossing. A hybrid "Tainung13" has been obtained from a cross between Smooth Cayenne x Tainung 8 (Chang, 1995). Its average plant height is 127.5 cm, leaf length 104cm, leaf width 5.1 cm and leaf margin spineless (base and tip have small spines). It bears cone shaped fruits, average weight 1.2 kg, flesh-firm, green yellow and slightly fibrous. It also has some degree of wilt tolerance in the field. Gopimony and Balakrishnan (1984) have tried to combine the large fruit size of Kew and good dessert quality of the Ripley Queen. Among the 16 hybrids obtained one (Hybrid-8) appeared to bear the desired fruit qualities. Sanewski and Giles (1997) have developed a hybrid resistant to blackheart. This inter group hybrid was without blackheart after storage at 10°C for 14 days followed by 20°C for 8 days. In Taiwan two dwarf and high yielding hybrids namely Typhone1 and Typhone 3 have been released. The fruit shape is cylindrical in Typhone 3.The development of a new hybrid through hybridization is a slow process.

CO-2: Developed as an inter specific cross between Pineapple Research Institute hybrid clone 58-1184 and Pineapple Research Institute (PRI) hybrid clone 59-443, and has been named hybrid CO-2. This hybrid clone generally resembles the parent varieties, but is distinguished there from in that this plant produces a sweeter, more vitamin C rich fruit, which is low in parasitic fruit diseases, and tolerant or resistant to the physiological disorder called internal browning. This plant has more uniform yellow color, good fruit appearance after refrigerated storage, and produces acceptable crop yields of saleable fresh pineapple (Bartholomew et al., 2010).

Out Crossing

The best breeding system will usually be the one that maintains genetic variability in the breeding population but allows the frequency of desirable genes and gene combinations to increase. Out crossing with simple recurrent selection in heterozygous species such as pineapple will shift the mean for characters under selection in the direction of selection. However, this shift will not change the heterozygosity as long as the breeding 98 population is not diluted with unimproved cultivars. The number of selections made in each generation will be a function of the frequency of the desirable genes (Allard, 1960).Such a system initially relies on a good representation of foundation cultivars. The ceiling of a simple recurrent selection program will be set by the most favourable combination of alleles within the foundation population. The PRI program has the longest documented history of pineapple breeding. This program started in the 1920s with a system of out crossing between the five cultivars 'Smooth Cayenne', 'Smooth Guatamala' (Monte Lirio), 'Perola', 'Queen' and 'Ruby' (Spanish) and then inter-crossing the F1 selections. Since all of the F1 population included 'Smooth Cayenne', always exploited as the female parent, the F2 generations were all half sib crosses. Other programs started in a similar fashion as the PRI program with the F1 population composed of crosses between basic Amerindian cultivars. The cultivars used were, in Thailand 'Smooth Cayenne' and 'Queen' (Sripaoraya, 2009), in CIRAD (France) 'Smooth Cayenne' and 'Perolera' (Cabot, 1987), in Brazil 'Smooth Cayenne' and 'Perola', 'Perolera' or 'Primavera' (Cabral *et al.*, 1993), and in Malaysia 'Red Spanish', 'Smooth Cayenne' and 'Queen'. The best parents in a simple recurrent selection program will usually be late generation selections. Currently, as a few late generation cultivars are available, the commencement of a program with only two unimproved parent combinations such as 'Queen' with 'Smooth Cayenne' is most likely a comparatively inefficient approach. The exception is where a widely segregating population is required to identify novel characters. Since simple recurrent selection relies on incremental improvements with each generation, it is important to quickly assess selections so the best can be used as parents in the next generation. In the PRI program it required at least 7 and up to about 13 years before selections were used as parents such that selections made in 1973 were the progeny of selections made in the late 1950s. Strategies to decrease the interval between generations are critical to making rapid progress. New approaches such as genome wide marker selection might be worth evaluating (Volz *et al.*, 2009).

Backcrossing

Backcrosses are often compatible with at least one of the parents. 'Smooth Cayenne' is often used as the recurrent parent in backcrosses. While 'Smooth Cayenne' is probably a good F1 parent, its continued use as a recurrent backcross parent in a simple recurrent selection program is questionable if alternative high yielding late generation cultivars are available. Backcrossing to an unimproved cultivar, like 'Smooth Cayenne', is likely to decrease the frequency of desirable alleles in breeding stocks of an advanced program. It will almost always be better to use current generation cultivars as parents.

SIB and Half SIB Crosses

While most programs have commenced without crossing between unrelated cultivars, subsequent generations usually encounter some mild inbreeding, usually in the form of sib or half sib crosses, because of the use of 'Smooth Cayenne' in most F1 crosses. Half sib crosses must therefore be common where F2 populations are generated. Mild forms of inbreeding such as half sib crosses which produce large populations of progeny should actually allow a greater chance of recombination of favourable alleles compared to severe inbreeding such as selfing, which often produces very few progeny (Allard, 1960). It might however be necessary to alternate out crossing with other foundation cultivars to identify more favourable gene combinations. Half sib crosses are often compatible (usually in one direction only), but might be accompanied by mild inbreeding depression and will require heavy selection pressure for vigour. Inbreeding coefficients (IBC) for different levels of inbreeding are shown in Table 4. The data show the successful cultivar '73-114' has an IBC approximately equivalent to a half sib cross. Sib crosses are often incompatible or encounter excessive inbreeding depression at 99% germination but are feasible with some parent combinations (Sanewski, 2009). Germination of seed and early growth of seedlings, especially where some inbreeding is involved, should be optimal to allow identification and discard of the weaker types.

Mutation Breeding

Somatic mutations occur naturally, and a large number of distinct clones within a cultivar can now be identified. Collins and Kerns (1938) identified 30 Cayenne mutants. Natural mutants of cv. Spanish, Queen, and Abacaxi have also been identified, and attempts to induce mutation have been made. Chemical mutagens or irradiation may be used to create artificial mutagens. Irradiation of the che cultivars Kew and Mauritius in Kerala resulted in growth retardation and premature suckering. Merz (1964) claimed that X-irradiating pollen during meiosis resulted in the induction of self fertile mutants. When using a $_{60}$Co source, there was an inverse relationship between irradiation dose and pollen viability and tube growth. Mutations have occurred in the Cayenne variety, giving rise to a series of new or changed forms; some are so profoundly changed that they no longer show the variety characters, others show only minute changes. These mutant forms are treated separately from the strains already discussed, principally because we know something about the origin of these mutations and we do not know how the strains originated. It is quite possible that the latter also started as somatic mutations. The mutations in Cayenne can be conveniently divided into six classes one of the characters which are altered. Four classes affect the fruit and two classes affect both fruit and plant characters. They influence one or a few characters and do not, for the most part are present profound genetic changes, even though the character alteration may be conspicuous. The mutant Crowning Beauty is an example of this kind of mutation.

Shoot Proliferation and Radio-Sensitivity

The experiment established LD_{50} to be 45 Gy. These results are consistent with those reported by other workers on mutation breeding in pineapple (Lapade *et al.*, 1995), and with the previous results on micropropagation and radio-sensitivity. Multiple Range Test (p=0.05) showed that concentrations of 10, 15, 20 and 25 uM significantly affected shoot proliferation. More shoot buds were formed with increasing concentration of BAP in the medium. In both cultivars, the optimal proliferation of shoot buds occurred with 20 uM BAP and rooting was achieved on medium supplemented with 1.5 uM IBA and 0.75 uM IAA. The proliferation of 'Smooth Cayenne' was better than that of 'Sugar Loaf.

Colchicine Induced Morphological Variants in Pineapple

The colchicine treated callus produced normal green plants on the regenerating media; however, phenotypic variation was also observed among the regenerated plantlets. The 133 calluses transferred, seven gave variant lines with a frequency of 5.26%. On the callus induction media (MS + 2,4- D at 4.52 μM) the albino shoots regularly callused. The colour and growth rate of variant-callus was a little different to that of the normal callus. When the variant callus was transferred to the shoot regeneration media, it developed into albino shoots. When variant shoot tips were cultured on regenerating media containing NAA (2.69 M), the newly generated multiple shoots turned out to be variants as well. In the following generation, no cases of green or chimaeras were discovered. The regenerated variant lines grew well alongside normal plants, and in vitro cloning for variant regenerants were simple; however, their growth was severely stunted when compared to normal plants. Although there was less variation in plant height in the first four weeks, the difference became remarkably large in the later stages of growth, with the variants averaging 1.84 0.08 cm of plant height compared to 4.58 0.17 cm in normal green plants. Leaf morphological variation was also noted among the regenerated progenies. In normal plants length/breadth (L/B) ratio of leaf was significantly higher compared to that of the variant leaf where L/B ratio was recorded to be 5.52. As regards roots, induction and further growth of it was severely inhibited in variant lines. In the latter roots attained a length of 0.88 cm while in normal plants the average root length was 10.9 cm in six months' cultures. Surprisingly, the variation range in the number of adventitious shoots was smaller between the variant lines and normal green plants. Chlorophyll analysis further demonstrated that in the variant lines, total chlorophyll was about 15 times less than that of normal plants. Similarly, chlorophyll a and chlorophyll b were significantly higher in normal plants compared to those in the variant lines (Table 3). In the present investigation isoenzyme variation of both normal and regenerated albinos were studied. In most cases, differences were noticed in the occurrence and absence of isoenzyme bands. Normal calluses exhibited a total of five isoenzymes of peroxidase bands compared to three in the albino type. Both the normal and albino had three common bands (Rf 0.05, 0.37 and 0.64). In other words, the normal type showed two additional bands of Rf 0.10 and 0.26 and a faint band of Rf 0.40 which was absent in the variant. Besides, distinct differences in band intensities were also observed. The regenerated variant grew very slowly; it grew for three years or more without any regenerative loss. The detailed evaluation of the variant lines may prove helpful in utilizing any of its novel trait in the improvement of this popular crop. Phenotypic variation particularly in the synthesis of pigment and growth pattern was noticed among the regenerated clones. The variants showed significantly low chlorophyll compared to that of normal plants. Besides, leaf size, the rate of shoot and root growth was substantially reduced in the variants. They grew with less vigour and produced identical clones without further segregation when cultured *in vitro*. The altered phenotype was stable and it maintained its usual high regenerative potential for over three years (Mujib, 2005).

Ploidy Breeding

Polyploidy is an organism which has more than two sets of chromosomes or genomes in its somatic cells. Some features of polyploidy plants include nuclei and larger cell contents; leaves and flowers which grow larger; and it may undergo changes in chemical compounds including an increase or change in the size or proportion of carbohydrates, proteins, vitamins, or alkaloids. Collins *et al.* (1931) stained and counted chromosomes in a number of varieties in 1930, discovering 50 in root cells and 25 in pollen. 75 was also discovered in the roots of Cayenne x Wild Brazil hybrids. The discovery of these triploids ignited great interest because it provided the possibility of improving variety by chromosome number manipulation. They tried the latter first with temperature treatments, then extended the cold temperature treatment of germ cells after discovering some giant pollen cells. Triploids were discovered in other hybrid populations, as well. It was determined that these came from unreduced egg cells based on genetic and morphological evidence. Tetraploids were later discovered in hydrid populations. With the introduction of colchicine in 1937, a large programme was introduced to cause auto tetraploids, which they hoped to cross with diploids to create triploids with the tetraploid parent's complete chromosome complement. Cameron *et al.* (1940), was loaned to the department by the Genetics Department of University of California in 1940. He improved chromosome staining and counting methods and conducted a survey of hundreds of breeding selections and Collins-Baker expedition accessions, resulting in the discovery of over 100 more triploid and tetraploid selections. Despite the fact that polyploids had a higher percentage of retention in the reselection screening process, no polyploids were ever considered for commercial use in these tests. They had lower brix, bigger but less eyes, smaller berries, and took longer to mature in general.

Resistance Breeding

Resistance to Fungal Diseases: Heart and root rot in pineapple is caused by Phytophthora species (Kamoun, 2001). Heart rot disease infects the basal young leaf tissues at the heart gradually leading to fruit rot while root rot leads to root necrosis which ultimately reduce yields. Previously many attempts were made for the introduction of antifungal genes such as chitinase (chi) and ap24 in to pineapple genome for obtaining resistance against *Phytophthora nicotianae* (Espinosa *et al.*, 2002). The chi gene helps in the degradation of the chitin which is the important constituent of the fungal cell wall and ap24 gene assist in production of antifungal proteins which destabilize the fungal proteins. Delivering the transgene rice actin I-chitinase in to the callus derived from leaf discs through *Agrobacterium tumefaciens* was done *by* Espinosa *et al.* (2002). Different plasmids like pHCA 58, like pHCA 59, pTOK233 carrying the trans gene were harboured with the bacterium. pHCA 58 carried class-I bean chi gene and tobacco ap24 gene. The trans genes were expressed in the plants obtained but their resistance to the pathogen was not reported.

Many studies revealed antimicrobial peptide conferring resistance to bacterial and fungal diseases Attempts were made to induce fungal resistance in pineapples. MSI-99 and Myp20 are the two antimicrobial peptides that have been utilized for their expression in transgenic crops for disease resistance (Li *et al.,* 2001). MSI-99 is the derivative of original magainin sequence with three substitutions. Introduction of antimicrobial peptide MSI-99 gene in to pineapple cv. Queen was done by *Agrobacterium tumefaciens* strain EHA 105 mediated transformation (Mhatre *et al.*, 2009). pMSI168 plasmid with MSI-99, a magainin analogue, was harboured by co-cultivation with Leaf bases collected from in vitro derived shoots of pineapple. Many studies revealed antimicrobial peptide conferring resistance to bacterial and fungal diseases. Mhatre *et al.* (2009) revealed that in their investigation, the MSI-99 gene had been stably integrated into pineapple genome and its expression was confirmed by RT-PCR analysis. However, resistance acquired in the transgenic plants need to be confirmed by further bioassay studies.

Yabor *et al.* (2019) developed transgenic plants of the pineapple cv. 'Serrana Smooth Cayenne' by the method described by Espinosa *et al.* (2002). In addition, micropropagated control plants that were not exposed to *Agrobacterium tumefaciens* or phosphinothricin (untransformed) were produced. Chitinase gene from *Phaseolus vulgaris*and a DNA sequence cloned from *Nicotiana tabacum* which is designated as AP24were integrated in to pineapple. AP24 codes for antifungal protein mainly has characteristic feature of disrupting the fungal cell wall. A selectable marker gene which provides resistance against herbicide phosphinothricin derived from actinobacteria was also included in the transfer. Transgenes were harboured by the constitutive promoters i.e, OCS-35S CaMV-rice actin I, 35S CaMV, and maize Ubi1 (Espinosa *et al.*, 2002).Long-term effects of transgenic pineapple plants were tested by growing them in the field for 8 years up to fourth vegetative generation. Four treatments were compared i.e., macropropagated control plants (untransformed),control plants (untransformed), and micropropagated genetically transformed plants (clones 90 and 98) coefficients of Variation (OCV) were calculated for different parameters. Fewer differences were observed in aspects of leaf and root histology, there were no significant differences among the treatments in aspects of average weight of fruit with crowns among the four treatments investigated. This proved that yield parameter was not altered among the fruits produced from the control and transgenic plants and genetic modification and micro propagation did not influence the yield (Yabor *et al.*, 2019).

Mealy bug wilt resistance: Mealy bug wilt of pineapple (MWP) is a prevalent disease all over the world mainly in tropical regions. The disease is characterised by severe dieback, downward curling of the leaves, turning of leaves in to reddish brown and final wilting of the plants (Hu *et al.*, 2005). The disease is mainly through virus particles which are spread through mealy bug vectors which are secondarily spread through ants. Pineapple mealybug wilt responsible viruses are species of the Ampelovirus genus of the Closteroviridae family (Kishore *et al.*, 2018). Control of this disease through conventional breeding is not possible mainly in case of cultivars like Smooth cayenne where sterility is prevalent. Transgenic approach was used in developing resistance to MBW by using gene constructs which engage RNA mediated resistance technology (Ko *et al.*, 2013). In a study conducted by Sether *et al.* (2005), coat protein (CP) gene of PMWaV-2 was constructed as an inverted-repeat in the pCAMBIA 1300 vector and by using gene transfer methods particle bombardment and *Agrobacterium*-mediated transformation, transgenic pineapple plants resistant to MBW were developed but the resistance didn't long for more time and plants gradually got infected and later, transformed plants with PMWaV-2 gene by engineering the gene in sense and inverted orientation in to pCAMBIA1300 vector by using particle bombardment and *Agrobacterium*-mediated transformation for inverted repeats and sense orientation respectively. Seven lines of transgenic lines resistant to pineapple mealy bug wilt were obtained and resistance persisted even after repeated trails.

Nematode resistance: Reniform nematode (*Rotylenchulus reniformis*) is one of the devastating pests causing damage in pineapple crop. They are no cultivars that have natural resistance towards nematodes until now. Nematode infection can cause losses up to 40 percent in the first crop and 80 percent in subsequent ratoon crops (Sipes and Schimitt, 1994). Genetic engineering paved a way in development of transgenic plants by harbouring nematode resistance genes. Root genes activated by root knot or reniform nematode are identified for using their promoters for utilizing in nematode resistance strategies. Sipes *et al.* (2002) transformed pineapple variety MD 1 resistant to nematodes. Stem segments and leaf bases were constituted with a modified crystatin protease inhibitor by using AGLO strain of *Agrobacterium tumefaciens*. Around 22 transgenic lines were produced and tested for growth and reproduction of nematodes in comparison with the wild plants. Nematode infection and reproduction was found less in the transgenic plants compared with the wild plants. But, range of nematode reproduction per plant was found greater compared with wild plants due to chimerism in the transformed pineapple plants (Sipes *et al.*, 2002).

Black Heart resistance in pineapple: Internal browning or Blackheart is a physiological disorder in the pineapple which is usually observed when day or night temperatures fall below 20°C and photoperiod is less than required during fruit development stage. It also arises when the fruits are stored in cooler temperatures for longer period. Watery spots start appearing at the bottom of the fruit, which gradually discolour the core along the length of the fruit and external damage is not visible. Ko *et al.* (2013), used transgenic technology utilizing particle bombardment technique to control black heart disease in pineapple cv. Smooth Cayenne. Buds excised from crowns were sterilized and cultured *in vitro* in M.S medium supplied with essential nutrients and then multiplied. PPO genes have been isolated using Polymerase Chain Reaction with abundant oligo nucleotides and cloning of cDNA collected from the black heart affected pineapple fruit has been done. To visualize the stable expression, Gus and gfp reporter genes were used .Two plasmids were constructed for expression of the nptII selectable marker gene and five plasmids were constructed that express gus, gfp or ppo genes. 15 individual transgenic lines with gus and gfp reporter genes and 22 ppo lines were obtained. Expression of these genes was confirmed by Southern blotting technique.

Herbicide tolerance: Pineapple is one of the stress tolerating plants but it cannot cope up with the weed flora. Most of the pineapple production and commercialization is limited due to the problems caused by the weed flora. Pineapple genes responsible for herbicide tolerance were not identified making it impossible for development of resistance cultivars. Only alternative to unravel this problem is genetic engineering. Sripaoraya *et al.* (2001) developed Thai pineapple (*Ananas comosus* `Phuket') which was resistant for the herbicide 'basta' having glufosinate ammonium as the active compound. Leaves of micro propagated shoots are transformed with plasmid AHC25, carrying the b-glucuronidase (gus) reporter gene and the bialaphos resistance (bar) gene for herbicide tolerance usingmicro projectile-mediated delivery. Regenerated plants sprayed with Basta2 containing glufosinate ammonium up to 1400 mg L^{-1} remained healthy and retained their pigmentation. Expression of the bar gene in regenerated plants was confirmed by Southern analysis and RT-PCR techniques.

Weed resistance: Weeds are the main cost factor for pineapple production as at least a half of production costs are related to weed control. All current commercial pineapple cultivars are sensitive or susceptible to biodegradable herbicides. This research project aimed to establish an F1 hybrid pineapple cultivar resistant to herbicide by conventional breeding. Direct and reciprocal crosses between genetically modified 'Phuket' pineapple (TP) carrying the bialaphos-resistant (bar) gene and the commercial pineapple cultivars 'Pattavia' (PV) and 'Phuket' (PK) were made in 2005 at the Faculty of Agriculture, Rajamangala University of Technology Srivijaya, Nakhonsithammarat, Thailand. The crosses resulted in 1,172 F1 seeds and 886 F1 seedlings were obtained from the first crossing of the two crosses. No seeds were obtained from selfing of TP, which shows that transforming 'Phuket' did not eliminate the incompatibility of TP×TP. The cross TP and PK also did not produce seeds but 14 seeds were obtained from the reciprocal cross of PK×TP showing that transformed 'Phuket' was at least somewhat compatible with the untransformed 'Phuket'. The 886 F1 plants were evaluated for agronomic characteristics, fruit quality and bialaphos herbicide resistance. There were 410 F1 plants that showed resistance to bialaphos herbicide. Furthermore, there were 18 F1 plants showing other good agronomic characters and good fruit quality. Three promising hybrids, (PV×TP)-51; (PV×TP)-46 and (PV×TP-34), showed resistant to biodegradation bialaphos herbicide, a °Brix of 18-22.2, a fragrant smell, crispy flesh and a smooth leaf with only the tip being spiny. In particular, (PV×TP)-34 hybrid plants had palepink bracts and fruit with a 22.2°Brix, cylindrical fruit shape, dark yellow flesh, and pleasant flavor. These potential hybrids will be propagated and evaluated to confirm the traits and subsequently released to farmers (Sripaoraya and Davey, 2011).

Salinity stress: Pineapple (*Ananas comosus),* one of the important horticultural crops in tropical and sub-tropical regions of the world possesses CAM photosynthesis (Chen *et al.,* 2002). While the effects of salinity on growth and metabolic processes are well documented in many plants, knowledge about NaCl induced changes in pineapple is still limited. The study to determine the level of salt concentrations influencing the mineral nutrient and pigment contents in pineapple may improve the understanding of tolerance mechanisms, to the extent they exist, in this species under salinity stress. This could play a role in improving management practices where pineapple grown in saline areas such as the 'bris' area in the east coast of Peninsular Malaysia. Therefore, a study was conducted with the aim of assessing the effect of NaCl concentrations on K, Ca, Mg, P and Na and chlorophyll and carotenoid contents in pineapple under tissue culture conditions. Under in vitro conditions, the cultures are free from deleterious effects of uncontrollable external environments, technically easy to handle and require minimal management cost as compared with those under field conditions. Moreover, the salt tolerance assessment under in vitro conditions using physiological traits could be more effective than the assessment performed in the field using yield components (Housmand *et al.,* 2005). The results could also provide an alternative means to formulate growth medium for preservation of pineapple germplasm under *in vitro* conditions.

References

As'adah, M., Rahayu, T., Hayati, A. 2016. Metode Pemberian Kolkisin Terhadap Research

Allard, R. W. 1960. Principles of Plant Breeding. John Wiley & Sons, Inc. 485p. Anderson, E.J. and Collins, J.L. 1949. *Progress in Breeding for Resistance to Heart Rot - Root Rot. Special Report of the PRI,* **16**:8-13.

Almeida, W. A. B., Matos, A. P. and De-Souza, A. 1995. Effects of Benzylamino purine (BAP) on in vitro proliferation of pineapple (*Ananas comosus* (L.) Merr.). *Newsletter of Pineapple Working Group of the International Society for Horticultural Science. Vol. II,* March 1996.

Bartholomew, D. P., Coppens de' Eeckenbrugge, G. and Chen, C. C, 2010. Fruit and nut register list No. 45. *HortScience,* **45**: 716–756.

Benega, R., Isidron, M., Arias, E., Daquinta, M., Dominquez, Q., Borroto, C. G. and Hidalgo, M. 1995. Comparative study between two pineapple (*Ananas comosus* L.) cultivars. *Newsletter of Pineapple Working Group of the International Society for Horticultural Science. Vol. II,* March 1996.

Bhomik, G. 1980. Selection of male parents on the basis of male gametophyte for pineapple breeding. *Indian Journal of Agricultural Research,* **50(10)**:753-756.

Cabot, C. 1987. Practice of pineapple breeding. *Acta Horticulturae,* **196**:25-36.

Cabral, J. R. S., de Matos, A. P. and de Cunha, G. A. P. 1993. Selection of pineapple cultivars resistant to fusariose. *Acta Horti*culturae, **334**:53-58.

Cabrala, J. R. S., de Matos, A. P. 2009. Imperial, a New Pineapple Cultivar Resistant to Fusariosis. *Acta Horti*culturae, **822.**

Cabrala, J. R. S., de Matos, A. P., Junghans, D. T. and Souza, F. V. D. 2002. Pineapple Genetic Improvement in Brazil. *Acta Horticulturae,* **822.**

Cameron, D. R. and Collins, J. L. 1940. A study of polyploidy in pine apple productions. *Pineapple Producers Cooperative Association Quarterly,* **1**:122-138.

Chan, Y. K. 2006. Hybridization and selection in pineapple improvement: the experience in Malaysia. *Acta Hort.* **702:** V International Pineapple Symposium.

Chan, Y. K. and Lee, H. K. 1999. Performance of F1 pineapple hybrids selected for early fruiting. *Journal of Tropical agriculture and Food Science,* **27**(1):1–8.

Chang, C. 1995. Tainung 13-pincapple. *Journal of Agricultural Research (China)*, **44**(3): 287–296.

Chen, H., Hu, B., Zhao, L., Shi, D., She, Z., Huang, X., Priyadarshani, S., Niu, X., Qin, Y. 2019. Differential expression analysis of reference genes in pineapple (*Ananas comosus* L.) during reproductive development and response to abiotic stress. *Trop Plant Biology,* **22**:1–11.

Chen, L. S., Lin, Q. and Nose, A. 2002. A comparative study on diurnal changes in metabolites levels in the leaves of three crassulacean acid metabolism (CAM) species, *Ananas comosus,* Kalanchoe daigremontiana and K. pinnata. *Journal of Experimantal Botany,* **53**:341-350.

Collins, J. L. 1960. The pineapple, Botany, Utilization, Cultivation. Leonard Hill Ltd, London, **pp**. 294.

Collins, J. L. 1960. The Pineapple: Botany, Cultivation and Utilization. Inter science Publishers, New York. **pp**. 301.

Collins, J. L. and Kerns, K. R. 1931. Genetic studies of pine apple. *Journal of Heredity,* **22**(5): 139-142.

Collins, J. L. and Kerns, K. R. 1938. Mutation in pineapple: A study of 30 inherited abnormalities in the Cayenne variety. *Journal of Heredity,* **29**:162-172.

Daquinta, M., Cisneros, A., Rodriguez, Y., Escalona, M., Perez, M. C., Luna, I. and Borroto, C. G. 1995. Somatic embryogenesis in pineapple. *Newsletter of Pineapple Working Group of the International Society for Horticultural Science. Vol II,* March 1996.

Dominquez, Q., Gonzalez, B. L., Borroto, C. G., Exposito, L. A., Gonzalez, R., Rodriguez, Y. and Gonzalez, J. L. 1995. Effect of magnetism application upon the shoot production of pineapple *in vitro* plants in the multiplication phase. *Newsletter of Pineapple Working Group of the International Society for Horticultural Science. Vol. II,* March 1996.

Dujardin, M. 1991. Cytogenetique de l'ananas. *Fruits.* **46**:376-379.

Espinosa, P., Lorenzo, J. C., Iglesias, A., Yabor, L., Menendez, E., Borroto, J., Herandez, L., Arencibia, A. 2002. Production of pineapple transgenic plants assisted by temporary immersion bio reactors. *Plant Cell Reports*, **21**:136-140.

Firoozabady, E., Nicholas, J. and Gutterson, N. 1995. *In vitro* plant regeneration and advanced propagation methods for pineapple. *Newsletter of Pineapple Working Group of the International Society for Horticultural Science. Vol II*, March 1996.

Fitchet, M. 1989. Observation on pineapple improvement in Taiwan Republic of China, *Information Bulletin of Subtropical, Fruit Research Inst*itute, **207**: 10-12.

Frank, R. 2003. Going for 'The Gold' Turns Pineapple World Upside Down. *Wall Street Journal*, Oct. 7, p. 1.

Gonzalez, R., Dominquez, Q., Exposito, L. A., Gonzalez, J. L., Martinez, T. and Hidalgo, M. 1995. Effectiveness of eight strains of Azotobacter in the adaptation of pineapple vitro plants (*Ananas comosus* (L.) Merr.) cv Smooth Cayenne. *Newsletter of Pineapple Working Group of the International Society for Horticultural Science. Vol II*, March 1996.

Gopimony, R. and Balakrishna, S. 1984 Features of inter varietal hybrids of pine apple *.South Indian Horticulture,* **32**(2):65-68.

Guillermo, A. P., Jumwa, A. M., Sagarra, F. and Carlos, E. A. B. 2011. Morphological and physiologicalcharacterization of two new pineapple somaclones derived from in vitro culture. *In Vitro Cellular and Developmental Biology - Plant.* **47**. 428-433. 10.1007/s11627-011-9342-y.

Houshmand, S., Arzani, A., Maibody, S. A. M. and Feizi. M. 2005. Evaluation of salt tolerant genotypes of durum wheat derived from in vitro and field experiments. *Field Crops Research,* **91**:345-354.

Hu, J. S., Sether, D. M., Metzer, M. J., Perez, E., Gonslaves, A., Karasev, A. V. and Nagai, C. 2005. Pineapple mealybug wilt associated virus and mealybug wilt of pineapple. *Acta Horticulturae,* **666**: 209-212.

Jai Prakash , Bhattacharyya, S. , Chattopadhyay, K., Roy, S., Das, S. P. and Singh, N. P. 2009. PQM-1: A newly developed superior clone of pineapple for north eastern India as evident through phenotype, fruit quality and DNA polymorphism. *Scientia Horticulturae,* **120** :288–291.

Kamoun, S. 2001. Non host resistance to Phytophthora: novel prospects for classical problem. *Current Opinion In Plant Biology,* **4**:295-300.

Kerns, K. R., Williams, D. F. and Smith, J. B. 1968. A review of seediness in pineapple. PRI Research Report No. 24.

Kishore, K. D., James, C. G., Michael, M., Wayne, B. and John, S. H. 2018. Mealybug Wilt of pineapple and associated viruses. *Horticulturae,* **4**(52).

Ko, L., Eccleston, K., O'Hare, T., Wong, L., Giles, J. and Smith, M. 2013. Field evaluation of transgenic pineapple (*Ananas comosus* (L.) Merr.) cv.'Smooth Cayenne'for resistance to blackheart under subtropical conditions. *Scientia Horticulturae,* **159**: 103–108.

Kumar, N. 2006. Breeding of Horticultural Crops: Principles and practices. New India Publishing Agency, India. **pp** 118-120.

Lapade, A. G., Veluz, A. M. S. and Santos, I. S. 1995. Genetic improvement of the Queen variety of pineapple through induced mutation and *in vitro* culture techniques. Proceedings, Joint IAEA/FAO *Internatl. Symp.* "Induced Mutations and Molecular Techniques for Crop Improvement", IAEA, Vienna, 684-687.

Li, Q., Lawrence, C. B., Xing, H. Y., Babbitt, R. A., Bass, W. T., Maiti, I. .B. and Everett, N. P. 2001. Enhanced disease resistance conferred by expression of an antimicrobial magainin analog in transgenic tobacco. *Planta.* **212**:635–639.

Loison-Cabot, C. 1990. From 1978 to 1990 Twelve Years of Research on the Genetic Improvement of Pineapple, *Fruits (Paris)*, **46**: 367–371.

Merz, G. S. 1964. Study of self-incompatibility in the pine apple. *Agricultural Resources Pretoria,* **2-6**:561-562.

Morton, J. 1987. Fruits of warm climates. *Creative Resource Systems, Inc. NC.* **1987**:18- 28

Mujib, A. 2005. Colchicine Induced Morphological Variants in Pineapple .*Plant Tissue Culture and Biotechnology,* **15(2)**: 127-133

Nath, V., Kumar, G., Pandey, S. and Pandey, S. 2019. Impact of climate change on tropical fruit production systems and its mitigation strategies. In: Sheraz Mahdi S (ed) Climate change and agriculture in India: impact and adaptation. *Springer, Berlin,* **pp.**129–146.

Pickersgill, B. 1976. Pineapple (In) Evolution of crop plants (Simmonds 0 Longman,) London, **pp**. 14–18.

Py, C., Lacoeuilhe, J. J. and Teisson, D. C. 1987. The Pineapple: Cultivation and Uses. G.P. Maisonneuve and Larose, Paris, **pp**. 127-137.

Remirez, O. D. 1970. Pineapple PR-1-67, Rev. Cafe, *Puerto Rico*, **25**: 117–120.

Robert, E Paull, Duane, P. Bartholomew and Ching, Cheng Che 2003. Pineapple breeding and production practices. Tropical Plant and Soil Sciences, University of Hawaii at Manoa, 3190 Maile Way, Honolulu, Hawaii, 96822, USA

Sanewski, G. M. 2009. The effect of different levels of inbreeding on self-incompatibility and inbreeding depression in pineapple. *Acta Horticulturae,* **822**:63-70.

Sether, D. M., Melzer, M. J., Busto, J., Zee, F. and Hu, J. S. 2005. Diversity and Mealybug Transmissibility of Ampeloviruses in Pineapple. *Plant Diseases*, **(89):** 450–456.

Sipes, B., Nagai, C., McPherson, M., Atkinson, H., Christopher, D., Hu, J., Paull, R., Rohrbach, K. , Moore, P., Oda, C.,Wood, P. and Conway, M. 2002. Pineapple genetically modified for resistance to plant –parasitic nematodes. *Phytopathology*, **92**: S-76.

Sipes, B. S. and Schimitt, D. P. (1994) Evaluation of pineapple, *Ananas comosus*, for host plant resistance and tolerance to *Rotylenchus reniformis* and *Melodogyne javanica. Nematropica*, **24(2)**:113-121.

Smith, M. K., Ko, H. L., Hamill, S. D. and Sanewski, G. M. 2002. Pineapple transformation: managing somaclo nal variation. *Acta Hort*iculturae, **575**: 69-74.

Sripaoraya, S. and Davey, M. R. 2011. F_1 Hybrid Pineapple Resistant to Bialaphos Herbicide. 7th International Pineapple Symposium. *Acta Horticulturae,* **902.**

Sripaoraya, S., Blackhall, N. W., Marchant, J., Power, J. B., Lowe, K. C. and Davey, M. R. 2001. Relationship in pineapple by random amplified polymorphic DNA (RAPD) analysis. *Plant Breeding,* **120**: 265-267

Sripaoraya, S., Keawsompong, S., Insupa, P., Power, J. B., Davey, M. R. and Srinivas, P. 2006. Genetically manipulated pineapple; transgene stability, gene expression and herbicide tolerance under field conditions. *Plant Breeding,* **125**:411-413

Subramanian, N., Iyer, C. P. A. and Singh, R. 1981. Surmounting self-incompatibility in pineapple (*Ananas comosus* L.) with pollen irradiation. *Indian Journal of Horticulture,* **38**:3-4.

Torres, Navarro. H., Lozoya, Saldana., H. and Uriza, Avila, D. 1989. Seleccion clonal de pina *Ananas comosus* L. Merr. con resistencia a la marchitez roja y de caracteristicas comerciales. *Revista Chapingo,* **13-16** (62-63): 156-160

Ventura, J. A., Costa, H.,Cabralc, J. R. S. and de Matos, A. P. 2009. 'Vitória': New Pineapple Cultivar Resistant to Fusariosis .*Acta Horticultuare,* **822.**

Ventura, J. A., Zambolim, L. and Chaves, G. M. 1995. Tissue culture techniques for rapid clonal propagation of pineapple cultivars. *Newsletter of Pineapple Working Group of the International Society for Horticultural Science. Vol.II*, March 1996

Volz, R. K. E., Rikkerink, E., Austin, P., Lawrence, T. and Bus, V. G. M. 2009). "Fastbreeding" in apple: a strategy to accelerate introgression of new traits into elite germplasm. *Acta Horticulturae*, **814**:163-168.

Wee, Y. C. 1974. The Masmerah pineapple. *Word Crops*, **26**:64-67.

Williams, D. D. F. and Fleisch, H. 1993. Historical review of pineapple breeding in Hawaii. *Acta Horticulturae,* **34**:67-76.

Williams, D. D. F. and Yamane, G. M. 1970. Open flowers induced on closed-flower varieties by gibberellin. *PRI News,* **18**:11-12

Yabor, L. , Perez, L., Gomez, D., Olivera, V. A., Mendoza, J. , Martínez J, Escalante, D., Garro, G., Hajari, E. and Lorenzo, F. J. 2020. Histological evaluation of pineapple transgenic plants following 8 years of field growth. *Euphytica,* **216**(2) :22-30.

Yapi, A, and Coppens D' Eeckenbrugge, G. 1997. Evaluation of the first hybrid clones (Cayenne x Perolera) at the end of the selection process in Cote d'Ivoire, *Acta Horticulturae,* **425**: 213-219.

Zhang, J., Liu, J. and Ming, R. 2014. Genomic analyses of the CAM plant pineapple. *Journal of Experimental Botany,* **65:**3395–3404.

10

Biotechnological Intervention in Pineapple Improvement

[1]Santhoshkumar G.M., [2]Bidyarani Devi Senjam
[3]Novin Chamling and [4]Nilesh Bhowmick

[1]Department of Fruit Science, J C Diwakar Reddy Horticultural College Tadipatri, Andhra Pradesh
[2]ICAR- National Research Centre for Orchids, Pakyong, Sikkim
[3]Senior Research Fellow, NICRA, Darjeeling KVK, Uttar Banga Krishi Viswavidyalaya, Kalimpong
[4]Department of Pomology & Postharvest Technology, Faculty of Horticulture, Uttar Banga Krishi Viswavidyalaya, Pundibari, Cooch Behar West Bengal, India

---

Pineapple is a short-stemmed herbaceous plant that necessitates a significant amount of planting material. For example, depending on the intensity of planting and the cultivar used, one hectare of cultivation needs 35,000 to 70,000 planting parts. Realizing the need for quick multiplication, Ma (1968) devised a method for quick pineapple multiplication using the decapitation technique. Later a technique for rapid in vitro propagation has been developed. Since then, numerous reports on crop micro-propagation using tissue culture techniques have been published. On supplemented MS medium, different types of explants, such as leaf base, shoot base, excised lateral buds and tips from crowns, young-leaf segments, and excised leaf buds, have been used to regenerate plants. Plantlets raised in this manner thrived in the field. In vitro mutation may also be induced using tissue culture techniques. Shoot tips from in vitro grown plantlets may be treated with chemical mutagens (EMS) or irradiated with gamma rays to produce mutants with superior traits such as spineless leaves, larger fruits, improved pulp consistency, or disease resistance. Similarly, somaclonal variants can be screened out early to improve field results

In vitro Culture

Pineapple cultivars Smooth Cayenne and Sugar Loaf were used. As explants, young basal suckers were collected from a commercial farm near Kwabenya. The explants were washed in tap water, then cleansed for 15 minutes in 7X detergent, rinsed in distilled water, and surface sterilised for 20 minutes in 20 percent Clorox containing 2-3 drops of 'Tween 80'surfactant. After that, four rinses in sterile distilled water were performed. The shoots were cut even further, Shoot tips were cultured on MS salts , 3.5 percent sucrose, 3 uM thiamine HC1, 3 jaM naphthalene acetic acid (NAA), and varying concentrations (0, 10, 15, 20, and 25 uM) of 6-benzylaminopurine (BAP). The pH of the medium was changed to 5.8 before sterilisation. The cultures were randomly assigned to three replicates for each treatment in a fully randomised design. After 8 weeks of culture, multiple shoot buds were formed in liquid media and agitated on a rotary shaker at 70 rpm. Strong MS medium supplemented with 1.5 uM indole-3-butyric acid (IBA), 0.75 uM indole-3-acetic acid (IAA), and 0.8 percent agar was used to induce rooting. All cultures were held at 27°C for 16 hours under 3000 lux cool-white fluorescent lighting. Tissue culture experiments were therefore conducted to develop rapid multiplication procedures for 'Yeppoon Gold'. Axillary buds from suckers were treated with 0.025% (w/v) mercuric chloride for five minutes and placed in different media. Rapid shoot initiation was found in MS medium supplemented with 1 mg L-1 BAP after four weeks incubation. Shoots were subcultured by bisectecting vertically and placing on MS or Bs medium containing 30.0 g L-1 sugar, 8.0 g L-1 agar and 0.0, 0.5, 1.0, 1.5, 2.0 or 2.5 mg L-1 BAP. The highest multiplication was obtained in Bs medium at 2 mg L-1 BAP. Repeated subculturing in the same medium for seven months resulted in root initiation. Thus, Bs medium with 2 mg L-1 BAP can serve as both multiplication and rooting medium. This protocol, which eliminates the extra step required for root initiation, has the potential to produce up to a million plants per year from about 10-20 suckers.

Radio-Sensitivity and Mutation Induction

Shoot tips produced in vitro were irradiated at a rate of 215 Gy/hr with 0, 15, 25, 45, 60, 80, 100, and 120 Gy gamma rays from a CO^{60} source. Each treatment had 50 explants, and each treatment was repeated three times. Irradiated shoot tips were transferred to a medium supplemented with 20 uM BAP and irradiated shoot tips were transferred to a medium supplemented with 20 uM BAP (based on results from in vitro culture). After 8 weeks of culture, the radiation response was measured in terms of *ex plant* survival and shoot proliferation. Surviving shoot tips were transferred to fresh medium and sub-cultured up to M1V4 in a sequential manner. Following the radio-sensitivity procedure, in vitro-grown shoots of both cultivars were irradiated with a 45 Gy dose and sub-cultured to M1V4 using the micro-propagation process.

Plantlets derived from in vitro irradiated material were transplanted to a greenhouse for the selection of high-temperature-tolerant variants. Plants were grown in loamy soil in 30x25 cm black polythene bags. Plants were gently washed in water to eliminate any traces of agar on the roots before being transplanted into soil. The potted plants were put in a heat chamber with low relative humidity and a day temperature of 45°C and a night temperature of 40°C.or two weeks, the plants were watered with a 6.6 mg/ml 'Raizal 400' macro-nutrient solution, then with daily water as required. Plants were tested for the existence of chlorophyll and spineless variants on a regular basis. The plants were kept in the greenhouse for a long time so that field testing could take place during the dry season. The weaned plants were moved to the field for further selection and evaluation after 5 months.

Haploids and Doubled Haploids

Limited success has been reported for the production of callus from both anthers and ovules of pineapple although there was a high genotype dependency (Benega *et al.,* 1996, 1997a, 1998). Irradiation of pineapple pollen has demonstrated a dose and genotype response (Benega *et al.*, 1997b) but the use of irradiated pollen to produce haploid seed has not been reported. A method for developing polyploid pineapple from colchicine treated callus was described by Nagatomi *et al.* (1997). This technique might have use in conjunction with callus derived from anthers or ovules. At this stage, the techniques for production of haploid and double haploid pineapple cultivars do not appear to be mature enough to be useful. Also, the viability of completely homozygous pineapple plants is questionable given the response to a loss of heterozygosity. Techniques to improve the success of selfing are likely to be more successful, easier and less expensive.

Genetic Engineering

Genetic engineering facilitates the improvement of human traits by targeting specific genes of interest. Genetic transformation studies in the sense of fruit trees have focused on developing lines with increased resistance to pathogens (viruses, fungi, and bacteria) as well as abiotic stresses such as drought, salinity, and cold (Gambino and Grib-audo, 2012). We based our efforts on developing transgenic pineapple with enhanced tolerance to a fungal pathogen in order to overcome the foregoing constraints. Since genetically modified plants must be tested at different stages to ensure efficacy of the added trait and the protection of the developed product, this is a long-term project. Our specific aim was to create pineapple plants that could withstand *Phytophthora nicotianae*var. *parasitica* infection (Kamoun 2001).Two genes were stacked to impart resistance, namely a chitinase gene from *Phaseolus vulgaris* and a DNA sequence cloned from *Nicotiana tabacum*, to increase the probability of success (designated AP24). The AP24 portion codes for an anti-Phytophthora protein that interrupts the fungi's cell membrane. A selectable marker gene (the bar gene from actinobacteria) was also included, which conferred resistance to the herbicide phosphinothricin (Thompson *et al.,* 1987).Constitutive promoters were used to drive all transgenes, including OCS-35S CaMV-rice actin I, 35S CaMV, and maize Ubi1 (Espinosa *et al.,* 2002). As part of the regulatory processes regulating the use of genetically modified plants, these plants must be thoroughly tested to ensure that the genetic engineering has not compromised the crop's protection (in terms of changes in allergenicity, toxicity, potential for increased weediness, etc.)As a result, both the expected and unintended consequences of genetic engineering must be examined. The intended function of the chitinase and AP24 genes is known and has been demonstrated in previous studies in the context of this study (Nagendra-Pradra *et al* 2007).However, in order to assess the protection of the modified seed, unintended effects that could occur as a result of the genetic modification must also be identified (Hossain 2016). Research on the characterization of genetically modified staple crops (e.g., maize, soybean, potato, etc.) is easily available (Della Penna 2015). However, similar information on unintended effects in genetically modified tropical crops, including pineapple are limited.

DNA molecular fingerprinting techniques: The genetic diversity and relationships among pineapple accessions through the use of DNA molecular fingerprinting techniques would assist breeders using pineapple cultivars for breeding programs. Polymerase chain reaction (PCR) based DNA markers provides an opportunity to characterize genotypes and measure genetic relationships more precisely than other markers (Vanijajiva, 2011). Among the various molecular marker techniques, random amplified polymorphic DNA (RAPD) and inter-simple sequence repeat (ISSR) are easily to be carried out, low cost, and do not require prior genetic information (Williams *et al.,*1990; Zietkiewicz *et al.*, 1994). The two techniques have been used effectively for identification and assessment of germplasm in pineapple (Prakash *et al.,* 2009). In Thailand the RAPD technique has been used to document the genetic diversity and constructed genetic relationships among Thai pineapple cultivars (Sripaoraya *et al.*, 2001, Popluechai *et al.,* 2007). Although RAPD could amplify the entire information of the genome, but RAPD-PCR has the disadvantage of exhibiting spurious bands due to the low stringency of the technique (Zhao *et al.,* 2009; Kumar *et al.,* 2010). ISSR-PCR has been described as an alternative approach for polymorphism studies (Zietkiewicz *et al.,* 1994).The high reproducibility with ISSR technique is attributable to the use of longer primers allowing for higher annealing temperatures than those of RAPDs (Pradeep *et al.,* 2002). Analysis of genomic DNA using the technique of ISSR is increasingly employed to determine genetic diversity and relationships in plants such as melon and cucumber (Parvathaneni *et al.*, 2011), orange (Kumar *et al.*, 2010), mango (Luo *et al.*, 2011), durian (Vanijajiva, 2012), including pineapple (Carlier *et al.*, 2004; Prakash *et al.*, 2009). Assessment of genetic diversity using ISSR techniques in Thai pineapple cultivars may give additional or more information possibly be used in improve breeding of pineapple. In present study, ISSR polymorphism is used to support the phenotypic identification and to distinguish Thai pineapple cultivars. Genetic diversity and relationships among commercially Thai pineapple varieties were assessed.

Volatile fingerprinting and sensory analysis: The key volatile compounds and aroma profle of six pineapple varieties grown in Malaysia were investigated by gas chromatography–olfactometry (GC-O), gas-chromatography–mass spectrometry and qualitative descriptive sensory analysis. A total of 59 compounds were determined by GC-O and aroma extract dilution analysis. Among these compounds, methyl-2-methylbutanoate, methyl hexanoate, methyl-3-(methylthiol)-propanoate, methyl octanoate, 2,5-dimethyl-4-methoxy-3(2H)-furanone, δ-octalactone, 2-methoxy-4-vinyl phenol, and δ-undecalactone contributed greatly to the aroma quality of the pineapple varieties, due to their high favour dilution factor. The aroma of the pineapples was described by seven sensory terms as sweet, foral, fruity, fresh, green, woody and apple-like.Inter-relationship between the aroma-active compounds and the pineapples revealed that 'Moris' and 'MD2' covaried majorly with the fruity esters, and the other varieties correlated with lesser numbers of the fruity esters. Hierarchical cluster analysis (HCA) was used to establish similarities among the pineapples and the results revealed three main groups of pineapples.

Micro RNAs (miRNAs) are a class of small, non-coding, endogenous RNAs that negatively regulates gene expression by cleaving or repressing the translational process of protein-coding mRNA transcripts. To determine if pineapple fruit development and ripening was mediated by miRNA, miRNA were detected and quantified at pre- and post-ripening stages using RT-qPCR. The expression profiles of 13 miRNAs namely miR156, miR157, miR159, miR162, miR164, miR165, miR167, miR168, miR169, miR170, miR171, miR172, and miR390 were detected in pineapple. Among these 13 miRNAs, only the genes (miR168, miR390, and miR164) displayed clear differential expression during pre-ripening and post-ripening stages of the fruit. During the post-ripening stage, miR168 and miR390 were up-regulated 1.79- and 2.64-fold, respectively, while expression of miR164 dropped to 0.07-fold compared to the pre-ripening stage. The discovery of miRNAs from this study together with its stage-regulated expression profiles are excellent candidates for further investigation into pineapple crop improvement through functional genomes.

DNA extraction and RAPD reaction: DNA extraction was carried out according to Ramessur and Ranghoo (2011). The RAPD-PCR reactions were carried out in 30 µL PCR tubes using a Biorad PCR machine. Each 30 µL PCR reaction volume consisted of 1xPCR buffer (160 mM $(NH_4)_2SO_4$, 670 mM tris-HCl and 0.1% Tween 20), 3 mM $MgCl_2$, 0.25 mM dNTP, 0.6 mM primer, 1 U *Taq* polymerase, 50 ng DNA template and 17.7 µL sterile water. Negative control lacking DNA template was also included in the PCR reaction. The cycling parameter set comprised of 1 cycle of 90 sec at 95°C, 40 cycles of 30 sec at 92°C, 1 min at 35°C and 3 min at 72°C. This was followed by an extension of 10 min at 72°C and a final delay of 5 min at 15°C. Reaction analysis was evaluated using electrophoresis on a 2% (w/v) agarose gel using 0.5xTBE buffer at 115 V. It was then stained with EtBr for 5 min and destained in running water (5 min). Finally, it was visualized under a UV transilluminator.

RAPD profile analysis: Sixty Operon primers were purchased from Inquaba Biotechnology Industries Ltd., South Africa and used to evaluate genetic diversity. All the screened primers that gave clear and distinct amplification products were used to run with all the six varieties. Each genotype was characterized by its banding pattern using the DNA hyperladder 2 (Bioline) as basepair ladder. The RAPD markers as viewed from the gels after electrophoresis analysis and staining that were converted into a matrix of binary data, where the presence of the band corresponded to value 1 and the absence to value 0. The statistical software NTSYS-PC (Rohlf, 2005) and DARwin 5 software (Perrier and Jacquemoud-Collet, 2006) were used to construct a UPGMA dendrogram using hierarchical clustering. Using NTSYS software, a dissimilarity matrix was calculated utilising Jaccard (1908) coefficient. The matrix was converted to a dissimilarity matrix corresponding to the complement (dissimilarity = 1-similarity). Cluster analysis based on the dissimilarity matrix, was performed using un-weighted pair group method arithmetic averages (UPGMA) (Sneath and Sokal, 1973) of the NTSYS-PC version 2.2 (Rohlf, 2005). Synthetic deoxyribo-nucleotides were used as primers for amplification of pineapple DNA. Among the 60 primers tested, 13 RAPD primers designated as OPA-03, OPA-04, OPA-10, OPA-13, OPD-02, OPD-03, OPD-07, OPD-13, OPC-11, OPC-15, RAPD-02, RAPD-03, RAPD-04 produced bands. Cluster analysis conducted on the RAPD markers produced a dendrogram of the genetic relationship of the 30 pineapple cultivars. The genetic distances among the cultivars analyzed ranged from 81% for closely related cultivars and 44% for those more distantly related. The results of the dendrogram separated the thirty cultivars into two major groups. Group I included five cultivars including 'Cayenne' introduced from Thailand, 'Thom Ha Noi', 'Thom BT23', 'ThomBT27', and 'Queen Tay Ninh'. Group II included the remaining twenty five cultivars, which were separated into two minor groups. Group II-1 consisted of the closely related cultivars 'ThomBT9', 'ThomBT29' and 'ThomBT39' and these cultivars were all derived from Ben Tre province. Group II-2.1 included the cultivars 'Cayenne UcI và' and 'Cayenne UcII' and these cultivars were introduced from Australia. Group II-2.2 included the closely related 'Cayenne Ivory Coast' and 'Cayenne Dai Loan BL' which had a close genetic relation to 'Cayenne Martinique-I'. 'Cayenne Ivory Coast' and 'Cayenne Martinique-I' were originally from the French. 'Cayenne Thai Lan MC', 'Cayenne Dai Loan I', 'Thom Nhat', 'Thom Phung', 'ThomBT28', 'ThomBT24', 'ThomBT22', 'ThomBT25', 'ThomBT32', 'ThomBT30', 'ThomBT31', 'Thom Tay Ninh', 'Cayenne Lam Dong', 'DN17', 'Cayenne Trung An', 'Khom Ben Luc', 'Khom Kien Giang' had similarity coefficients of about 70% .

Molecular marker-assisted selection is a breeding method which provides the potential for increasing selection efficiency by allowing for earlier selection and reducing the plant population size required for field selection (Khawale *et al.*, 2007). RAPD is one type of molecular marker and the basis of this technique is using short random oligonucleotide sequences for differential PCR amplification of genomic DNA (Williams *et al.,* 1991; Agarwal *et al.,* 2008). The main advantages of RAPD are that it is simple, cheap and there is no requirement for DNA sequence information. However, its reproducibility is lower and it is also difficult to analyze approximate fragments (Tingey and Tufo, 1993; Schlötterer, 2004). In the present study, the average genetic similarities between EMS treated plants and the control decreased as the concentration of EMS increased. The proximity analysis also showed that higher concentrations of EMS induced plants that were more distinct from the control. It seems that the genetic dissimilarities of EMS treated plants gradually increased when the EMS concentration increased. This indicates that EMS concentration plays a critical role in the efficiency of mutation induction in pineapples and RAPD is able to detect the difference in EMS induced mutants. Some induced morphological changes of EMS treated pineapple progenies were found in the present study such as changes in chlorophyll and anthocyanin contents in leaves, elongation and swelling of roots, leaf length and width, and spine patterns on leaf margin, etc.. These mutagen-derived morphologic traits might be useful for further gene function studies. In addition, those EMS treated plants produced in this study have been transplanted in the field for further evaluation and possible selection. Flowering and fruit characteristics will be examined.

Simple Sequence Repeats (SSRs): Simple Sequence Repeats (SSRs), also known as microsatellites or Short Tandem Repeats (STRs), are ubiquitous, hypervariable and abundant in many eukaryotes' genomes. They are generally thought to differ in size due to the effect of replication error and unequal recombination. Their distribution varies across genic, intergenic regions and non-coding regions of the DNA. SSRs may be predominant in certain regions which may reflect the tendency of mutational processes to create certain repeat patterns under selective pressure. The expansion of repeat motifs in the coding regions of the gene may alter gene expression. This paper reports the discovery of tandem repeat motifs and their frequency from a fully-ripe pineapple fruit transcriptome generated from contigs derived from paired-end Solexa mRNA sequencing. Paired-end sequences, with lengths of 75 bp each and with insert sizes of 200 bp, were assembled using Velvet which generated 28,728 contigs. These contigs were then used to mine for di-, tri,- and tetranucleotide SSR motifs, which are present in the fruit, by using SynaRex (Synamatix). Primers flanking the SSR loci were then designed by using the online software PRIMER 3. Mining of the regenerated contigs showed that only 3.5% of the fully-ripe pineapple fruit transcriptome contained SSRs. Dinucleotides were the most abundant and they accounted for 498 (49.6%) contigs containing SSR followed by trinucleotides and tetranucleotides with 467 (46.5%) and 40 (3.9%) contigs, respectively. Among the dinucleotide motifs, (TC)n and (GA) n were the most abundant with 45 and 41.6% occurrences, respectively. For trinucleotide motifs, (GGA)n, (CTC)n, (AGA)n and (TTC)n showed higher occurrences compared to other trinucleotide motifs with 20.3, 16.5, 13.1 and 9.2%, respectively. Of all the contigs containing SSR, only 26.6% were suitable for designing flanking primers for PCR. Of these, 66 contigs were dinucleotide, 190 contigs were trinucleotide and 11 contigs were tetranucleotides. Further validation of the PCR primers will provide wider application in biomarker identification, genome mapping and characterization, phenotype mapping, marker assisted selection and finally in diversity studies of pineapples (Onga and Kumar 2011).

Total RNA Extraction and Small RNA Purification: Total RNA was isolated from the fruit tissues using an optimized method originally developed by Li *et al.* (2005). Briefly, approximately 3 g of frozen tissue was pulverized with a mortar and pestle. About 6 ml lysis buffer (10 mM Tris-HCl, pH 8.0, 20 mM EDTA, 400 mM NaCl, 1% SDS and 2% β-mercaptoethanol) was immediately added and the mixture was vortexed vigorously for 1 min and then centrifuged at 10,000× g for 10 min at 4°C. Next, the supernatant was transferred into a fresh tube and 0.3 volume of saturated NaCl solution was added and mixed well. After that, an equal volume of phenol/chloroform mix (5:1, pH 4.5) was added and again vortexed vigorously for 1 min and followed by centrifugation at the same settings. The resulting upper layer was re-extracted with an equal volume of chloroform/isoamyl alcohol mix (24:1). After centrifugation, the upper layer was transferred to a new tube and the total RNA was precipitated with the addition of an equal volume of cold isopropanol and left overnight at -20°C. The total RNA was collected by centrifugation at 15,000× g for 20 min at 4°C. The total RNA was then re-suspended in 50 µl RNAse-free water. Large 179 RNA molecules (with >300 nucleotides) were then removed by using the mirVana microRNA isolation kit (Ambion). This was done to enrich the sample with small RNAs which contains the miRNAs.

Construction of Small RNA Library: The small RNAs (sRNAs) were electrophoresed in a 15%/8 M urea denaturing polyacrylamide gel along with a microRNA ladder (NEB) as a size standard. The gel was stained with 1x SYBR Gold solution and the regions corresponding to 17-25 nt were excised. The small RNAs were eluted overnight at 4°C with 0.3 M NaCl solution. Next, the small RNAs were precipitated with 20 µg of glycogen and 3 volumes of absolute ethanol. The small RNAs were then ligated sequentially to the 3'-linker (5'-rAppCTGTA GGCACCATCAAT-NH2-3') and 5'-linker (5'-ATCGTrArGrGrCrArCrCrUrGrArArA-3'), with T4 RNA ligase 2 (truncated) and T4 RNA ligase (NEB), respectively (Lau *et al.*, 2001). The underlined sequence indicates the position of the BanI restriction endonuclease site. The end-ligated small RNAs were then reverse transcribed with 200 U of SuperScriptIII to obtain cDNAs. The sRNA-cDNAs were then PCR-amplified with Phusion High Fidelity DNA polymerase (Fynnzyme) using the conditions suggested by the manufacturer. Next, the PCR products, which were approximately 50-60 bp, were digested with BanI. The digested fragments, which were about 40-50 bp, were then gel-purifed. These small fragments containing sticky ends at both terminals were concatamerized to form long strands of sRNA-cDNAs by T4 DNA ligase (NEB). The free terminals of these concatamers were then filled-in with Taq DNA polymerase (Promega) before it was ligated to pGEM-T Easy vector (Promega) and transformed into JM109 competent cells, according to the manufacturer's protocols. A total of 150 white clones were then randomly selected for Sanger sequencing.

Analyses with Bioinformatics Tools: Small RNAs were isolated from the concatamer sequence by removing the concatamer connector sequence (5'-CTGTAGGCACCTGAAA-3'). All sRNAs were screened for the presence of conserved miRNAs by searching the online miRBase database (http://www.mirbase.org). The precursor-miRNAs were identified by BLASTing the small RNA sequences with available pineapple ESTs. Next, MFold was used to predict the stem-loop structures for putative precursor-miRNA sequences. To predict for the target genes, the antisense-miRNA sequences were BLAST with pineapple ESTs using the tool available at GenBank. To identify differentially expressed profiles between pineapple and tomato miRNAs, newly discovered conserved and putative pineapple-miRNAs were used to screen and compare with the online Tomato Functional Genetics Database (TFGD). The potential target genes were also screened using the tomato ESTs database at TFGD.

Delayed Flower Initiation

Flowering is important physiological process in the plant ontogeny. It is influenced by many endogenous and exogenous factors acting through specific plant hormones. In pineapple industry, natural flowering is problematic to some extent as it is not synchronised with induced flowering. Conditions like short days and low temperatures increase the concentration of ethylene in the plant which stimulate flowering. Fruit maturity occurs 4 -6 weeks prior to the normal summer harvest in natural flowered pineapple plant resulting in need for an extra harvest also sometimes lead to losses in yield (Trusov and Botella, 2006). Silencing of the host gene ACC synthase gene (ACACS2) was done in 'Smooth Cayenne' pineapple by co-suppression mechanisms. ACACS2 gene is a main gene responsible for flowering in pineapples and suppression of its expression will help in controlling the natural flowering. ACC synthase gene is expressed in meristems and suppression of ACACS2 gene gene is done by methylation which delayed the flowering in transgenic plants. (Trusov and Botella, 2006). Methylation is the main mechanism resulting in suppression of gene and thus silencing the gene associated, which is also associated with transformation with foreign DNA and transposon inactivation (Jaenisch and Bird, 2003).

Transgenic Pink Fleshed Pineapple

Genetically modified pineapple for pink flesh was developed by US company Del Monte Fresh Produce Company. This is the only approved genetically modified variety in pineapple named as 'Extra sweet pink flesh pineapple'. This variety is a modification of the very popular export variety MD-2.'Extra sweet pink flesh pineapple' was patented by Firoozbady and Young in the year 2015, under the name 'Rose' and is commonly denoted as the GM EF2-114 event. Modification of the fruit resulted in production of fruit with red to pink flesh due to accumulation of carotenoid pigment. Characteristic pink flesh colour is attained was mainly due to accumulation of red lycopene pigment in the concentration of 21 ppm which similar to that of present in other red coloured fruits such as watermelon, grapefruit and papaya. This transgenic pineapple was developed through *A. tumefaciens*-mediated transformation. Two T-DNA constructs were formed separately and introduced into the GV3101 (pMP90) *A. tumefaciens* strain. The first plasmid, pHCW.T-7 is a combination of four cassettes. First is the phytoene synthase (*psy*) gene from tangerine (*Citrus unshiu*), second is a partial coding fragment of the β-lycopene cyclase (*Lcyb*), third is the ε-lycopene cyclase (*Lcye*) gene and fourth is mutant acetolactate synthase (ALS) gene from tobacco (*SuRBHra*), conferring chlorsulfuron resistance The second plasmid, pHCWflACC3'-2, contained a partial coding fragment of the pineapple meristem *ACS* gene (flACC3'in the sense and antisense orientation.. Anti-sense sequences help in RNA interference process and gene silencing of the gene which in turn caused expression of the LCYB and LCYE and accumulation of lycopene. The sense and anti-sense fragments, promoters, terminators and flanking sequences, all were originated from pineapple. Thus, introduced DNAs in to the plant constitute only DNA's from the edible plants and only one gene was taken from tobacco. MD-2 shoots were initially cultured in vitro and the separated leaf base sections were co-cultured withbacterial strain, which resulted in the co-delivery of T-DNA fragments from both the pHCW.T-7 and pHCWflACC3'-2 plasmids.

Source: www.isaaa.org

Nutrient Accumulation in Pineapple (Biofortification)

First report on the transformation of pineapple with nutrients like iron and zinc in the transformed plants was done by Mhatre *et al.* (2011). They transformed pineapple cultivar "queen" with soybean ferritin cDNA using *Agrobacterium tumefaciens* strain EHA 105mediated gene transfer. Leaf bases were used for the transfer of soybean ferritin cDNA in the pSF and pEFESF plasmids. Co-cultivated leaf bases with *Agrobacterium tumefaciens* were shifted to Murashige and Skoog's medium supplemented required nutrient medium required for shoot proliferation. After 6 weeks of culturing in in the invitro conditions, multiple shoots were produced from which 1-2 cm shoots were selected and further multiplied. Multiplied shoots of 5cm length were further shifted to rooting medium supplemented with cefotaxime 400 mg/l and kanamycin 100 mg/l. Rooted plants when analysed through PCR, southern analysis and reverse transcription PCR conformed the expression of ferritin gene in transformed plants. Transformed plants rt5t5t5t5t56667showed around, 5.03-fold increase in iron content and 2.44-fold increase in zinc content in the leaves of pSF-transformed plant in the analysis carried by Atomic absorption Spectrometry. Nearly, iron content increased 3.65 times and zinc level to 2.05 times in pEFE-SF-transformed plants (Mhatre *et al.*, 2011).

References

Agarwal, M., Shrivastava, N. and Padh, H. 2008. Advances in molecular marker techniques and their applications in plant sciences. Plant Cell Reports, 27: 617-631.

Benega, R., Isidron, M., Arias, E., Cisneros, A., Martinez, J., Companioni, L. and Borroto, C. G. 1997a. Plant regeneration from pineapple ovules (Ananas comosus (L.) Merr.). Acta Horticulturae, 425:247-250.

Benega, R., Isidron, M., Arias, E., Daquinta, M., Dominquez, Q., Borroto, C. G. and Hidalgo, M. 1995b. Comparative study between two pineapple (Ananas comosus L.) cultivars. Newsletter of Pineapple Working Group of the International Society for Horticultural Science. Vol. II, March 1996

Benega, R., Isidron, M., Cisneros, A., Arias, E., Daquinta, M., Companioni, L. and Martinez, J. 1996. Induction of callus in pineapple anthers. Cultivos Tropicales 17:72- 74.

Benega, R., Viciedo, L., Martinez, J., Castillo, E., Cisneros, A., Romero, M., Isidron, M., Fernandez, J. and Arias, E. (1997b) Effect of gamma irradiations on pineapple pollen germination and tube growth. Fruits, 51:425-428.

Carlier, J. D., Reis, A., Duval, M. F., d'Eeckenbrugge, G. C. and Leitao, J. M. 2004. Genetic maps of RAPD, AFLP and ISSR markers in Ananas bracteatus and A. comosus using the pseudo-testcross strategy. Plant Breeding, 123:186-192.

Della Penna, D. 2015. Plant metabolic engineering. Plant Physiology, 125:160–163.

Espinosa, P., Lorenzo, J. C., Iglesias, A., Yabor, L., Menendez, E., Borroto, J., Herandez, L. and Arencibia, A. 2002. Production of pineapple transgenic plants assisted by temporary immersion bio reactors. Plant Cell Reports. 21:136-140.

Gambino, G. and Gribaudo, I. 2012. Genetic transformation of fruit trees: current status and remaining challenges. Transgenic Research, 21:1163–1181.

Hossain, M. F. 2016. World pineapple production: an overview. African Journal of Food and Agriculture, 16:11443–11456.

Jaccard, P. 1908. Nouvelles recherches sur la distribution florale [New research on the floral distribution]. Bulletin De La Societe Vaudoise Des Science Naturelles, 44: 223-270.

Jaenisch, R. and Bird, A. 2003. Epigenetic regulation of gene expression: how the genome integrates intrinsic and environmental signals. Nature Genetics, 33:245-254.

Kamoun, S. 2001. Non host resistance to Phytophthora: novel prospects for classical problem. Current Opinion in Plant Biology, 4:295-300.

Khawale, R. N., Yerramilli, V. and Singh, S. K. 2007. Molecular marker-assisted selection of in vitro chemical mutagen-induced grapevine mutants. Current Science, 92: 1056-1060.

Kumar, S., Jena, S. N. and Nair, N. K. 2010. ISSR polymorphism in Indian wild orange (Citrus indica Tanaka, Rutaceae) and related wild species in North-east India. Scientia Horticulturae, 123: 350– 359.

Lau, N. C., Lim, L. P., Weinstein, E. G. and Bartel, D. P. 2001. An abundant class of tiny RNAs with probable regulatory roles in Caenorhabditis elegans. Science, 294:858- 862.

Li, Q., Lawrence, C. B., Xing, H. Y., Babbitt, R. A., Bass, W. T., Maiti, I. B. and Everett, N. P. 2001. Enhanced disease resistance conferred by expression of an antimicrobial magainin analog in transgenic tobacco. Planta, 212:635–639.

Luo, C., He, X. H., Chen, H., Ou, S. J., Gao, M. P., Brown, J. S., Tondo, C. T. and Schnell, R. J. 2011. Genetic diversity of mango cultivars estimated using SCoT and ISSR markers. Biochemical Systematic and Ecology, 39: 676-684.

Ma, S. 1968. The decapitation method for rapid multiplication of pine apple Journal of Hoerticulture Society of China, 14(1-2):1-5

Mhatre, M., Srinivas, L. and Ganapathi, T. 2011. Enhanced Iron and Zinc Accumulation in Genetically Engineered Pineapple Plants Using Soybean Ferritin Gene. Biological Trace Elemental Research, 144. 12-19

Nagatomi, S., Degi, K. and Ikemiya, H. 1997. Development of polyploidy induction and discrimination methods in pineapple. Technical News-Institute of Radiation Breeding, pp. 2.

Nagendra-Pradra, D., Sudhakar, N., Murugesan, K. and Mohan, N. 2007. Application of ozone on induction of resistance in Vigna unguiculata cv. Co 6, against Fusarium wilt. Archives of Phytopathology and Plant Protection, 3:1–10.

Onga, W. D. and Kumar, S. V. 2011. Frequency and Distribution of Simple Sequence Repeats (SSRs) in Pineapple Fruit Transcriptome.7th International Pineapple Symposium. Acta Horticulturae, 902.

Parvathaneni, R. K., Natesan, S., Devaraj, A. A., Muthuraja, R., Venkatachalam, R., Subramani, A. P. and Laxmanan, P. 2011. Fingerprinting in Cucumber and Melon (Cucumis spp.) genotypes using morphological and ISSR markers. Journal of Crop Science and Biotechnology, 14: 39-43.

Perrier, X. and Jacquemoud-Collet, J. P. 2006. DARwin software. Genetic Improvement of Vegetatively Propagated Crops. http://darwin.cirad.fr/Home.php.

Popluechai, S., Onto, S. and Eungwanichayapant, P. D. 2007. Relationships between some Thai cultivars of pineapple (Ananas comosus) revealed by RAPD analysis Songklanakarin, Journal of Science and Technology, 29: 1491-1497.

Pradeep, R. M., Sarla, N. and Siddiq, E. A. 2002. Inter simple sequence repeat (ISSR) polymorphism and its application in plant breeding. Euphytica, 128: 9–17.

Prakash, J., Bhattacharyya, S., Chattopadhyay, K., Roy, S., Das, S. P. and Singh, N.P. 2009. PQM-1: A newly developed superior clone of pineapple for north eastern India as evident through phenotype, fruit quality and DNA polymorphism. Scientia Horticulturae, 120: 288–291.

Ramessur, A. D. and Ranghoo-Sanmukhiya, V. M. 2011. RAPD marker-assisted identification of genetic diversity among mango varieties in Mauritius. International Journal of Agriculture and Biology, 13: 167-173.

Rohlf, F. J. 2005. NTSYSpc (Numerical Taxonomy and Multivariate Analysis System). Version 2.2, Exeter Software, Applied Biostatistics Inc., New York, USA.

Schlötterer, C. 2004. Opinion: The evolution of molecular markers - just a matter of fashion. Nature Review on Genetics, 5: 63-69.

Sneath, P. H. A. and Sokal, R. R. 1973. Numerical Taxonomy. 1st Edn., W.H. Freeman and Co., San Francisco, USA., ISBN-10: 0716706970, P: 573.

Sripaoraya, S., Blackhall, N. W., Marchant, J., Power, J. B., Lowe, K. C. and Davey, M. R. 2001. Relationship in pineapple by random amplified polymorphic DNA (RAPD) analysis. Plant Breeding, 120: 265-267.

Tingey, S. V. and Tufo, J. P. 1993. Genetic analysis with random amplified polymorphic DNA markers. Plant Physiology, 101: 349-352.

Trusov, Y. and Botella, J. R. 2006. Delayed flowering in pineapples (Ananas comosus (l.) merr.) caused by co-suppression of the acacs2 gene. Acta Horticulture, 702: 29–36.

Trusov, Y. And Botella J. 2006. Silencing of the ACC synthase gene AcACS2 causes delayed flowering in pineapple [Ananas comosus (L.) Merr.]. Journal of Experimental Botany, 57: 3953–3960.

Vanijajiva, O. 2011. Genetic variability among durian (Durio zibethinus Murr.) cultivars in the Nonthaburi province, Thailand detected by RAPD analysis. Journal of Agriculture and Technology, 1105-1114.

Williams, J. G. K., Kubelik, A. R., Livak, K. J., Rafalski, J. A. and Tingey, V. S. 1990. DNA polymorphism amplified by arbitrary primers are useful as genetic markers. Nucleic Acids Research, 18: 6531- 6535.

Zhao, G. H., Li, J., Zou, F. C., Mo, X. H., Yuan, Z. G., Lin, R. Q., Weng, Y. B. And Zhu, X. Q. 2009. ISSR, an effective molecular approach for studying genetic variability among Schistosoma japonicum isolates from different provinces in mainland China. Infection, Genetics and Evolution, 9: 903– 907.

Zietkiewicz, E., Rafalski, A. and Labuda, D. 1994. Genome fingerprinting by simple sequence repeat (SSR)-anchored polymerase chain reaction amplification. Genomics, 20: 176–183.

11

Edaphic and Climatic Requirement of Pineapple

[1]M. Premi Devi, [2]Prahlad Deb and [3]U. K. Behera

[1]College of Agriculture, Central Agricultural University, Kyrdemkulai Meghalaya, India
[2]Department of Horticulture & Postharvest Technology, Institute of Agriculture, Visva-Bharati, Sriniketan, West Bengal, India
[3]College of Agriculture, Central Agricultural University, Kyrdemkulai Meghalaya, India

Successful cultivation of a crop is governed by various factors including the edaphic and climatic conditions of the region, owing to which the crop responds in terms of quality and productivity. Pineapple (*Ananas comosus* L. Merr.) is one of the major crops cultivated across the tropical and sub-tropical countries worldwide. Adaptability of the crop under different climatic conditions, topography, temperature, photo-periodicity, rainfall and edaphic conditions will be discussed in this chapter.

Climatic zone: Pineapple is a tropical fruit crop though it can thrive well in the sub-tropical condition as well.Pineapple grows well inwarm and humid climate ranging from mild coastal climate up to an altitude of about 1000 meters, provided the area is free from frost (Singh, 2014).Pineapple is commercially grown over a wide range of latitudes from 30° N (Assam, India) in the northern hemisphere to 33°S in the Southern hemisphere (South Africa) (Malezieux *et al*., 2003; Purseglove, 1972).

Topography: Pineapples are easy to cultivatesince they thrive naturally in a variety of places with different elevation, especially in tropical countries (Kaudo, 2014). Though the plant survives at different altitude, the fruit quality and size may alter at different elevation. When pineapple is grown at altitudes greater than 1000 m, smaller fruits with elevated tartness with less attractive pulp colour (Purseglove, 1968).Moreover, fruit flavour can also vary, wherein the fruit tends to bemore acidic and sour when cultivated above 1800 meters, but will be sweet and tasty under 1350 meters (Kaudo, 2014).

Temperature: On an average, the temperatures required for growing pineapples range within 18-30°C (Kaudo, 2014) andmore specifically at 23 - 24 ° C (Neild and Boshell, 1976; Py *et al*., 1987). The optimum day and night temperatures for pineapple were reported to be about 30°C and 20 °C, respectively (Neild and Boshell, 1976). Bartholomew and Kadzimin (1977), however have reported that leaves and roots grow best at 32°C and 29°C respectively and the growth practically ceases below 20°C and above 36°C. Py and Tisuucau (1965) have observed 25°C to be optimal for fruit ripening.

This crop cannot withstand extreme cold, frost or temperature near 0°C for a prolonged duration. Temperatures below 20°C can cause chlorotic discoloration. This fruit can resist cold nights but for short periods of time, otherwise it will slow down the growth, delay maturity, and it lowers the quality of the fruit by making it more acidic, reducing the sweetness or even freezing it. It's ideal for the location of pineapples plantations to have a range 10 of cold nights and sunny days so it doesn't freeze nor get sunburned. A high temperature at night is deleterious and a difference of at least 4°C must be maintained between day and night as it is clear from an experiment made in phytotron (Bartholomew and Kadzimin, 1977). There are large differences in average annual temperature for various areas where pineapple is grown on a commercial scale. A mildly cold winter gives fruit of a better quality. An absolute maximum temperature has not been established, but it is known that overheated fruits suffer from the 'sun scald'. With exposure to temperatures well over 30°C heat damage may occur due to increased respiration rate and metabolism and impaired nutrient absorption (Bartolomew and Kadzimann, 1987). The countries, where temperature extremes detrimental to pineapple are most common., are South Africa, Australia and Taiwan, where both sunburn and frost damage can occur.

Climatic variations have potential implications for challenging agricultural production. The observed relationship between temperature and yield implies pineapple production is highly sensitive to temperature and is a major climatic factor to consider for pineapple production (Williams *et al*., 2017). The contribution of minimum temperature to yield show variations up to 82%, supporting the view ofRainha *et al*. (2013) that minimum temperature is a critical climate variable in pineapple production. Increases in minimum temperature above 26 °C are known to hinder flower induction and this contributes to loss of yield and quality (Cunha, 2005). Even moderate elevation of temperatures poses possibly significant reduction in flowering of crops (Sthapit *et al*., 2012).

Photo-periodicity: Sunlight is a very important factor for the cultivation of pineapple as the amount of sunlight received determines the fruit weight

(Engebos, 2012). In general, pineapple needs a sunny climate, though there are no exact figures on hours of solar radiation required. Collins (1960) stated that pineapple show almost no photo-periodicity but Py and Tisscau (1965) observed that Smooth Cayenne was a short-day plant and succession of a long dark period was able to induce bloom. Other cultivars react only weakly to day length but may flower when the temperature falls.During periods of intense sunlight and high temperature above about 35°C, the fruit is susceptible to sunburn damage. The damage caused by sunburn in pineapple production areas in the world where this fruit is harvested can be very important in periods when solar radiation is high. Percentages of reject fruit because of this damage can be over 40% in some months of the year if preventative actions are not taken (Gonzalez and Espinoza, 2008). Hence partial shade is desirable in areas receiving high intensity of sunlight.

The best aspect is north-east, followed by north and north-west. Steeper north-west slopes are to be avoided to reduce the risk of sunburn. The worst aspects are southerly ones because they do not get warm in winter.

Rainfall: Pineapple cultivation requires well distributed rainfall. It is suitable in areas having a relatively high atmospheric humidity and an optimal rainfall of 100-150 cm (Bartholomew and Kadzimin, 1977). Almeida *et al.* (2002) stated that in general pineapple requires a minimum monthly rainfall total of 80–100 mm. Where the annual rainfall is less than 500 mm, irrigation is essential. If it is above this threshold, irrigation is still needed if there are three consecutive months with monthly rainfall of less than 15 mm, or four consecutive months with less than 25 mm month^{-1}, or five months with less than 40 mm month^{-1}.

In India, pineapple is grown commercially in Assam, West Bengal, Bihar, Coastal Andhra Pradesh, Kerela, Karnataka, Orissa, Goa, Tripura, Meghalaya and Tamil Nadu. A great deal of variation exists among the agro-climatic situations of these areas. In Kerela, where the rainfall is heavy, it is grown as an intercrop in banana and coconut. Strong sunshine is not desirable for pineapple. Similarly, complete shade is also not suitable, but in India it is sometimes grown in partial shade. It grows well in Golaghat in Assam with annual precipitation of 75 cm and in the Khasi hills with annual rainfall of 500 cm. Optimal rainfall ranges from 1000 to 1500 mm (Bartholomew and Kadzimin, 1977). Successful production in low rainfall areas requires good annual distribution of rainfall and water-conserving management practices such as use of plant litter/plastic mulches. In high rainfall areas, good drainage is important because the pineapple root system is very sensitive to water logging. These appear to be the best guidelines available to pineapple growers contemplating whetheror not to invest in irrigation (other than for crop establishment). There is a similar lack of detailed advice on the yield and quality benefits/penalties that result fromthe application of good/poor water management practices.

Rainfall variation, however, has less significant correlation with yield variation. A study showed that high and low rainfall variations influence particularly the vegetative and yield formation stages of pineapple production stretching the growth stages, increasing the production costs, affecting the quantity and quality of fruits produced and ultimately reducing fruit productivity.

According to FAO Water ("Crop Water Information: Pineapple," 2015) this fruit crop is very sensitive to water deficits because it retards grow, fruiting, and flowering specially in the vegetative phase of the plant's growing process. Pineapples have the ability to resist a wide range of rainfall conditions depending on the location and humidity (best between 70-80%) because of their CAM photosynthesis. The best precipitation would be 1,100mm, even though it can range from 650-3,800mm. It can resist long periods of drought due to their ability to store water on its leaves (Engebos, 2012).

In other cases, high water use efficiency (WUE) of pineapples, making it tolerable to dry conditions, showed insignificant effect of precipitation to yield (Carr, 2012). Meanwhile, Williams *et al.* (2017) stated that the contribution of rainfall to yield variations was up to 45% though comparison between standardized yield anomalies and rainfall anomalies indicated influences of rainfall on yield with the highest frequency. Rainfall influences on yield have also been attributed to the interaction between the cumulative growing period rainfall variations and rainfall during various growth stages (Omoyo *et al.*, 2015). In addition, other studies have also indicated that dry periods during a growing season especially when accompanied by high temperature have the tendency to affect production (Ayanlade *et al.*, 2009; Thornton *et al.*, 2009).

Edaphic condition: The best soils for pineapple production are well drained non-compacted loams, sandy loams and clay loams, without heavy clay or rock within one meter of the surface. The flavour and quality of fruit grown on light soils is considered to be superior. Acidic soil rich in organic matter and potassium is desirable to pineapple cultivation ["Pineapple-Land requirements for growing pineapple", 2014]. The soils required for growing this fruit should be low in magnesium and calcium in order for the pH to stay constantly low. Therefore, acid soils are good for the crop to let chlorosis occur in the plant's leaves. The best soil pH varies between 4.5-5.6 ("Land Requirements for Growing Pineapple," 2013).

Soil temperature influence both biological and chemical processes in the soil, for example, rates of decomposition and mineralization of soil organic matter and release of CO_2. Soil temperature also affects plant growth directly (through its effect on physiological activity) and indirectly (through its effect on soil nutrient availability) (Paul *et al.*, 2002; Paul *et al.*, 2004). Soil temperature controls the rate of evaporation and aeration (Waker, 1971; Awe *et al.*, 2015). The optimum soil temperature for pineapples grow is 20-30°C (Bartholomew*et al.*, 2003). Soil temperature in bare land and 1 month after planting give highest soil temperature than other because the soil not shaded by the plant canopy. Peak drought occurs in October 2015, with the highest soil temperature, reached 54.5°C on soil surface. With irrigation can decrease soil temperature until 14% (Astuti *et al.*, 2016).

The plant cannot resist waterlogging or subsoils, "Pineapples-Cultural practices",2014). Good drainage is fundamental for the root system to be strong and to avoid heart rot diseases. On the hills it can be grown successfully provided the soil depth is at least 45 to 60 cm ("Pineapple-Soil and Climate", 2014). Slopes between 2-6% are best for pineapple growing. Below this range, it may be hard to achieve adequate drainage (Bartolomew and Kadzimann, 1987). Above this range, operational and erosion control costs escalate.

Conclusion: Pineapple can be cultivated in a wide range of edaphic and climatic conditions, making it one of the most suitable fruit crop for cultivation in different parts of the world, especially tropical and sub-tropical countries. The fruit yield, quality and performance may differ according to the variations in elevation, temperature, photoperiodicity, rainfall and soil quality of different locations. In order to improve and maintain the yield and quality performance of pineapple, cultural practices may be specifically practiced according to the requirements and needs of the varied factors prevalent in the region.

References

Anonymous. 2014a. Pineapple-Land requirements for growing pineapple. Department of Agriculture and Fisheries. Queensland Government. https://www.daff.qld.gov.au/plants/fruit-and-vegetables/fruit-andnuts/pineapples/land- requirements-pineapple. Accessed on 31 October, 2014.

Anonymous. 2014b. Pineapples-Cultural practices Information system of tropical crops in Hawaii Department of Tropical Plant & Soil Sciences University of Hawaii at Manoa. http://www.ctahr.hawaii.edu/fb/pineappl/pineappl.htm Accessed on 8 November 2014.

Anonymous. 2014c. Pineapple-Soil and Climate http://nhb.gov.in/fruits/pineapple/pin012.pdf. Accessed on 12 December 2014.

Anonymous. 2016. Land Requirements for Growing Pineapple. (2013, October 31). Retrieved November 10, 2016, from https://www.daf.qld.gov.au/plants/fruit-and-vegetables/fruit-andnuts/pineapples/land-requirements-pineapple

Astuti, N. K., Cahyono, P. and Rahmat, A. 2016. Soil temperature on various conditions of pineapple (Ananas comusus L. Merr.) in red acid soil, Lampung, Indonesia. Proceeding of International Conference on Climate Change 2016, 190-195 DOI: 10.15608/iccc.y2016.564

Awe, G. O., Reichert, J. M. and Wendroth, O. O. 2015. Temporal variability and covariance structures of soil temperature in a sugarcane field under different management practices in Southern Brazil. Soil and Tillage Research, 150: 93-106.

Ayanlade, A., Odekunle, T. O., Orinmogunje, O. I. and Adeoye, N. O. 2009. Inter-annual climate variability and crop yields anomalies in the middle belt of Nigeria. Advances in Natural and Applied Science, 3(3):452–65.

Almeida, O. A., DeSouza, L. F. DaReinhardt, D. H. and Caldas, R. C. 2002. Influência da irrigação no ciclo do abacaxizeiro cv. Pérolaemárea de tabuleirocosteiro da Bahia. Revista Brasileira Fruticultura, 24(2):431– 435.

Bartholomew, D. P., Paull, R. E. And Rohrbach, K. G. 2003. The Pineapple–Botany, Production and Uses, Wallingford, CABI, 2003. pp, 69- 108.

Bartholomew, D. P. and Kadzimin, S. B. 1977. Pineapple. Chapter 5. In: Ecophysiology of Tropical Crops, (Eds P. de T. Alvim and T. T. Kozlowski). London: Academic Press. pp, 113–156

Bartolomew, D. P. and Kadzimann, S. B. 1987. Ecophysiology of tropical crops. In: P.T. Alvin and T.T. Kozlowski (Eds). New York. Academic Press.1987. pp, 502.

Carr, M. K. 2012. The water relations and irrigation requirements of pineapple (Ananascomosus var. comosus): A Review. Experimental Agriculture, 48(04): 488–501.

Cunha, G. A. 2005. Applied aspects of pineapple flowering. Bragantia, 64(4):499–516.

Engebos, M. 2012. Pineapple Adaptation. Retrieved November 10, 2016, from http://zotero.org/support/quick_start_guide

Gonzalez, A. C. and Espinoza, F. R. 2008. Pineapple News, Saran Shade Cloth to Prevent Sunburn in Pineapple Fruit. Newsletter of the Pineapple Working Group, International Society for Horticultural Science June, 2008; 15: 21-23.

Kaudo, H. O. 2014. Pineapple Production [Blog]. Retrieved November 10, 2016, from http://hellenomondikaudo.blogspot.com.co/2014/12/pineapple-production.html

Malézieux, E., Côte, F. and Bartholomew, D. P. 2003. Crop environment, plant growth and physiology. Chapter 5. In: The Pineapple, Botany, Production and Uses, (Eds D. P. Bartholomew, R. E. Pauli and K. G. Rohrbach). Wallingford, UK: CAB International. pp, 69–107

Neild, R. E. and Boshell, F. 1976. An agro-climatic procedure and survey of the pineapple production potential of Colombia. Agricultural Meteorology, 17: 81-92.

Omoyo, N. N., Wakhungu, J. and Oteng'i, S. 2015. Effects of climate variability on maize yield in the arid and semi-arid lands of lower eastern Kenya. Agriculture & Food Security, 4(1):1.

Paul, K. I., Polglase, P. J., O'Connell, A. M., Carlyle, J. C. and Khanna, P. K. 2004. Soil temperature under forest: a simple model for predicting soil temperature of forest types. Agriculture and Forest Meteorology, 121 (3-4): 167–182.

Paul, K. I., Polglase, P. J., O'Connell, A. M., Carlyle, J. C. and Smethurst, P. J. 2002. Soil nitrogen availability predictor (SNAP): a simple model for predicting mineralisation of nitrogen in forest soils. Australian Journal of Soil Research, 40: 1011–1026.

Purseglove, J. W. 1968. Tropical crops: Monocotyledons, Vol.1, John Wiley & Sons. New York.1968. pp, 334.

Purseglove, J. W. 1972. Tropical Crops: Monocotyledons. London: Longman.

Py, C., Lacoeuilhe, J. J. and Teisson, C. 1998. The pineapple, cultivation and uses. Paris: G.P. Maisonneuve& Larose. pp, 568.

Rainha, N., DeMedeiros, V. P., Rodrigues, A. C., Simas, A., Arruda, C., Pacheco, A. S., Cruz, C. and Ponte, D. 2013. News from the Azores. In: Newsletter of the pineapple working group. International Society for Horticultural Science, 20: 9.

Singh, V. B. 2014. Fruits of North East Region: Package of Practice for Pineapple http://megapib.nic.in/pppineapple.htm.

Sthapit, B. R., Rao, V. R. and Sthapit, S. 2012. Tropical fruit tree species and climate change. Biodiversity International (New Delhi). pp, 15–26.

Thornton, P. K., Jones, P. G., Gopal, A. and Andresen, J. 2009. Spatial crop yield response to climate change in East Africa. Global Environmental Change, 19:54–65.

Waker, J. M. 1971. Temperature depression while aerating moist soils and the resultant effect on maize seedling growth. Plant and Soil, 34 (1): 487-497.

Williams, A., Crespo, O., Atkinson, C. J. and Essegbey, G. O. 2017. Impact of climate variability on pineapple production in Ghana. Portia Agric& Food Security, 6:26. DOI 10.1186/s40066-017-0104-x.

12

Propagation and Quality Planting Material Production of Pineapple

Ranjit Pal

Department of Fruit Science, College of Horticulture & Forestry, Rani Laxmibai Central Agricultural University, Gwalior Road, Jhansi, India

Sexual Propagation in Pineapple

Pineapple occupy a gametophytic self-incompatibility system so that commercial cultivated pineapple rarely produces very few seeds in the field (Coppens d'Eeckenbrugge *et al.* 1992). However, if two varieties are grown close together and flowering time is same, it is chance for insect-aided crosspollination to set seeds. The vectors for natural cross-pollination vary with the location. Where humming birds are present, they are likely to be a primary pollinator. Where these birds are absent, honeybees and ants would be the main pollinators. The small corolla tube and the fact that pollen is not shed rule out wind as a pollinator. Since seedy fruit is not acceptable for either fresh or processing, it is rare to find more than one cultivar on large plantations. If more than one cultivaris grown in a farm, spatial separation and careful management of time of forcing will prevent seed production. However, cross-pollination between cultivars is used by pineapple breeders to generate new hybrids like "MD-2" and "CO-2" (Williams and Fleisch 1993). Sexual propagation is important for the generation of new varieties and is successfully used in pineapple breeding programs (Reinhardt *et al.*, 2018). Pineapple Seeds are placed in shallow trays on moistened sand beds under controlled environmental conditions for germination(Junghans *et al.*, 2015).Seed germination come up mostly from two weeks after sowing and rates are high, usually 80% or more (Chang *et al.*, 2003). The new growing seedlings have 6 to 8 leaves stage within about three months after sowing and are then transplanted to pots with agricultural substrate and are grown under greenhouse or screen house conditions. After an additional 9 to 15 months of growth stage, during which the seedlings height reach 25 cm, which is considered adequate for planting in the field, the plants are finally acclimatized in an open nursery. Cultural practices applied are similar to those used for growing conventional planting material, but growth is usually slower and the first sexual cycle longer, taking about 2 years under tropical conditions. The mostly high heterozygosity of the parental cultivars leads to a high segregation in the majority of the important characters observed in the young plants (Cabral *et al.,* 1993).

Vegetative Propagation in Pineapple

The propagation methods of Pineapple exclusively done by vegetatively from a number of planting materials produced by the plant: the crown, born on the fruit apex; slips, born on the fruit peduncle; and suckers, lateral and basal shoots born on the plant stem (Py, 1979). Suckers and slip generally preferred for commercial propagation methods in most of the pineapple growing countries. These planting materials preferred for planting as they come to bearing earlier than the crown and produce larger fruits. Crowns are also used with varying degrees of success. Suckers and slip removal process from mother plant are very expensive and laborious. Slips would be the first choice for planting purpose. However, some species of pineapple produce very less or no slips and MD-2, Smooth Cayenne do not produce slips in the hot equatorial environments. Where slips are not available suckers are the only choice, and in equatorial climates suckers are mostly produced after fruit harvest (Yan Cheng *et al.,* 2018). Crown grows on the top of the fruit. It is the vegetative growth at the top of the fruit, attached to the central core of the fruit. It is called crown as it adorns the apex of the fruit. Suckers arise and grow from buds below the ground level and sparsely produced. Shoots are leaf-branches arising from the buds in the axils of leaves above the ground level. Both of them look-like daughter plants by the side of the mother-plant. They emerge out as the mother plant flowers, and can be seen growing fast when fruit develops. After the fruit is harvested from the mother-plant, these grow still vigorously and bear fruit in the forthcoming season. This is known as the ratoon crop (Chadha *et al.,* 1998).Slips emerge on the peduncle the differentiation of lateral buds during fruit initiation. However, they are quite different in shape because slips grow outward and then upward around the base of the fruit, resulting in a sharp curvature at the base of their stem (Hepton, 2003).Out of three planting materials, crown is the smallest, generally weight from 200-300g; slips are intermediate in size between the smaller crown and larger suckers, generally weight 300-400g; suckers are largest planting material among the others, generally weight 400 to 500g. Besides these, others planting materials are discs, hapas, butts or stumps. Fruit-stalks cut into bits known as 'discs' can also be used for propagation, but their use in field is very limited. 'Hapas' are shoots produced at the base of the peduncle. They are intermediate in size between shoots and slips. 'butts' or 'stumps' are the stem parts of the plant that have borne fruits. These consist of entire plants after the fruits have been harvested and from which the base of the stem, roots, leaves and peduncle have been removed. The older they are, the less suitable they will be as a plant (Chadha *et al.,* 1998).

For commercial pineapple cultivation, characterized by vigour, growth rate, time taken for bearing and uniformity of fruit production and quality attributes varies with the type of material used for planting in pineapple. The time taken for fruit initiation is depends on use of kind of planting materials during planting. The basal suckers produced fruit within day after planting about 12-14 months, while, in stem suckers take about 16-18 months. Planting material crown will take more than two years for production of fruits (Anonymous, 2000). In, West Africa (Benin) during planting time, pineapple producers often mix different types and weights of planting materials, depending on their availability. It is well-known that larger or heavier planting material shows more vigorous growth than smaller or lighter planting material and would produce more vigorous plants at flowering induction time than smaller or lighter planting material (Norman, 1976; Reinhardt *et al.*, 2003).Planting materials from larger size of slips produced more yield and gave good quality for Sugarloaf pineapple cultivar under Benin conditions in Africa (Fassinou Hotegni *et al.,* 2015).Vegetative propagules, slipes and suckers, weighting around 350g and 450g, were performing the best in term of fruit production and as well as quality parameters for Kew variety under Croog climatic condition in Karnataka (Singh *et al.*, 1978).In Assam climatic condition (Jorhat) slips weight 301-400g and suckers' weight 501-750g were found to be best planting material (AICFIP Cell I, 1982), while suckers weighting about 501-1000 g were the ideal planting material for Thrissur, Kerala (Verkey *et al.*, 1984).

Micro-Propagation in Pineapple

The advance propagation methods like micro-propagation or tissue culture techniques for obtained quality planting materials as commercially scale have been one of the biotechnological advancements with momentous influence on horticultural crops in last few years. Micro-propagation techniques has made massive contributions to the vegetative reproduction of many horticultural crop species, accelerating propagation of new important genotypes and producing a maximum number of quality planting materials in a very short time and often of better quality than that of other conventional methods. The fruitful application of these methods depends, on number of factors that need to be controlled.

Rationale for micro-propagation of pineapple: A single sucker is normally left in place to make the subsequent ratoon crop. Conventional methods of propagation include crowns, harvesting slips at the base of the fruit, and harvesting lateral and basal shoots, each of which will produce a plant. Each of these propagules requires a different amount of time from planting until flowering, creating a problem in field harvest if plants in the field come from different types of propagules.

There are several problems with using the methods described above for the vegetative propagation of pineapple. Generally, pineapple planting spacing is very less so planting material required more at time of planting. In, India most of the varieties planted in high density planting system and approximately 60000 to 70000 plants per ha. In some areas where spiny cultivars are cultivated and a density of 15000 to 35000 plants per ha is common. When such a large number of quality planting materials are needed, propagule material becomes a limiting factor in expansion, even with the use of chlorflurenol-methyl ester. The problem is compounded when a promising new sport is identified or a new cultivar is developed. Further, vegetative propagation may maintain and multiply somatic mutations.

An efficient system of in vitro propagation might allow for a much faster increase in planting material in pineapple if the micro propagated plants are true to type. On the other hand, genetic improvement of pineapple by conventional methods is difficult. Most improved types are mutants that have been selected and propagated. If plants produced in vitro are highly variable, tissue culture could be a source of genetic variability that might be of use to pineapple breeders (Moore *et al*., 1992).

Summary of previous research on micro-propagation of pineapple: The tissue cultural research activity was done on pineapple has focused on micropropagation techniques for production of planting material. Direct planting material production from terminal buds was first achieved by Aghion and Beauchesen in 1960. A single plant was produced from individual shoot tip cultured on solid medium (Lakshmi Sita *et al.,* 1974).

Mapes (1973) first mentioned multiple shoot production directly from axillary buds. Several research reports have investigated the factors involved in the establishment of axillary buds in culture (Wakasa *et al,*. 1978; Mathews and Rangan 1979; Drew 1980; Zepeda and Sagawa 1981). In almost all these reports, the basal medium of Murashige and Skoog (1962) (MS medium) has been used. Various growth regulators have been tested for the initiation and maintenance of proliferating cultures, with most researchers using a combination of auxin and cytokinin. Complex additives such as coconut water and casein hydrolysate have been added to the medium (Aghion and Beauchesne 1960; Zepeda and Sagawa 1981), but are not necessary if hormones are present (Wakasa *et al.,* 1978).Liquid culture medium was superior to solid medium for shoot proliferation from axillary bud cultures and shaken liquid cultures were superior to stationary liquid cultures (Mathews *et al*., 1976; Mathews and Rangan 1979).

Pineapple shoots have also been regenerated from other explants. Wakasa *et al*. (1978) produced shoots from protocorm-like calli derived from immature syncarps, whole immature crowns, and an immature slip. Immature leaf explants have also been used for shoot proliferation (Mathews and Rangan, 1979).Callus arising from axillary bud cultures on MS medium supplemented with 10 mg/l NAA, 15% coconut water, and 400 mg/l casein hydrolysate gave rise to shoots when transferred to a medium with lower hormone concentrations (Mathews and Rangan, 1981).Shoot proliferation from hybrid embryo-derived callus has also been reported (Srinivasa Rao *et al.,* 1981).All of these reports have focused primarily on the establishment of cultures. Although some researchers have stated that their systems are highly efficient (Pannetier and Lanaud, 1976; Drew 1980), data on culture efficiency and numbers of plants regenerated was not presented, with the exception of Wakasa *et al*. (1978) who reported the establishment of 400 regenerated plants in soil.

Micro Propagation Protocol

Explant preparation: Commercial grown of pineapple is called clones because they are vegetatively propagated and are self-sterile but cross easily with plants outside their varietal group (Evans *et al*., 2002). There are many commercial clones (Table 1) classed in different groups including 'Cayenne', Spanish', 'Queen' and 'Pernambuco' which may represent botanical varieties. The most popular variety for micropropagation in many parts of the world is cv. Smooth Cayenne Serrana (Daquinta and Benegas, 1997) but some others such as, cvs. Queen, Kew, Giant Kew and Mauritius are also widely cultivated in other growing regions.

Table 1: Commercially grown pineapple varieties in different regions.

Name of variety	Growing region	Remarks
Smooth Cayenne	Thailand, Philippines, Hawaiian Islands	Most commonly available
Queen	South America, Australia, India	Smaller, little drier and less sweet than Smooth Cayenne
Red Spanish	Caribbean	Smaller, little drier and less sweet than Smooth Cayenne
Pernambuco		Medium sized
Sugarloaf		Large and heavy
Baby		Very sweet
MD2 ('Extra Sweet' or 'Golden Ripe')	Europe	Sold as fresh fruit

(*Source*: Mhatre, 2007)

Types of explants: Dormant axillary buds from crowns: Crowns from harvested ripe pineapple fruits can be used as source of dormant axillary buds for culture initiation. Conventionally, after removal of all the leaves the base of the crown is sectioned and each quarter is then planted to yield a single plant. Beneath each leaf of the crown is a 1–2 mm sized dormant bud (nearly 12–15 buds per crown) which remains dormant in this cultivation method. However, under in vitro conditions it is possible to scoop out this bud and culture it on defined medium so that it can be made to sprout and yield numerous tiny shoots. Each shoot can be subsequently rooted to obtain a complete plant. However, of the 12 to 15 buds present in each crown, only the larger 8–10 buds in the middle portion of the crown have the potential to sprout in culture (Soneji *et al.,* 2002a).

Leaf bases of *in vitro* shoots: The basal ends of in vitro produced shoots isolated from stock cultures can also be used as explants after peeling them and placing them with their ventral sides in contact with the defined multiplication medium. The basal section can be cultured (one per tube) in the medium for the induction of multiple shoots. This process would take 6 weeks and the shoots thus obtained can be multiplied and maintained on the same medium. Elongation of (Soneji *et al.,* 2002b) shoots can be done on hormone-free medium and rooted on liquid rooting medium.

Disinfection of plant material: Dormant axillary buds (one at the base of each leaf of the crown) of pineapple can be carefully excised with minimum unavoidable surrounding tissue (Figure 1, step 1; Figure 2A) and collected in a 5 × 5 cm pre-sterilized damp (dampened with tap water) muslin cloth, tied into a bundle with a string and placed into a pre-sterilized conical flask and stoppered. The bundle containing the excised axillary buds can then be washed thoroughly with soap and tap water. Following this, surface sterilization of the buds should be carried out in a laminar air flow unit. The axillary buds can be sterilized in 0.1% mercuric chloride for 2 min and rinsed several times (4–5 washes) in sterile tap water (Figure 1, step 2). Each axillary bud (one per culture tube) can be placed vertically for sprouting with its basal end slightly embedded in the bud sprouting medium (Figure 2B).

Shoot regeneration and maintenance: ***Media:*** The surface sterilized dormant axillary buds of pineapple should be cultured on 'bud sprouting media' (BSM) comprised of: hormone-free N (Nitsch's, 1951) or MS (Murashige & Skoog, 1962)basal medium, containing BAP (4.44 µM). The buds will sprout and give rise to 1–1.5 cm, 1–2 shoots (Figure 2C) after 4–6 weeks in culture (Figure 1, step-3). A single shoot should then be transferred to 'shoot multiplication medium' (SMM) – comprised of MS basal medium supplemented with NAA (9.67 µM), IBA (9.84 µM) and Kn (9.29 µM) for further proliferation (Figure 2D). In this medium each shoot can be made to proliferate to give rise to 10– 20 shoots. SMM can be used also as a maintenance medium on which a tuft of 4–5 (3–4 mm) shoots can be sub-cultured and maintained as a cyclic proliferating stock culture leading to 50–60 shoots per culture (Figure 2D). SMM can also be used as liquid medium (without addition of phytagel) in flasks for further proliferation of shoots (Figure 2E). However, shoots produced in flasks tend to be elongated and can be directly rooted on rooting medium. A tuft of 4–5 (1–2 cm) shoots proliferated on SMM solid medium, can be sub-cultured on 'hormone-free MS' basal medium (HFMS) for shoot elongation. It is necessary for each shoot to attain a height of 4–5 cm before they can be isolated and induced to root. Shoot elongation would take 4–5 weeks on HFMS medium (Table 2).

Table 2: Nutrient media compositions used for pineapple micropropagation

Medium	Macro elements	Micro elements	Carbon source (%)	Growth adjuvants (µM)
BSM	1	1	2	BAP 4.44
SMM	1	1	2	NAA 9.67 IBA 9,84 Kn 9.29
HFMS	1	1	2	-
RM	1	1	1	NAA 0.54 IBA 1.97

BSM: Nitsch's (1951) or MS (1962) macro- and microelements, vitamins; SMM: MS (1962) macro- and microelements, vitamins; HFMS: MS (1962) macro- and microelements, vitamins; RM: White's (1954) macro- and microelements, vitamins; Gelation of solid media was with gelrite (0.2%); Fe EDTA was added separately according to Murashige and Skoog (1962) medium and the pH was adjusted to 5.8 prior to autoclaving; NAA, naphthalene acetic acid; IBA, indole butyric acid; Kn, kinetin; BAP, benzyl amino purine.

Culture vessels and culture conditions: All culture vessels (test tubes and flasks) used in the protocol should be made of Borosil glass, 25 × 150 mm size. Non-absorbent cotton should be used for plugging the tubes prior to autoclaving at 121°C, 1 kg/cm for 15 min. All cultures should be incubated at 25 ± 2°C in 16/8 h dark/illumination (12.1 μMm–2s –1, Philips) cycle. Phytagel (Sigma) 0.2% is appropriate for solidifying culture media – BSM, SMM and HFMS. RM is devoid of phytagel and therefore the shoots (3–4 cm) have to be placed over filter paper bridges for support and for the induction of rooting. Carbon source used was sucrose (1–2%) as described in the text and Table 2. The pH of all media should be adjusted to 5.8 using 1N NaOH or 1N HCl, prior to autoclaving. Soilrite (Soilrite mix, TC is a mixture of 75% Irish peat moss and 25% expanded Perlite, obtained from M/s Chougule Industries Ltd., Mumbai) or autoclaved fine soil may be used to harden the in vitro raised plantlets (Mhatre, 2007).

Rooting of shoots, recovery of plantlets and acclimatization: Each elongated shoot can then be isolated and sub-cultured on liquid 'rooting medium' (RM) comprised of White's (1954) basal medium supplemented with NAA (0.54 μM) and IBA (1.97 μM) and 1% sucrose (Table 2), for the induction of roots (Figure 2F). Plantlets with actively growing roots can be carefully removed from culture tubes and washed with tap water and transferred to paper cups containing either autoclaved soilrite or soil (Figure 2G). All plantlets were transferred to soil in polythene bags after 4 weeks, and maintained in the green house for hardening. In 30–45 days, these plantlets can be transferred to the field for further growth, flowering and fruiting (Figure 2H). Fruiting begins at the end of 14 months from the time of field transfer and can be harvested in 4–5 weeks (Figure 2I,J).

1st Step	Pineapple – Physiological mature fruits – crowns – excise dormant axillary buds (1 at the base of each leaf)
2nd Step	Surface sterilization of dormant axillary buds (washing with soap and water – treat with 0.1% $HgCl_2$, 2 min – rinsing with autoclaved tap water, 4–5 times)
3rd Step	Culture buds in Medium I (BSM = hormone free Nitsch 1951/ Murashige & Skoog's 1962 solid basal medium + BAP)
	Buds sprout – (3–8 shoots produced in 4–6 weeks)
4th Step	Culture in Medium II (SMM = Murashige & Skoog's 1962 solid basal medium + NAA+ IBA + Kinetin)
	Formation of multiple shoots (tuft of 10–20 tiny 2 mm sized shoots, 4 weeks)
5th Step	Tuft of 4–5 shoots sub-cultured on Medium II (maintained as stock cultures)
	Proliferation – 50–60 shoots (in 2–3 months)
6th Step	Subculture tuft of 4–5 shoots (1–2 cm) in Medium III (HFMS = hormone free Murashige & Skoog's 1962 basal medium) (shoot elongation, 4 cm, 4–5 weeks)

7th Step	Isolate each shoot and sub-cultured for rooting in Medium IV (RM = White's 1954 liquid basal medium + NAA + IBA and 1% sucrose) over filter paper bridges
8th Step	Acclimatization of rooted plantlets in soilrite/soil in the greenhouse, 30–45 days
9th Step	Transfer the hardened plants to the field

Fig. 1: Pineapple micro propagation protocol

Fig. 2: Pineapple, Serial Micropropagation: scooping of dormant axillary bud from crowns **A)**, Culture of bud on BSM solid medium **B)**, sprouting of dormant axillary bud **C)**, Proliferation in solid SMM **D)**, Proliferation in SMM liquid medium **E)**, Rooting of isolated shoot in RM **F)**, transfer to papercups with Soilrite **G)**, Field transfer of in vitro produced plants **H)**, Fruiting **I)**, harvested fruits **J)** (*Source*: Mhatre, 2007).

Mass Multiplication in Bioreactors

The method of mass propagation of pineapple (cultivar 'Smooth Cayenne') plants using a temporary immersion bioreactor has been described by Escalona *et al.* (1999).This method is automated and involves three distinct phases: shooting, bud differentiation and elongation. The initial explant used in this protocol is shoots from previously established cultures in a 'shooting medium' formulated by Daquinta and Benegas (1997).

In this protocol the bioreactor system should consist of two 1000-ml vessels – one to act as a reservoir for the liquid medium (volume = 1000 ml) and the other (volume ranging from 25 to 100 ml) for growing the plants. The two vessels were connected by silicone and glass tubes and the airflow can be sterilized during passage through 0.2-µm hydrophobic filters (Escalona *et al.,* 1999). Using an air compressor, sterile medium can be pressured to flow from the reservoir to completely immerse five explants (2–3 cm shoots) placed in the other vessel (plant vessel). The medium can be drained back into the first vessel by controlled reversion of air flow. The frequency and length of immersion can be electronically controlled. Exposure to the medium for 2 min every 3 h at 25°C under cool-white fluorescent tube (80 µmol · m–2 · s –1 (3.3 µM) in the immersion medium promotes shoot growth and induces axillary bud proliferation but avoids unnecessary leaf development during the shoot-formation stage (Escalona *et al.,* 1999). Shoot multiplication period ranged from 4–7 weeks. After this a change in medium composition to MS supplemented with BAP (2.2 µM) and GA_3 (2.9 µM) for 7 days (step 1) and later to MS supplemented with GA_3 (2.9µM) – step 3, can achieve shoot elongation. All shoots produced by this method can be rooted in the green house ex vitro by transplantation onto a substrate consisting of compacted red ferralitic salt mixed in a 1:1 ratio with sugarcane mill baggasse. The temporary immersion system combines the advantages of solid and liquid medium and exhibit high multiplication rates.

Synthetic Seed Preparation in Pineapple

Synthetic seeds can be prepared using tiny shoots (2–5 mm), from in vitro grown stock cultures, as encapsulation propagules (Soneji *et al.,* 2002c). Isolated shoots should be mixed in an alginate matrix [3% sodium alginate (Sigma) prepared in MS basal medium or SMM and autoclaved] for encapsulation. Each shoot thus coated with alginate can be picked up using a pair of forceps and gently dropped into an autoclaved solution of $CaCl_2{\cdot}2H_2O$ (1.36 g/150 ml). The coated buds should be allowed to remain in this solution for 30 min for complexation to occur. At the end of 30 min, the $CaCl_2{\cdot}2H_2O$ solution can be carefully decanted off and the encapsulated shoots washed 3–4 times with sterile tap water and blot dried on sterilized filter paper to remove the excess solution. Synthetic seeds are ready to be either cultured on solidified nutrient media for germination or stored in sealed Petri plates at 4°C in the refrigerator.

Pre-treatment and encapsulation matrix: Prior to encapsulation the shoots should be soaked in liquid White's (1954) basal medium supplemented with sucrose (1%), NAA (10.8 μM) and IBA (39.4 μM) and agitated at 100 rpm (on a gyratory shaker) for 48 h. The shoots can be blot dried on a filter paper, mixed in a gel of 3% sodium alginate (Sigma) prepared in MS basal medium or SMM and encapsulated using $CaCl_2 \cdot 2H_2O$ (1.36g / 150 ml)solution.

Storage and synthetic seeds: Synthetic seeds of pineapple can be stored in sealed petriplates at 4°C for a period of 2 months without sprouting. Whenever required, they can be cultured on HFMS medium to recover shoots and subsequently rooted plantlets for multiplication.

References

Aghion, D. and Beauchesne, G. 1960. Utilisation de la technique de culture sterile d'organes pour obtenir des clones d'ananas. Fruits, 15: 464-466

AICFIP. 1982. Standardization of tipe and size of planting material in 'Kew' pineapple. Research Reports of the All-India Co-ordinate Fruit Improvement Project (Cell-I), 1982, pp. 81-87

Anonymous. 2000. Pineapple, Department of Agriculture, Sri Lanka.

Cabral, J. R. S., Matos, A. P. de, Cunha, G. A. P. 1993. Selection of pineapple cultivars resistant to fusariose. Acta Horticulturae, 334: 53–58.

Chadha, K. L., Reddy, B. M. C. and Shikhamany, S. D. 1998. Pineapple, Indian council of agricultural research, New Delhi, 27-38

Chang, Y. K. Coppens D'Eeckenbrugge, G. And Sanewski, G. M. 2003. Breeding and variety development. In: Bartholomew, D.P.; Paull, R.E.; Rohrbach, K.G. (Ed.). The pineapple: botany, production and uses. New York: CABI Publishing. pp, 33-55.

Cheng, Y., Bartholomew, D., and Qin, Y. 2018. Biology of the Pineapple Plant. In: R. Ming (ed.), Genetics and Genomics of Pineapple, Plant Genetics and Genomics: Crops and Models 22, Springer Nature Switzerland. pp, 27-40. https://doi.org/10.1007/978-3-030-00614-3_2

Coppens D'Eeckenbrugge, G, Duval, M. F. And Van Miegroet, F. 1992. Fertility and self-incompatibility in the genus Ananas I International Pineapple Symposium. Acta Horticulturae, 334:45–51

Daquinta, M. and Benegas, R. 1997. Brief review of tissue culture of pineapple. Pineapple Newsletter, 3: 7–9.

Domingo Haroldo, Reinhardt, R. C., Bartholomew, D. P., D'Souza, F. V., D'Carvalho, A. C. P. P., D'Pádua, T. R. P., Junghans, D. T. and D'Matos, A. P. 2017. Advances in pineapple plant propagation. Revista Brassilia Fruticola, 40 (6): e-302.

Drew, R. 1980. Pineapple tissue culture unequalled for rapid multiplication. Queensland Agricultural Journal, 106: 447-451

Escalona, M., Lorenzo, J. C., Gonzalez, B., Daquinta, M., Gonzalez, J. L., Desjardins, Y. and Borroto, C. G. 1999. Pineapple (Ananas comosus L., Merr.) micropropogation in temporary immersion systems. Plant Cell, 18: 743–748.

Evans, D. O., Sanford, W. G. and Bartholomew, D. P. 2002. Growing pineapple. Co-op. Extn. Serv., Fruits and Nuts – 7, CTAHR, 2002(10): 4–8.

Hepton, A. 2003. Cultural system, In: Bartholomew, D. P., Paull, R. E., and Rohrbach, K. G., The pineapple-Botany, Production and Uses, CABI Publishing, 6: 109-142.

Hotegni, V. N. F., Willemien, J. M. Lommenl, E. K. A. and Paul, C. S. 2015. Influence of weight and type of planting material on fruit quality and its heterogeneity in pineapple [Ananas comosus (L.) Merrill], Frontiers in Plant Science, 5(798): 1-16. doi: 10.3389/fpls.2014.00798.

Junghans, T. G., Junghans, D. T., Matos, E. M., Batista, E. A., Mielke, M. S. and Ledo, C. A. S. 2015. Seed germination of three pineapple progenies in different temperature regimes. Perspectiva Erechim, 39(147): 61-67.

Lakshmi Sita, G., Singh, R. And Iyer, C. 1974. Plantlets through shoot-tip cultures in pineapple. Current Science, 43: 724-725.

Mapes, M. 1973. Tissue culture of Bromeliads. Journal of International Plant Propagations Society, 23:47-55.

Mathews, H., Rangan, T. and Niuayanaswamy, S. 1976. Micropropagation of Ananas sativus in vitro. Plant Physiology, 79: 450-454.

Mathews, H., Rangan, T. S. 1979. Multiple plantlets in lateral bud and leaf explant in vitro cultures of pineapple. Scientia Horticulturae, 11: 319-328.

Mathews, H. and Rangan, T. S. (1981) Growth and regeneration of plantlets in callus cultures of pineapple. Scientia Horticulturae, 14: 227-234.

Mhatre, M (2007) Micropropagation of pineapple, Aananas comosus (l.) merr. In: S.M. Jain and H. Häggman (eds.), Protocols for Micropropagation of Woody Trees and Fruits, Springer, 45, 499–508.

Moore, G. A., DeWald, M. G. and Evans, M. H. 1992. Micropropagation of Pineapple (Ananas comosus L.). In: Bajaj Y.P.S. (eds) High-Tech and Micropropagation II. Biotechnology in Agriculture and Forestry, vol 18. Springer, Berlin, Heidelberg. https://doi.org/10.1007/978-3-642-76422-6_24

Murashige, T. and Skoog, F. 1962. A revised medium for rapid growth and bioassays with tobacco tissue cultures. Physiologia Plantarum, 15: 473–497.

Nitsch, J. P. 1951. Application of different macro and micro nutrients in the study for a balanced basal medium. Physiologia Plantarum, 15, 473–497.

Norman, J. C. 1976. Influence of slip size, deslipping and decrowning on the "Sugarloaf" pineapple. Scientia Horticulturae, 5: 321–329. doi: 10.1016/0304- 4238(76)90127-8

Pannetier, C. and Lanaud, C. 1976. Divers aspects de l'utilisation possible des cultures in vitro pour la multiplication vegetative de /' Ananas comosus 1. Merr, variete Cayenne lisse. Fruits, 31: 739-750.

Py, C. 1979. Production acceleree de materiel vegetal de plantation. Fruits, 34: 107-116.

Reinhardt, D. H., Souza, A. P. M., Caldas, R. C., Alcântara, J. D. P., and Almeida, A. A. D. 2003. Management of slips and its effect on growth and production of Pérola pineapple plants. Revista Brassilia Fruticola, 25: 248–252. doi: 10.1590/S0100- 29452003000200016

Singh, H. P., Dass, H. C. and Gangapathy, K. M. 1978. Studies on different types and size of planting material in Kew pineapple. Research Report and Project proposal on banana, pineapple and papaya presented at Fruit Research Workshop, Bangalore, 31.7.1978 to 2.8.1978. pp, 134-135.

Soneji, J. R., Rao, P. S. and Mhatre, M. 2002a. Somaclonal variation in micropropagated dormant axillary buds of pineapple (Ananas comosus L., Merr.). Journal of Horticultural Science and Biotechnology, 77: 28–327.

Soneji, J.R., Rao, P.S. and Mhatre, M. 2002b. In vitro regeneration from leaf explants of pineapple (Ananas comosus L., Merr.). Journal of Plant Biochemistry and Biotechnology, 11: 117–119.

Soneji, J. R., Rao, P. S. and Mhatre, M. 2002c. Germination of synthetic seeds of pineapple (Ananas comosus L., Merr.). Plant Cell Reproduction, 20: 891–894.

Srinivasa Rao, N., Doreswamy, R. and Chacko, E. 1981. Differentiation of plantlets in hybrid embryo callus of pineapple. Scientia Horticulturae, 15:235-238.

Verkey, P. A., Nair, M. N. C., George, T. E., Aipe, K. C, and Leela Mathew. 1984. Growth, flowering duration and growth characters as influenced by size of propagules in pineapple (Ananas comosus Merr.) cv. Kew. South Indian Horticulture, 32: 327-329

Wakasa, K., Koga, Y. and Kudo, M. 1978. Differentiation from in vitro culture of Ananas comosus. Japan Journal of Breeding, 28: 113-121.

White, P. R. 1954. The Cultivation of Animal and Plant Cells. Ronald Press, New York. pp, 74–75.

Williams, D. D. F. and Fleisch, H. 1993. Historical review of pineapple breeding in Hawaii. Acta Horticulturae, 334:67–76

Zepeda, C. and Sagawa, Y. 1981. In vitro propagation of pineapple. HortScience, 16:495

13

Planting System of Pineapple

Tanmoy Mondal and Fatik Kumar Bauri

Department of Fruit Science, Bidhan Chandra Krishi Viswavidyalaya Mohanpur, Nadia, West Bengal, India

Pineapples are commonly propagated through vegetative means, as the crop sets very few or no seeds and seeds possess slow germination process and low survival rate of seedlings under natural conditions. Vegetative parts such as stem suckers, peduncle slips, and fruit crowns are commonly utilized for vegetative multiplication of pineapple.

Selection of planting materials: Slips or hapa suckers are commonly preferred for commercial multiplication because they may produce fruit within 14-16 months after planting. Crowns are not commonly used as they may take up to 24 months after planting to harvest. Ground suckers (ratoon) and side suckers (side shoots) produce fruit 12-14 months and 18-20 months after flowering, respectively.(Pineapple Technical Group, 1999).

The fresh suckers should not be used for planting. Pineapple suckers should be stored for a few months to allow them to dry before planting, as they may begin to decay if planted in moist soil. Suckers are graded according to vigour and size, such as small, medium, and large. Planting the same graded suckers in the same plot helps to avoid uneven growth of the plant, uneven flowering times, and harvesting times. Additionally, using uniform planting material makes cultural practices easier, increases labour efficiency and other inputs, ultimately reducing the cost of cultivation.

Chadha *et al.* (1974) observed the superiority of slips and suckers as planting materials compared to crowns. In both slips and suckers, larger planting materials resulted in more vigorous plants. Medium-sized slips or suckers led to more flowering, while larger grade material resulted in a more staggered flowering period. Smaller slips or suckers yielded fruit earlier. Slips produced better quality fruits, and the yields were comparable to the highest yield obtained with the largest suckers.

Preparation of planting materials: Suckers should be treated with Imidachloprid 1 ml/L of water and Carbendazim @2g/L of water solution to protect mealy bug and crown rot respectively.Some lower scaly leaves of the suckers should be removed (2-5cm) for easy root formation, 4-5 days before planting. The basal end of the suckers should be kept in metalaxyl 8% + mancozeb 64% @ 2.5 g per L of water to avoid heart-rot disease.

Planting time: Planting time depends on climate which is varies region to region and the variety used. The time of planting is very important, since its coincidence with natural flowering time which variesfrom region to region. Planting time determines the season of harvesting of the plant crop. In the natural flowering time, if plant dose not attends physiological maturity, either it will be escaping flowering or flowering will be next season, or if induced for flowering in the same season, plant will be bearing small fruit. So, ideal planting time should be 12-15 months before natural flowering time of this region which varies from December to March (Chadha *et al.,* 2013). Gandgi (1949) recommended that planting time of western India during September to November. In Assam, ideal planting is during August to October and in Kerala during April to May. In north Bengal districts, ideal planting time during October -November and other district of West Bengal is June -July.

Land preparation: The land for planting is ploughed and pulverized to a fine tilth by giving four to five tillages with sufficient intervals. The land should be properly levelled to avoid water logging. Adequate drainage facility should also be provided. The field should be free from weed.

Planting systems: The system of planting depends on topography of land and rainfall. There are four planting system practices viz. flat-bed planting system, furrow planting, trench planting and contour planting (Chadha *et al*., 2013).

1. **Flat-bed planting system:** This is commonly vogue in West Bengal, mostly in plain where there is no issue in soil erosion. Planting is done by dibbing the pit in 15-20 cm deep. Channel is made for good drainage of excess rain fall. Flatbed planting system is also followed in irrigated area in western India.

2. **Furrow planting system:** This ispractised in the coastal area of Kerala, where natural precipitation is moderate, fairly distributed but not intensive and crop is given without irrigation. In this system planting is done 30-45cm deep furrow alternating with 1.2-1.5m broad ridges.

3. **Trench planting:** Trench planting involves creating side anchorages for planting. Dhareswar (1950) suggested that growing pineapple in a trench depth of 45.72cm is suitable for soil moisture conservation. In

this planting system, trenches 30-45 cm deep are opened at a distance of 1-1.5m. There are two types of trench planting systems: single row and double row planting. Additionally, triple row and four-row planting systems are occasionally seen. The single-row trench system is commonly followed in hillside plantations of Kerala, Assam, and Tripura, where the soil slope is gentle and soil erosion is a problem due to moderately heavy rainfall. The double-row trench system is widely practiced and popular in Karnataka, Kerala, Assam, Tripura, West Bengal, and Goa, and wherever plantations are on plains, whether crops grow with or without irrigation. The field is laid out with trenches alternating with mounds, made across the slope. In each trench, two shallow furrows about 10-15cm deep and 15cm inside from the edge of the trench are opened, and suckers or slips are planted in these furrows. In the two-row trench, plants are arranged within the trench so that they are not exactly opposite to each other.

The advantage of treble-row trench and four-row trench over double row trench is that more number of plants can be accommodated per unit area. Experiments at the Indian Institute of Horticultural Research have revealed that the double-row trench system is better than three-row and four-row trench system in view of the convenience in farm operations. The depth of the trench should be about 22.5-30.0cm (Reddy and Prakash, 1982; Radha *et al*,1990), and this is important for plant-crop that ratoon crops.

4. **Contour planting:**This system is frequently observed in hillside plantations of Assam and Tripura, where rainfall is abundant and soil erosion is a concern. Planting is carried out in terraces created by digging contour trenches, which capture runoff water and sediment, also functioning as drainage channels. The spacing between two parallel lines of plants and the terrace width are determined by the slope gradient.

Designs of Planting

1. **Single row planting: R**ow to row distance is 60 cm and plants are spaced at 30 cm in this design of planting. This design is mostly uneconomical and sunburn in fruits is very common due to penetration of profuse sunlight.

2. **Double row planting:** This is most commonly followed planting design and suitable for north eastern Indian condition. This design of planting allows maximum utilization of land and other resources as well as fewer infestations of weeds, protection from lodging and sunburn of fruits, due to mutual support of plants. Two rows of pineapple suckers are planted

at 45 to 60 cm and plants are spaced at 30 to 45 cm within the row. The planting of suckers in pair of rows are done in triangular or zig-zag fashion. The pairs of rows in a bed are spaced at 90 cm which are separated by trenches. The soil from the trenches is used to raise the beds. This system accommodates 40,000-64,000 plants per hectare.

3. **Triple row planting:** This planting design almost similar to the previous design except having three rows in a single raised bed. The planting of suckers are also done in pair of rows in triangular method or zig-zag fashion in this method. This design of planting is followed mostly in dwarf or small canopied varies of pineapple like as Queen. Most disadvantage of this planting system is the low yield.

Depth of planting: Pineapple is a shallow feeder crop. It should not be planted very deep in soil. Suckers or slip should be planted about 10-12cm in depth. Deep planting should also be avoided to prevent soil from entering the plant's core, which could result in its death. During planting, it's crucial to firmly secure the sucker/slip by compacting the soil around it and applying pressure.In a study by Mohan and Ahmed (1984), various trench planting depths (15 cm, 21.5 cm, or 30 cm) and planting densities (30 x 45 x 105 cm or 25 x 60 x 90 cm) were investigated. Although plant growth and fruit quality did not exhibit significant differences across the treatments, there were variations in fruit yield. The highest yield (54.32 q/ha) was achieved with a trench depth of 21.5 cm and a spacing of 25 x 60 x 90 cm, while the lowest yield (44.87 q/ha) was recorded with a trench depth of 30 cm and the same spacing. Dhareswar (1950) suggested growing pineapple under partial shade of coconut and young mango trees in plains of south India, where the sun is severe and rainfall is scanty.

Spacing and planting density: Planting density depend on growth of the planting materials and system of planting. Adaptation of low planting density is a major constrain in India which increase the cost of production per unit area. The ideal planting density of sub-tropical and mid humid condition is 63,400 plant /ha (25.5cm×60cm×75cm), whereas for hot and humid condition a plant density 53,300 plant/ha (25cm×60cm×90cm) gave maximum yield. The recommended plant density of rain fed, high fertile and hilly area of north eastern state is 43,500plant/ha (30cm×60cm×90cm).However close planting increased the number of fruits per unit area, weight and yield due to more number of plant accommodation per unit area but reduced the fruit size. Higher plant population showed reduction of the fruit quality by increasing the fruit acidity and reduction in total sugar content. Laishram *et al.* (2012) experimented with six different spacing and found that the spacing 100x100cm recorded maximum fruit weight with and without crown at 1.59kg and 1.31kg and maximum fruit length and diameter at 15.50cm and 11.67cm, respectively. However, the spacing 25 x 50 x 80 cm recorded maximum fruit yield with and with crown in with 45.23t/ha and 38.77t/ha, respectively. Fruit weight slightly decreased with increase in plant population, however, total yield/h increased significantly. Roy *et al.* (1981) found that triangular system of planting showed high percentage of flowering and increased yield compared to rectangular system of planting. An experimental trial conducted by Chadha *et al.* (1974) at Indian Institute of Horticulture Research, Bangalore have proved conclusively that dense planting of pineapple results in high yield without effecting the quality or size of fruits. In this experiment, the planting density of 63,758 resulted in the yield of 118.8 tonnes of fruits with uniform maturity and cylindrical shaped in comparison to a yield of 87 tonnes with 43036 plant population per hectare. Dass *et al.* (1978) in their trials observed the best result in a spacing of 25cm, 60cm and 90cm between plants, rows and beds respectively, in a two-row planting system having a plant population of 53333 per hectare. Choairy and Fernandes (1983) adopted yield 90-94 tonnes/ha by using 90900 plant /ha.A field experiment was conducted at Bidhan Chandra Krishi Viswavidyalaya reported that plant density of 64000 with a spacing of 25cm×35 cm×90 cm in a double row system was most suitable under North Bengal condition of West Bengal (Bose, 1985). Further increase in plant density up to 72945 per hectare can profitably be used by judicious manuring and irrigation and application of growth substance, if necessary, to increase the percentage of flowering (Das Biswas *et al.*, 1987) (Table 1).

Planting density was a determination factor for yield especially for the mother plant crop (Fig. 1), which was evident in the positive linear response between these two variables ($r^2 = 0.66$, $P < 0.001$). The ratoon crop also showed a positive relation, but with a higher level of variability ($r^2 = 0.24$, $P = 0.01$) (Leon *et al.*, 2012).

Method of calculation of plant population in specific planting density: Calculation of the required number of planting materials in the spacing of 30cm×60cm×90cm.

Required area for a single plant

$$= \frac{\text{Plant to plant spacing} \times (\text{Row to row spacing} + \text{Spacing between pair of rows})}{2}$$

$=0.3\times (0.6+0.9)/2 \text{ m}^2$

$=0.3\times 0.75 \text{ m}^2$

$=0.225 \text{ m}^2$

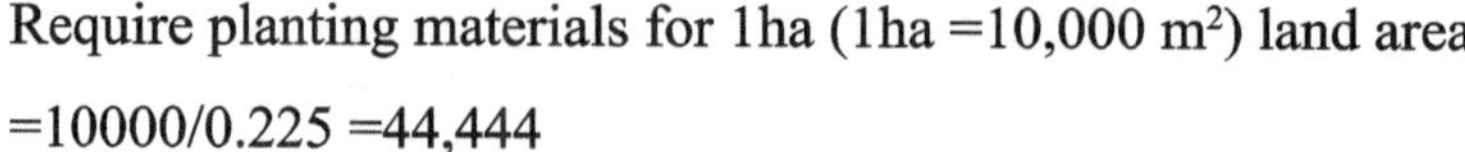

Require planting materials for 1ha (1ha =10,000 m^2) land area

=10000/0.225 =44,444

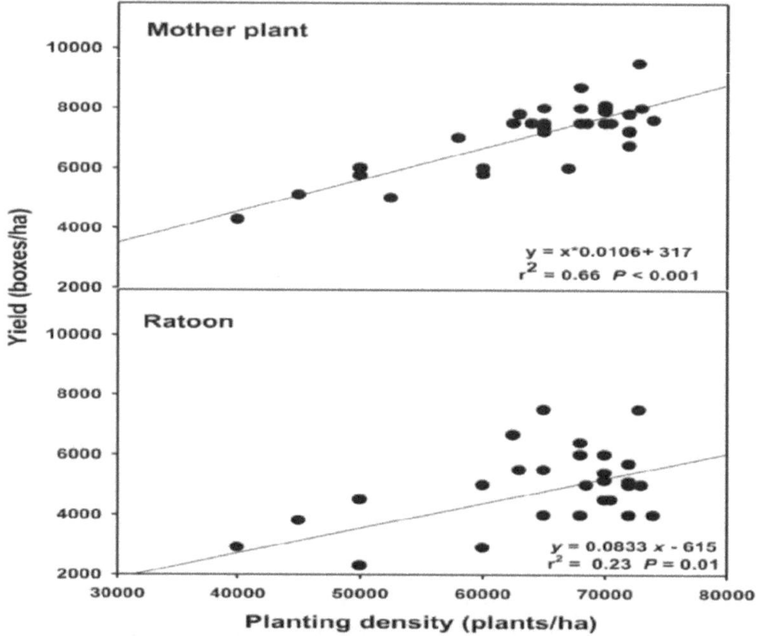

Fig. 1: Marketable fruit yield for the mother plant (33 farms) and ratoon crops (27 farms) as affected by planting density on 'MD-2' pineapple farms in Costa Rica in 2011. The data were best fit using polynomial linear equations; 1 ha = 2.4711 acres, 1 box = 25 lb = 11.3 kg, 1 box/ha = 0.4047 box/acre (Leon *et al.*, 2012).

Table 1: Fruit weight, yield and quality of fruits in pineapple under different plant densities

Plant Density (Per hectare)	Percentage of Flowering	Fruit weight (Kg)	Yield/ha (Tonnes)	TSS (Brix)	Total Sugar (% fresh weight)	Acidity (%)
27777	91.6	2.14	59.4	13.6	12.4	0.64
35714	90.8	2.10	71.6	13.4	11.8	0.64
49378	90.8	2.00	75.6	12.8	11.8	0.68
59259	90.4	1.97	85.2	12.8	11.8	0.72
72945	81.3	1.92	102.2	12.6	11.2	0.72
76923	76.8	1.88	98.1	12.6	9.2	0.78
86956	72.6	1.62	92.8	10.8	8.8	0.78
100000	70.4	1.58	88.9	10.2	8.8	0.81
S. E±	2.93	0.08	3.94	0.99	0.126	0.03
C.D. at 5%	8.79	0.23	11.82	N.S.	N.S.	0.09

Source: Das Biswas *et al.* (1987) (N.S. = Non-Significant)

Table 2: Spacing for different plant population per hectare (*Source*: Chadha *et al.*, 2013)

Spacing (cm)			Plant Population
Plant to Plant	Row to Row	Trench to Trench	
30	60	120	35,000
30	60	90	43,500
25	60	90	53,330
22.5	45	90	63,500
25	30	91.5	65,900
25	30	83.5	70,500

Table 3: Population density followed by different country

Country	Population density
Bangladesh	38,345
Brazil	90,900
Hawaii	40,000-45,000
India	40,000-53,333
Kenya	47,800-53,800
Malaysia	71,758
Philippines	62,000
South Africa	41,250
Taiwan	45,000

Source: Chadha *et al.*, 2013

Table 4: Recommended population density followed by different regions of India

Different region of India	Population density
Bangalore and Coorg	53,333
Jorhat, Assam	53,333
Kerala	40,404
Konkan area	56,437
Maharashtra	53,333
Meghalaya	43,500
West Bengal	43,500-49,300

Source: Chadha *et al.*, 2013

High-density planting (HDP): The traditional system of planting pineapple accommodates not more than 15,000 to 20,000 plants per hectare. High-density planting system is recommended for commercial cultivation to accommodate a greater number of plants per unit area. Besides increased productivity, it provides shade and reduces evaporation loss by overlapping of the basal leaves. Other advantage of high-density planting is reduced weed infestation. According to Chadha *et al.* (1974), the vigorously growing leaves tend to twist and grow upright due to crowing of the plants in high density panting, which provide the fruit natural covering to prevent the sun scald and result in uniformly coloured and lustrous fruits.In Assam, about 20-25% fruits are affected in sun burn in absence of shade. In high-density planting, a microclimate with high humidity favours fruit development. It is also possible to obtain at least two ratoon crops in dense planting.In a double row system (25x35x90cm), HDP is adopted in almost all pineapple growing region of India, accommodating approximately 64000 plants and yielding about 100-120t/ha of fruit.Neri *et al.* (2021) observed thatincreasing plantation densities (up to 55500 plants ha−1) did not have significant effects on the weight (crown) or diameter of pineapple fruits. The TSS contents and pH values of the harvested fruits varied due to the interaction of the cultivars and the sowing density, but in general, the fruits met the standards required for the market. Djido *et al.* (2021) showed that planting density had no effect on the crop development variables (such as the number of functional leaves and D-leaf length) at the moment of flowering induction. They also observed that the planting density had no effect on the total weight per fruit, infructescence weight, total fruit length, infructescence length, crown length, or fruit shelf life as perceived by traders. The yield increased from 54.9–69.1 up to 90.1 t/ha with an increase in the planting density.Prasanna and Bhowmick (2018) revealed that when different spacing's ranging from 64000 plants/ha to 29,630 plants/ha wereused, there was significant difference among 7 treatments for yield and quality. Highest fruit yield withcrown (52.59t/ha) was observed in the treatment with higher spacing of 64,000 plants/ha (T1) and highestfruit weight with crown (1290g) was observed in the treatment with lower spacing of 29,630 plants/ha (T7).

Cropping-system: Mauritius can be grown as a pure crop in garden land, reclaimed lowlands and wetlands. It can be used as an intercrop in arecanut, coconut and newly planted rubber plantations. In rubber plantation, it can be grown at the young stage of plantation for the first 3-4 years only.

The majority (63%) of the small-scale cultivators (less than 5 ha) grow the pineapple as monocrop while the majority of the medium (74%) (5 to 10ha) and large (77%) (more than 10ha) scale cultivators grow the pineapple as an intern crop on new or young rubber plantations. High productivity of pineapple cultivation in converted paddy fields is the most influential factor that led to the conversion of paddy fields into pineapple fields. Non- profitability of paddy cultivation and high cost of paddy workers are the other reasons reported in this regard. There can be three-paired rows in between two rows of coconut. Intercropping in rubber plantations. System of planting is in paired rows at 45 x 30 cm. There will be only one paired row of pineapple in between two rows of rubber.

Strategies for Year-Around Production and Improvement of Fruit Quality: The following strategies can be taken for year round production of pineapple production (Prakash *et al.*, 2010)

1. **Diversification of planting materials and planting time:** Pineapples are planted year-round to achieve continuous production by using different sizes of suckers and slips. Slips or suckers of pineapple should be planted from July to December, followed by application of etherl (0.25 ml/L) or calcium carbide (20 g/L), between 335 and 365 days after planting.
2. **Diversification of cultivars**: To ensure year-round fruit harvesting, various cultivars can be planted in the same field. For an example, 'Queen' normally matures in the month of April, May and June, whereas 'Kew' matures during July, August and September.
3. **Flower inducing chemicals**: Flower inducing chemicals should be applied to the heart of the plant between 5:00 to 6:00 pm in the evening and only when plants have at least 32-38 leaves. The flower emergence will come out after 55-60 days chemical spraying. Application of chemicals at an interval of 7 days from April to November can ensure steady harvest of the fruits throughout the year, if plants have at least 32-38 leaves at the time of chemical spraying.
4. **Removal of slips**: Slips should be removed immediately after emergence, if the number of slips will increase then fruit size and quality will decrease.
5. **Pinching**: Removal of the innermost whorl of leaflets along with the growing tips after 40 days of flowering was found best to get better fruit size, shape and quality. It enhances the fruit size and weight up to 15-20% in comparison to non-pinched.

References

Bose, T. K., Mitra, S. K. and Sanyal, D. 2001. Fruits: tropical and subtropical. Volume 1 (3rd Ed.). Naya Udyog, Kolkata.

Bose, T. K. 1985. Pineapple (by S.K. Sen). (In): Fruits of India Tropical and Sub- tropical. Naya Prokash, Calcutta.

Chadha, K. L., Melanta, K. R. and Shikhamany, S. D. 1974. Effect of the Type and Size of Planting Material on the Vigour of the Subsequent Plants, Yield and Quality in Kew Pineapple (Ananas Comosus (L.) Merr.). Indian Journal of Horticulture, 31(1): 9-15.

Chadha, K. L., Reddy, B. M. C. and Shikhamany, S. D. 2013. Pineapple, Directorate of Knowledge Management in Agriculture, ICAR, Pusa, New Delhi.

Das Biswas, S., Mitra, S. K. and Bose, T. K. 1987. Response of Kew pineapple to plant densities and calcium carbide. Scientific Cultivation, 53: 120-121.

Dass, H. C., Ganapathy, K. M., Singh, H. P., Reddy, B. M. C. and Prakash, G. S. 1978. Spacing and population density studies in pineapple var. Kew. Paper presented at the All-Indian Co-ordinated Fruit Improvement Project, held at Bangalore. From Research Report and Project Proposals on Banana, Pineapple, Papaya, pp. 130-32

Dhareswar, S. R. 1950. Trench cultivation of pineapple. Dharwar Agricultural College Magazine, 4: 9-11.

Djido, U., FassinouHotegni, N. V., Lommen, W. J., Hounhouigan, J. D., Achigan-Dako, E. G. andStruik, P. C. 2021. Effect of Planting Density and K_2O: N Ratio on the Yield, External Quality, and Traders' Perceived Shelf Life of Pineapple Fruits in Benin. Frontiers in Plant Science, 12: 627808.

Gandhi, S.R. 1949. Pineapple Culture in Western India. Poona Agricultural College Magazine, 40(3): 33-39.

Leon, R. G., and Kellon, D. 2012. Characterization of 'MD-2'pineapple planting density and fertilization using a grower survey. Horticulture Technology, 22(5): 644-650.

Mohan, N. K. and Ahmed, F. 1984. Effect of depth of trench and plant density on growth, yield and quality of pineapple var. Kew. Journal of Research, Assam Agricultural University, 5(1): 17-19.

Neri, J. C., Meléndez Mori, J. B., Vilca Valqui, N. C., Huaman Huaman, E., Collazos Silva, R. and Oliva, M. 2021. Effect of planting density on the agronomic performance and fruit quality of three pineapple cultivars (Ananas comosus L. Merr.). International Journal of Agronomy, 2021: 1-9.

Pineapple Technical Group. 1999. Pineapple production practices. Department of Horticulture, National Agriculture Research Institute, Mon Repos, East Coast Demerara, Guyana.

Prasanna, V. S. S. V. and Bhowmick, N. 2018. Effect of planting densities on fruiting characteristics of Pineapple [Ananas comosus (L.) Merr.] cv. Mauritius. International Journal of Chimerical Studies, 6: 1773-1776.

Radha, T., Baby, L.M. and Rajamony, L. 1990. Effect of depth of trenches on high density planting of Kew pineapple. South Indian Horticulture, 38: 183-88.

Reddy, B. M. C. and Prakash, G. S. 1982. Standardization of optimum depth of trench of planting Kew pineapple. Annual Report, 1982. Indian Institute of Horticultural Research. pp, 19.

Roy, A. 1981. Ph. D. Thesis, submitted to the Bidhan Chandra Krishi Viswa Vidyalaya, Kalyani.

Prakash, J., Suresh, C. P., Das, A., Chandra Das, S. and Bhattacharjee, T. 2010. Pineapple cultivation in hilly Tripura with year around production: improving livelihood opportunities in rural areas of Tripura. In VII International Pineapple Symposium. pp, 291-298.

14

Nutrient Management in Pineapple

Prerna Baraily

Department of Pomology and Postharvest Technology, College of Agriculture (Extended Campus), Uttar Banga Krishi Viswavidyalaya, Majhian D. Dinajpur, West Bengal, India

Pineapple is considered as one of the important tropical fruits which are extremely demanded in the international market. This is due to its various quality characteristics like sweeter taste, blemish-free flesh, a cylinder shape with golden yellow pulp, a very pleasant aroma when ripe, low acidity, and a longer shelf life compared to other varieties (Banful *et al.*, 2011; Malezieux *et al.*, 2003). Pineapple can grow well in low as well as high fertile soil but the best production is obtained with high fertile soils. Souza *et al.* (2007) found that a fruit's weight is directly correlated with the plant vigour at the time of floral differentiation with a minimum D-leaf length. Therefore, it is proved by Souza *et al.* (2007) that nutrients play animportant role in pineapple cultivation to increase the growth and yield of the crop. It is also noted that the pineapple plants have a large nutrient uptake demand, especially for potassium, followed by nitrogen, calcium, magnesium, sulphur, and phosphorus.

The focus on pineapple production should not only aim for stable volumes but also sustainable growth while maintaining quality standards amidst a growing population. The 'Green Revolution' with high inputs has plateaued, emphasizing the need for a natural balance to sustain life and property. Effective plant nutrient management is crucial for environmental, economic, and social reasons. Pineapple's nutrient requirements, high in nitrogen (N), potassium (K), and iron (Fe), with lower needs for phosphorus (P) and calcium (Ca), make soil organic matter desirable. Repeated cultivation exhausts soil nutrient reserves, leading to decreased yields. Heavy synthetic fertilizer use may cause soil impaction and acidification. Combining organic manure with chemical fertilizers is recommended for profitable and sustainable crop production while preserving soil fertility. However, excessive nitrogen application poses health risks. As farmers transition to organic farming, the negative impact of inorganic fertilizers on pineapple size, yield, and quality is evident. The role of various nutrients in pineapple, studied through nutritional experiments with graded doses, is discussed in this chapter (Spironello *et al.*, 2004; Ward, 2009; Wolfe and Patz, 2002).

Nutrient Management: Nutrient management is the science and art directed to link soil, crop, weather, and hydrologic factors with cultural, irrigation, and soil and water conservation practices to achieve the goals of optimizing nutrient use efficiency, yields, crop quality, and economic returns, while reducing off-site transport of nutrients that may impact the environment (Delgado and Lemunyon, 2006). Sustainable nutrient management combines organic and chemical sources to improve soil health and productivity, aiming to prevent on-site degradation and minimize off-site nutrient impact, reflecting a growing momentum towards conservation.Some important factors that need to be considered when managing nutrients include (a) the application of nutrients considering crop quality and achievable optimum yields; (b) application, and timing and management, of nutrients using a budget based on all sources and sinks active at the site; and (c) the management of soil, water, and crop to minimize the off-site transport of nutrients from nutrient leaching out of the root zone, surface runoff, and volatilization (or other gas exchanges) (Ganeshamurthy *et al.*, 2015).

Hazarika and Sharma in 2017 had stated that, the nutrient management is an important component in orchard management for high efficiency and high fruit quality. The Integrated Nutrient Management (INM) method calculates total nutrient demand based on target fruit yield, recommending NPK fertilizers considering soil fertility index and nutrient requirements at different growth stages. (Souza and Reinhardt, 2007).

Plant analysis to assess nutritional status: Fertilizer requirements are determined by assessing crop needs, considering factors like climate, soil conditions, water supply, and pest control. The nutritional status of fruit crops, particularly pineapples, is evaluated through leaf nutrient analysis, with the basal white portion of “D” leaves crucial for assessment in the Hawaiian Technique in India. Nitrogen (N) and potassium (K) are essential for plant growth, with the DRIS approach providing precision in nutrient balance evaluation. The Diagnosis and Recommendation Integrated System (DRIS) utilize dual nutrient ratios to address nutrient deficiencies and ensure a balanced nutritional status in plants, considering changes with plant age. This method proves efficient in assessing plant nutrition, overcoming issues related to concentration or dilution effects on nutrients in plants (Devadas, 2005; Beaufils, 1973; Walworth and Sumner, 1987; Singh *et al.*, 2000).

Integrated nutrient management: Integrated Nutrient Management (INM) involves optimizing soil fertility and plant nutrient supply for sustained crop productivity. This holistic approach considers the specific nutrient needs of plants, their combinations, timings, and integration methods to achieve the highest efficiency within environmentally acceptable limits (Lal and Dayal, 2014). INM intelligently utilizes organic, inorganic, and online biological resources to enhance yields, maintain soil properties, and provide economically viable and environmentally safe crop nutrition (Tandon, 1990). Integrated Plant Nutrient Management (IPNM) optimizes sustained crop productivity using a combination of resources like chemical fertilizers, amendments, organic manures, crop residues, N-fixing crops, and bio-fertilizers, tailored to location-specific conditions. Angers (1997) outlines five major concepts in simplified INM, including capital and annual investments in plant nutrients, losses, and outputs. Mineral fertilizers indirectly enhance soil biological activity, with nitrogen (N) fertilization causing soil acidification. Long-term experiments reveal that balanced applications of N, phosphorus (P), and potassium (K) promote microbial biomass and diversity (Zhang and Wang, 2005). Integrated nutrient management helps restore and sustain fertility, addressing emerging nutrient deficiencies beyond N, P, and K. The combined use of organic materials and inorganic nitrogenous fertilizers is essential for economic and ecological sustainability (Kumar *et al.*, 2007). Specific recommendations for fertilizer application, especially nitrogen, ammonium sulfate, and potash, are provided for different planting conditions. Studies show positive effects on flowering percentage, fruit yield, and biochemical parameters in pineapple with the application of recommended doses of NPK along with vermicompost and biofertilizer (Bhowmick *et al.*, 2017; Baraily and Deb, 2018). The combination of organic manure, inorganic fertilizer, and biofertilizer is found to yield the best results in terms of fruit length, weight, and total estimated yield of medium to high-quality fruits (Bhowmick *et al.*, 2017).

Components of INM

Organic sources of nutrients: Manures are organic materials derived from animal, plant, and organic wastes, providing nutrients and improving soil quality. Examples of organic manures include well-rotted farmyard manure (FYM), compost, and green manures. FYM and compost are produced from decomposed agricultural by-products, while green manures consist of undecomposed plant tissues that break down in the soil. Organic matter in manures enhances soil physical properties and can chelate toxic elements, regulating the availability of metals like iron, manganese, and aluminium in the soil. Many factors influence fruit quality, and one of the main factors is the fertilization system. The quality and quantity of applied fertilizers are key factors affecting the growth, yield, and quality of crops (Chang *et al*., 2010). However, as said earlier that continuous use of chemical fertilization leads to the deterioration of soil characteristics and fertility and might lead to the accumulation of heavy metals in plant tissues, affecting the fruit nutritional value and edibility (Shimbo *et al*., 2001). The application of organic fertilizers, which are made from animal excreta or other agricultural wastes, contributes to the sustainability of agriculture systems and is usually used to improve the soil structure and stability in addition to enhancing the yield and in particular, the quality of plants (Tejada and Gonzalez, 2003; Abou El- Magd *et al*., 2006;). According to Lui and Lui (2012) findings, Pineapple fruits ripened in winter showed enhanced quality, total relative contents of ester aroma components ,including increased TSS and total sugar levels which were treated with 200 ml peanut-press pulp liquid fertilizer + 20g compound fertilizer per plant. Daramola *et al*. (2013) reported that the pineapple plants received higher rates of poultry manure can have bigger fruit sizes and higher numbers of fruits. This is because chicken manure is a good source of plant nutrients. The manure contains not only macro elements but also trace elements (which may frequently be deficient in tropical soils) required by plants for their growth and development.

The return of residue to the field is a useful cultural practice to improve both soil fertility and soil organic carbohydrate storage (Aynehband *et al*., 2010; Lou *et al*., 2011). With the increasing demand for organic fruits and vegetables, the return of composter residue to fields as organic manure has recently attracted attention from farmers and scientists due to the positive effects of soil amendment while reducing the use of synthetic fertilizer (Chikae *et al*., 2006; Gabhane *et al*., 2012). Therefore, manygrowers compost the pineapple residue before returning it to the soil. In addition to this Liu *et al*. (2013) has reported that the Composted pineapple residue return (CPRR) decreased the bulk density of the soil and increased the organic matter, available N, available P, and available K contents. CPRR increased the abundance of bacteria, fungi, and actinomycetes and the activities of urease, catalase, acid phosphatase, and invertase in the soil. The chlorophyll, soluble sugar, and soluble protein contents of the leaves and root vigor were all enhanced by this practice. CPRR also promoted the growth and heightened the yield of pineapple fruits.

Darnaudery *et al*. (2018) also stated that Organic fertilization, combining *M. pruriens* green manure and local sugarcane vinasse, allowed in organic farming, provided very cheering results even if the yield was lower and the production cycle longer: fruit weight remained within the export standards and, most importantly, the organoleptic quality and the resistance to Leather Pocket fungal disease were the best mulching and mineral fertilization positively influenced fruit physico-chemical and organoleptic characteristics for sugarloaf pineapple, and consequently enhanced the percentage of fruit (Sossa *et al.*, 2017).

Farmyard Manure: Farmyard manure enhances soil by increasing water holding capacity for optimal water use, improving chemical and biological conditions. It boosts cation exchange capacity, provides essential elements, vitamins, hormones, and organic acids crucial for soil aggregation and beneficial microorganisms involved in biochemical processes and nutrient release. With nitrogen, phosphorus, and potassium content (0.5-1.5%, 0.4-0.8%, 0.5-0.9% respectively), FYM directly contributes to increased crop yields by accelerating respiratory processes, enhancing cell permeability through hormone growth action, and supplying nutrients through biological decomposition. It also indirectly improves soil physical properties, increasing permeability and water holding capacity.

Vermicompost: Vermicompost is the excreta of earthworms, which are rich in humus, macro and micronutrients. It helps to improve soil health status, increase crop production, and improve the physical properties of the soil (Adhikary, 2012; Azarm *et al.*, 2008. It has high porosity, aeration, drainage, water holding capacity and microbial activity, which helps to build an outstanding soil conditioner. There are many reports which indicate that vermicompost contains plant growth regulating hormones, such as humic acids, auxins, gibberellins and cytokinins that are liable for increasing plant growth and yield of many crops (Singh *et al.*, 2010). The addition to this vermicompost is very beneficial, as they helps increase the soil organic matter composition, which in turn aids in improving soil aeration, maintaining good soil aggregation, protecting against soil erosion, and increasing nutrient availability (Roy, 2006; Baligar *et al.*, 2001). Other than that, vermicompost also contains various plant nutrients such as N, P, K, Ca, Mg, S, Fe, Mn, Zn, Cu, and B, which improve the nutrient content of different plant components such as the roots, shoots, and fruits (Theunissen *et al.*, 2010; Chaudhuri *et al.*, 2016) found that the application of vermicompost in the tropical soils of pineapple agro-ecosystem promotes plant growth and yield of fruits through increase in density and biomass of earthworms and physicochemical properties of soil. The P, K, and Zn concentrations in the D-leaves of plants supplied with both vermicompost and chemical fertilizer were considered adequate for the crop.In addition to that they stated that the crop yield is very much related to the particular concentration of vermicompost, beyond the level of which production declines and increase in vegetative growth, fruit weight and fruiting percentage of pineapple are strongly linked with the soil pH, av. P, av. K, clay content and the earthworm density of soils. Mahmud *et al.* (2018) reported that the fruits produced were considerably larger in size with smaller crowns, but contained slightly lower TSS with the application of vermin-compost. Field application of vermicompost showed strong relationship with pineapple plant growth and yield of fruits.

Biofertilizers: Biofertilizer is commonly referred to as the fertilizer that contains living micro-organisms and it is expected that their activities will influence the soil ecosystem and produce supplementary substance for the plants (Ngampimol and Kunathigan, 2008). The application of bio-fertilizers is required to maintain and improve soil quality and productivity levels at low input costs. Combination of biofertilizer with organic sources of nutrients helps to conserve the soil health by maintaining the balance of organic matter and soil microflora eventuallyserving to improve physical, chemical and biological properties of the soil (Walia and Kler, 2009). Bio fertilizers are microbial inoculants of selective micro organisms like bacteria, algae, fungi which already have been existing in nature. They may help for improving soil fertility by the way of accelerating biological nitrogen fixation from atmosphere, solubilization of the insoluble nutrient in soil, generate little quantity of PGRs decomposing plant residues and stimulating plant growth and production (Patel *et al.*, 2018). Opinions vary greatly about the use of bifertilizers as a part of sustainable agriculture.

After the investigation Gonzalez *et al.* (2011) came out with an result that the bacteria *Azotobacter chroococcum* (INIFAT5 strain) significantly provoked changes in the plant fresh weight, the thickness of the leaf abaxial and adaxial cuticles, and the root exoderm width. Contrastingly, *A. chroococcum* did not affect the thickness of the leaf photosynthetic parenchyma. Baldotto *et al.*, (2010) also reported that the Humic acid from vermicompost and diazotrophic bacteria of the genus *Burkholderia*, separately and even more when combined, promoted the growth of the aerial part, root system and contents of N, P, K, Ca and Mg in pineapple plants 'Vitoria' cultivated *in vitro*, in the acclimatization phase. He also highlighted that the combined use of humic acid from vermicompost and *Burkholderia* did not affect the population of endophytic diazotrophic bacteria in the roots of pineapple plantlets.

Renewed and intensified efforts are being made to grow different crops using various microbial inoculations, ever since depleting soil fertility attained serious dimensions, more so, with increased cropping intensity coupled with heavy use of chemical fertilizers (Kohli *et al.*, 1996) triggering the menace of nutrient mining (NM) defined as the quantum of nutrient added – quantum of nutrients removed (if obtained on negative side, it denotes the NM). As *Azospirillum* was previously known as *Spirillum lipoferum*, it was only after its rediscovery by Dobereiner *et al.* (1972), that the bacterium gained the reputation of being the most studied plant-associated bacterium. *Azospirillum* spp. is ubiquitously distributed in many parts of the world with tropical, sub-tropical, and temperate climate conditions (Tilak *et al.*, 2005). The rhizosphere supports large and active microbial populations capable of exerting beneficial, neutral, or detrimental effects of plant growth (Orhan *et al.*, 2006). PGPR was first described by Kloepper *et al.* (1989), as soil bacteria that colonize the roots of plants following inoculation onto seeds and that enhance plant growth. Later, Bashan and Holguin (1998) proposed two new terms, biocontrol plant growth promoting bacteria and plant growth promoting bacteria. *Azospirillum* and *Pantoea* are defined as free living, plant-growth-promoting bacteria, capable of affecting the growth and yield in numerous plant species, many of agronomic and ecological significance (Bashan *et al.*, 2004). Later, Herman *et al.* (2008) suggested *Bacillus* (*B. subtilis* and *B. amyloliquefaciens*)-based PGPR for simultaneously improved production in bell pepper and reduced aphid infestation in peach. These PGPR have no preference for crop plants or weeds, or for annual or perennial plants, and can be successfully applied to plants that have no previous history of PGPR in their roots (Dobbelaere *et al.*, 2003). PGPR have been reported to enhance plant growth directly by a variety of mechanisms; fixation of atmospheric N that is transferred to the plant, production of siderophores that chelate Fe and make it available to the plant root, solubilization of minerals such as P, and synthesis of phytohormones (Dobbelaere *et al.*, 2001). *Azospirillum* spp. are known mainly for their ability to produce plant hormones as well as polyamines and amino acids in culture (Thuler *et al.*, 2003), but they are also involved in the biological fixation of N and the increased activity of glutamate dehydrogenase and glutamine synthetase (Ribaudo *et al.*, 2001). *A. brasilense* produces high quantities of extracellular indole-3-acetic acid (IAA), increasing root elongation, root surface area, and root dry matter (Molla *et al.*, 2001).

Inorganic fertilizers: Nutrients are normally supplied through chemical fertilizers. They are chemical compounds containing one or more of the primary nutrients and are generally produced by chemical reactions. Whatever may be the chemical compounds, its most important ingredient for plant growth is the nutrient content. The primary nutrients are nitrogen, phosphorus and potassium; however, their concentration in a chemical fertilizer is expressed as a percentage of total nitrogen (N), available phosphate (P_2O_5) and soluble (K_2O). According to FAO (2000), pineapple requires nutrients such as nitrogen

(N), phosphorus (P) and potassium (K) for fertility maintenance and crop production. These nutrients are specific in function and must be supplied in sufficient quantity to plant at the right time (Sinclair, 1993). Spironello *et al.* (2004), fruit yield of pineapple is very responsive to NPK fertilization. In fruit crops, it is known that applications of NPK fertilizer are vital for fruit yield and quality (Vho, 1999).

Many plants with high NO_3 and K, failing to undergo floral differentiation, continued in the vegetative stage from three to six months thereafter. Such plants produced longer and thicker stems, but shorter and sclerotic leaves than similar plants with fruits. Equiionic ratio of potassium to nitrate absorbed by the different cultures has been ranged from 2.63 to 0.73. High calcium concentrations interfered less with the absorption of potassium than high potassium concentrations with calcium. Nitrate concentrations in the stem and in the non-chlorophyllous basal tissues of the leaves increased in proportion to the concentrations in the substratum. Nitrate concentrations in the non-chlorophylls tissues were increased or decreased at subsequent growth intervals by the rates of absorption by the roots and translocation to the stem and leaves, and those of assimilation by the chlorophyllous tissues of the leaves and of plant volume expansion by growth. Nitrate undergoes little or no change in pineapple roots, in the stem and in the non-chlorophyllous basal tissues of the leaves (Sideris *et al.*,1938; Sideris *et al.*, 1947), but in the chlorophyllous tissues it is converted to protein by means of intermediary reactions which have not as yet been experimentally elucidated. Ammonia, contrasted to nitrate, is assimilated in the roots, stem and nonchlorophyllous or chlorophyllous leaf tissues.Nitrogen and potassium fertilizations have a high impact on pineapple fruit yield, organoleptic and sanitary quality (Spironello *et al.*, 2004). Potassium plays an important role in fruit quality (Hepton and Bartholomew, 2003), and plants and fruits require large quantities of K. In addition, the K/N ratio is also essential to both yield and organoleptic quality build-up (Owusu-Bennoah *et al.*, 1997; Sossa, 2017) N-K fertilization also enhanced TSS, glucose, fructose and malate contents, as well as pH, fruit weight, in frutescence length, proportion of exportable fruits and mean weight of exportable fruit. High values were globally obtained with treatments which associated higher nitrogen and potassium doses.

Spironello in 2004 revealed in his case study, that quadratic effects of N and K on yield and a maximum of 72 t/ha of fresh fruit has reached with the rates of 498 kg/ha N and 394 kg/ha K_2O. In order to reach the maximum fruit size, and to improve the percentage of first class fruit (mass greater than 2.6 kg), were necessary rates of N and K respectively 11 and 43% higher than those for maximum yield. No effect of P rates was observed on pineapple plant growth, despite the low availability of this nutrient in the soil. Nitrogen fertilizer generally increases fruit size (Paula *et al.*, 1991), juice content while decreases total soluble solids (TSS) and vitamin C (Teisson *et al.*, 1979; Paula *et al.*, 1991). Teixeira *et al.* (2002) observed that late application of N had anoptimistic effect on fruit yield but decreased TSS. Potassium also has a positive effect on fruit size, TSS and total acidity (Paula *et al.*, 1991). The effect of P rates on fruit quality is normally lower than that observed for N and K.Nitrogen and phosphorus had positive effects on growth, yield and quality of pineapple. The increased rates of N and P could be adduced to high energy synthesis and translocation activities stimulated by N and P application. According to Sakimin *et al.* (2017) NPK provided easily nutrients to the soil as compared to controlled-release fertilizer (Kamila). This supports that the SRI formulation enhanced nutrients uptake by the roots. The combination of NPK and SRI fertilizers showed the best performance on the growth of pineapple plants from beginning until 10 week after treatment (WAT) while for leaf length controlled release fertilizer Kamila showed the best performance until 10 WAT.

Macro Nutrients

Nitrogen (N): Nitrogen is an important constituent of proteins and their precursors, which form about 7% of the drymatter of the pineapple-plant. Nitrogen is required second after potassium by pineapple. Fruit yield components (branching, fruit length, fruit diameter,core diameter and flowering) increased, but fruit quality [Total soluble solid (TSS), percentage acidity and Vitamin C] decreased with increasing rates of N application. Nitrogen increases the chlorophyll contents of the leaves, thereby promote the photosynthetic capacity of the plant, plays a part in the manufacture of proteins and responsible for high fruit yield in pineapple (Mohamed and Ahmad, 2004). Higher doses of N, for example, increase productivity, size, and fresh weight of pineapple MD2 (Spironello *et al.*, 2004) and Vitoria cultivars (Silva *et al.*, 2012), The longest leaf as a result of N application is indicative of the role of N in promoting vigorous foliage growth and more intense physiological activities in the plant which favored the synthesis of more assimilates for fruit development.(Zhao, 2009) also justified that the soil nitrogen fertilization and foliar urea applications increased the content of almost all nutrients in leaves. According to Malavolta, (2006), The yellow coloration is related to the production of chlorophyll, since nutritional disturbance of N cause changes in the formation of chloroplasts (Malavolta, 2006), more specifically, chlorosis is associated with decreased chlorophyll content and activity of Rubisco, which provides low rates of photosynthesis in plants (Herman *et al.*, 2008).

Omotoso and Akinrinde (2013) also revealed that the effects of excess N is delay in food production and beyond the optimum level, toxicity could be expected.An excess of N also contributes to a reduction in fruit pulp firmness and an increase in translucency and, in hot periods, can also increase the risk of the appearance of a symptom known as "jaune" (ripening of the pulp, while the skin of the fruit remains green), according to Py *et al.* (1987).

Phosphorous (P): Despite the low demand for phosphorus (P), it plays a key role in plant productivity. Found in various organic and inorganic forms, phosphorus actively participates in the plant's energy transport mechanism and is essential for carbohydrate synthesis and decomposition.. According to the Py *et al.* (1975a) heavy dressing of phosphorous accelerate the fruit formation and ripening. According to Py *et al.* (1987), the symptoms of P deficiency include dark, bluish-green foliage, which is more pronounced in the presence of excess N, and leaves that completely dry out at the tips, starting from the older ones. Brownish-red tips and brown transverse striations and the borders of these leaves turn yellow from the tips. Plants suffering from P deficiency also have more erect, long and narrow leaves. Malezieux and Bartholomew (2003) pointed out that P deficiency causes a reduction in the growth of all parts of the plant. They called attention to the fact that visual deficiency symptoms are often not seen and are not very specific, and hence could be mistaken for symptoms resulting from damage to the root system caused by water deficiency, nematodes or pests.

Potassium (K): Souza and Reinhardt (2007) mentioned that for pineapples, potassium (K) is the nutrient which is required in greatest quantity. Potassium, the nutrient that accumulates in the largest amount in the plant, also influences productivity, although to a lesser extent than that of N. Potassium is found in inorganic forms everywhere in plant, particularly in young leaves and meristimatic tissues. Formation of sugar and starch, of organic acids, and of hardened tissues, reduction of nitrates and consequently synthesis of proteins are largely depends on the quantities of potassium present. Potassium is indispensible in transportation of photoassimilates from leaves to organ of storage and organs to utilization in physical process. K involved in stomatal regulation and water economy also. Pineapple has highest potash requirement soon after planting, and its requirement is highest at the time of fruit formation and ripening. K is reported as a factor controlling fruit quality (Py *et al.*, 1957a). Teisson *et al.* (1979) described the benefit of increasing amounts of K to an increase in ascorbic acid in the fruit pulp and, in consequence contributing to a reduction of internal fruit darkening. The lower fruit weight with high rates of K might be related to the reduced uptake of NH_4-N and other cations, as there is an interrelationship between the plant NH_4, Ca, Mg and K levels (Ohnoand Grunes, 1985; Hannaway, 1980).

On the other hand Razzaquea and Hanafib (2001) revealed that Potassium applications at high rates tended to depress most of the growth and yield attributes of cv. Gandul pineapple. (Ohno and Grunes, 1985; Wilkinson, 1983; Sideris and Young, 1945) also reported that high dressing of K had antagonistic effect on the uptake of Mg and NH_4-N. Diameter of fruit stalk and the number of fruits standing erect on the plant were found more than when the proportion of potassium was high in the total cations applied. K deficiency leads to the appearance of brown spot on green parts of leaves and withering of leaf tips. Theleaves are also shorter and narrow than normally nourished leaves (Py *et al.*, 1956.

Calcium (Ca): Like other plants, pineapple also needs calcium as it is indispensable for cell-wall formation and for cell division. Accumulation of calcium in eaves is characteristic of aging and maturation. Though calcium needs of pineapple is small, but an excess of cations in the soil leads to calcium chlorosis. Calcium seems to play an important part in the flower-bud differentiation and in fruit formation (Sideris and young, 1950)

Plants with calcium deficiency presented polyploidy nuclei, binucleate cells, amitotic divisions that cause cell death and paralyze plant growth, especially root, with subsequent vegetable tissue necrosis. In a controlled environment Ca deficiency led to the death of the apex, with the development of lateral shoots, which have similar symptoms. In calcium deficiency leaves were soft, flat and dark green and fruits were tasteless. Fruits from calcium deficiency have blackish spots internally with a gelatinous appearance. Calcium deficiency occurs rarely, except in very poor soils.

Magnesium (Mg): Magnesium (Mg) is vital for chlorophyll, essential for green plants like pineapple. Although pineapple's Mg requirements are modest, deficiency limits chlorophyll, reducing photosynthesis and enzyme activity. Magnesium also balances the high potassium needs of pineapple. Plants with Mg deficiency have yellow old leaves, except where shaded by younger leaves when they stay green, yellow spots which turn brown in a controlled environment; drying of older leaves, which have not finished growing before the deficiency appears; fruits without acidity, low in sugar content and without any flavor. Such deficiencies are very common on soils low in Mg, especially when intensely fertilized with K under conditions of high solar irradiation (Py *et al.*, 1987). Calcareous dolomites serve as the primary magnesium (Mg) source for pineapples, applicable in solid or liquid form spray.

Sulphur (S): Sulphur, being a constituent of some amino acids and all protein, is very much needed for plant. On the 154th day after starting the treatment, it was possible to identify the symptoms of sulphur deficiency by the occurrence of a light brown colour in the centre of the leaf blade and by the appearance of a light green color along the banks of the upper leaves Viegas *et al.*, 2014). Py *et al.* (1987) described the following symptoms associated with a deficiency of sulfur: pale yellow to gold colour foliage; pinkish leaf borders, especially on older leaves; normal plant stature; very small fruit. Sulphur supply is normally done by fertilizers that are at the same time sources of some of the primary macro-nutrients, like ammonium sulfate (23 to 24% S), potassium sulphate (17 to 18% S) and single superphosphate (10 to 12% S). In selecting which fertilizer to use, it is important to be certain of the amount of S they contain. This is especially so where the crop is intensely cultivated and on soils poor in organic matter, in order to prevent possible S deficiency.

Micro Nutrients

According to Su (1975), the micro-nutrients that are most important for pineapples in many parts of the world, are iron (Fe), zinc (Zn), copper (Cu) and boron (B). Normally, soils with little organic matter and available nutrients that have been exhausted by intensive cropping, or those with high pH (above 6.5) are likely to be also short of micro-nutrients and they require special attention in relation to micro-nutrients. Micronutrients also help in the uptake of major nutrients and play an active role in the plant metabolism process starting from cell wall development to respiration, photosynthesis, chlorophyll formation, enzyme activity, hormone synthesis, nitrogen fixation and reduction (Viegas *et al.*, 2014).According to Maeda *et al.*, (2011) the B and Zn sources did not affect the total soluble solids contents, total titratable acidity, mean fruit diameter, fruit length without crown, and maturity index. Application of zinc sulphate @ 0.5% and borax @ 0.5% is best with respect to improving the yield and quality of pineapple cv. Mauritius (Kumari and Deb, 2018).

Iron (Fe): Iron, being a constituent of several enzymes, plays an important as redox catalyst, participating in oxidation of carbohydrates and in reduction of sulphates and nitrates. Though iron is not a component of chlorophyll, its deficiency leads to reduced synthesis of the pigment. Development of chlorosis, starting in young leaves; the leaves are generally flaccid, wide, and yellow with a green 'web' of conductor veins. The fruit is red with a chlorotic crown. Iron deficiency can occur under a range of conditions, for example, on soil with high pH, soils rich in manganese (Mn/Fe = 2), compacted soils, areas of termite infestation and attack by pests both of which rapidly decrease root activity, drought, as well as other factors.

Zinc (Zn): Zinc is required for synthesis of tryptophan, which is a precursor for indole acetic acid (IAA) synthesis. It is involved in the activation of several enzymes such as alcohol dehydrogenase, lactic acid dehydrogenase, glutamic dehydrogenase and carbonic anhydrase. Zinc deficieny is often found in peaty and sandy soils which have high leaching rate or are strongly acidic with low p^H. Smaller or Larger transparent yellow spots on the leaf-lamina are characterictics of Zinc deficiency Theunissen *et al.,* 2010). Crook-neck disease in pineapple is also caused by the deficiency of Zinc.

Boron (B): Boron serves as a regulator of plasma permeability, promoting cation absorption and inhibiting anion absorption; insufficient boron results in abnormal root formation and deformities in fruits prone to damage during harvest and transport. Various symptoms linked to boron deficiency include discoloration, leaf stunting, curling, chlorosis with reddened borders, and anomalies in fruit development across different scenarios and regions. High pH soils often exhibit these symptoms, exacerbated by low soil B during drought. Severe cases result in small, cracked fruits, and boron deficiency leads to collapsed conducting vessels, hindering the transport of assimilates. Inadequate boron supply causes an accumulation of assimilation products like reducing sugars and soluble nitrogen compounds in leaves.

Copper (Cu): High concentration of copper to an extent of 70% occurs in leaf-chloroplast. Copper activates several enzymes, Viz. tryosinase, cytochrome oxide, polyphenol oxidase, ascorbic acid oxidase and phenolase. Light green narrow leaves with wavy borders, a pronounced U shape in cross-section and a small number of tricomes; leaf tips curved downward; old leaves with purple-red coloration at the fold; short roots with reduced hairs; stunted plant. Severe copper deficiency causes death of the plant. These symptoms are relatively common, but their descriptions are often imprecise.

References

Abou El- Magd, M. M., El-Bassiony, A. M. and Fawzy, Z. F. 2006. Effect of organic manure with or without chemical fertilizers on growth, yield and quality of some varieties of broccoli plants. Journal of Applied Sciences Research. 2: 791-798.

Abu, T. 2015. Effect of organic fertilizer water hyacinth on the growth and production of Taro [Colocasia esculenta (L.)Schott].Journal of Environment and Earth Science, 5: 70-74.

Adhikary, S. 2012. Vermicompost, the story of organic gold: A review. Agricultural Science, 3: 905-917.

Agbangba, C. E., Dagbenonbakin, G. D., Djogbenou, C. P., Houssou, P., Assea, D. E., Sossa, E. L., Kotomale, U. A., Ahotonou, P., Ndiaga, C. and Akpo, L. E. 2015. Influence de la fertilisation mineralesur la qualitephysico-chimiqueetorganoleptique du jus d'ananas Cayenne lisse au Benin. International Journal of Biological and Chemical Science, 9: 1277-1288.

Agyarko, K. and Asante, J. S. 2005. Nematode dynamics in a soil amended with neem leaves and poultry manure. Asian Journal of Plant Sciences, 4(4): 426-428.

Ahmed, O. H. and Husni, M. H. A. 2010.Exploring the nature of the relationships among total, extractable and solution phosphorus in cultivated organic soils.International Journal of Agricultural Research, 5: 746-756.

Aiyelaagbe, I. O. O., Oshuniyi, A. A. and Adegoke, J. O. 2012. Respone of 'smooth cayenne' pineapple to organic fertilizer in south western nigeria. International Society for Horticultural Science (ISHS): Leuven, Belgium, pp, 261-264.

Angeles , D. E., Sumner, M. E. and Barbour, N. W. 1990. Preliminary nitrogen, phosphorus, and potassium dris norms for pineapple. HortScience, 25(6): 652-655.

Anger, D. A. 1997.Imapact of tillage practices on organic carbon and nitrogen storage in humid soil of eastern Canada. Soil Tillage Research, 41: 191-201.

Arshad, A. M. and Armanto, M. E. 2012. Effect of nitrogen on growth and yield of pineapple grown on bris soil. Journal of Environmental Science and Engineering, 1: 1285-1289.

Aynehband, A., Tehrani, M. and Nabati, D. A. 2010.Effects of residue management and N-splitting methods on yield and biological and chemical characters of canola ecosystem. Journal of Food, Agriculture and Environment, 8(2): 317-324.

Baldotto, L. E. B., Baldotto, M. A., Canellas, L. P., Bressan-Smith, R. and Olivares, F. L. 2010. Growth promotion of pineapple 'vitoria' by humic acids and burkholderia spp. during acclimatization. Revista Brasileira de Ciencia do Solo, 34(5): 1806-9657.

Banful, B., Adjei, P. Y., Idun, I. A. 2011.The influence of two pre and post-planting fungicides on the growth and fruiting of MD2 pineapple [Ananas comosus L. (merr)].Agriculture and Biology Journal of North America, 2: 876-879.

Baraily, P. and Deb, P. 2018. Influence of integrated nutrient management on yield and bio-chemical parameters of pineapple (cv. Kew). Journal of Pharmacognosy and Phytochemistry, 7(5): 1339-1342.

Bashan, Y., Holguin, G. and de-Bashan, L. E. 2004. Azospirillum-plant relationships: Physiological, molecular, agricultural and environmental advances (1997-2003). Canadian Journal of Microbiology, 50: 521-577.

Bashan, Y. and Holguin, G. 1998. Proposal for the division of Plant Growth-Promoting Rhizobacteria into two classifications: biocontrol-PGPB (Plant Growth-Promoting Bacteria) and PGPB. Soil Biology and Biochemistry (in press).

Beaufils, E. R. 1973. Diagnosis and recommendation integrated system (DRIS). A general scheme of experimentation and calibration based on principles developed from research in plant nutrition. South Africa: University of Natal, Pietermaritzburg, Soil Science Bulletin, pp, 132.

Beverly, R. B. 1987. Modified DRIS method for simplified nutrient diagnosis of Valenica oranges by DRIS. Journal of Plant Nutrition, 10: 1401-1408.

Bhowmick, N., Ghosh, S. K., Munsi, P. S., Deb, P. and Ghosh, A. 2017. Effect of integrated nutrient management on flowering and fruiting characteristics of pineapple cv. Mauritius. Journal of Crop and Weed, 13(2): 144-156.

Chadha, K. L., Subramanian, T. R. and Melanta, K. R., Effect of varying levels of N, P and K on growth, yield and quality in variety Queen. Indian Journal of Horticulture, 33: 224-226.

Chang, K. H., Wu, R. Y., Chuang, K. C., Hsieh, T. F. and Chung, R.S. 2010. Effects of chemical and organic fertilizers on the growth, flower quality and nutrient uptake of Anthurium andreanum, cultivated for cut flower production. Scientia Horticulturae, 125: 434-441.

Chaudhuri, P. S., Paul, T. K., Dey, A., Datta, M. and. Dey, S. K. 2016. Effects of rubber leaf litter vermicompost on earthworm population and yield of pineapple (Ananas comosus) in West Tripura, India. International Journal of Recycling Organic Waste Agriculture, 5: 93-103.

Chikae, M., Ikeda, R., Kerman, K., Morita, Y. and Tamiya, E. 2006. Estimation of maturity of compost from food wastes and agro-residues by multiple regression analysis. Bioresource Technology, 97: 1979-1985.

Chundawat, B. S. 2001. Integrated nutrient management in troipical and sub-tropical fruits. Indian Journal of Horticulture, 58(1&2): 59-69.

Daramola, F. Y., Afolami, S. O., Idowu, A. A. and Odeyemi, I. S. 2013. Effects of poultry manure and carbofuran soil amendments on soil nematode population and yield of pineapple. International Journal of Agriculture Science, 3(4): 298-307.

Darnaudery, M., Fournier, P.and Lechaudel, M. 2018. Low-input pineapple crops with high quality fruit: promising impacts of locally integrated and organic fertilisation compared to chemical fertilisers. Experimental Agriculture, 54(2): 286-302.

Delgado, A. and Gomez, J. A. 2016. The soil. Physical, chemical and biological properties.In Principles of Agronomy for Sustainable Agriculture; Springer: Cham, Switzerland, pp, 15-26.

Delgado, J. A. and Lemunyon, J. 2006. Nutrient Management. In Encyclopedia of Soil Science Rattan Lal (Ed.). pp, 1157-1160.

Devadas, V. S. 2005. Standardization of leaf sampling for nutrient analysis in pineapple var. Mauritius.IV International Pineapple Symposium, Acta Horticulturea, 666: 191-192.

Devadas, V. S. and Kuriakose, K. P. 2005. Evaluation of different organic manures on yield and quality of pineapple var. Mauritius. Acta Horticulturea, 666: 185-189.

Dobbelaere, S., Croonenborghs, A., Thys, A., Ptacek, D., Vanderleyden, J. Dutta, P. Lanbandera-Gonzalez, C., Caballero-Mellado, J., Aguirre, J.F., Kapulnik, Y., Brener, S. Burdman, S. Kadouri, D. Sarig, S. and Okon, Y. 2001. Responses of agronomically important crops to inoculation with Azospirillum. Australian Journal of plant Physiology, 28: 871-879.

Dobbelaere S, Vanderleyden J, Okon Y. Plant growth-promoting effects of diazotrophs in the rhizosphere. CRC Crit. Revista Plant Science, 22:107–149.

Dobereiner, J., Day. J. M.and Dart, P. J. 1972. Printed in Great Britain Nitrogenase Activity and Oxygen Sensitivity of the Paspalurnnotatum-Azotobacter paspa Ii Association. Journal of General Microbiology, 71: 103-116.

Edwards, S. 1998. Openness, productivity and growth: what do we really know?.The Economic Journal, 108(447): 383–398.

FAO. 2000. Guide lines for on farm plant nutrition and soil management trials and demonstrations. Rome, Italy.

Gabhane, J., William, S. P. M. P., Bidyadhar, R., Bhilawe, P., Anand, D., Vaidya, A. N. and Wate, S. R. 2012. Additives aided composting of green waste: Effects on organic matter degradation, compost maturity, and quality of the finished compost. Bioresource Technology, 114: 382-388.

Gaind, S. and Gaur, A. C. 2003. Quality assessment of compost prepared from fly ash and crop residue. Bioresource Technology, 87: 125-127.

Ganeshamurthy, A. N., Kalaivanan, D., Selvakumar, G. and Panneerselvam, P. 2015. Nutrient Management in Horticultural Crops. Indian Journal of Fertilisers, 11(12): 30-42.

Gonzalez, R., Laudat, T. and Arzola, 2011. M. Effect of Azotobacter chroococcum on in vitro pineapple plants growth during acclimatization. In Vitro Cellular and Developmental Biology-Plant, 47: 387–390.

Hannaway, D. B., Bush, L. P. and Leggett, J. E. 1980. Plant nutrition: magnesium and hypomagnesium in animals, Kentucky Agriculture Experiment Station Bulletin, pp. 716.

Hartz, T. K., Mitchell, J. P. and Glannini, C. 2000. Nitrogen and carbon mineralization dynamics of manures and composts. HortScience, 35: 209-212.

Havlin, J. L., Beaton, J. D., Tisdale, S. L. and Nelson, W. L. 2005. Soil Fertility and Fertilizers; An Introduction to Nutrient Management. Pearson Education, Inc. New Jersey.

Hazarika, V. and Sarma, A. K. 2017. Advance in integrated nutrient Management (INM) in the Production of High Value Crops. Training Manual Entrepreneurship and Skill Development on Market Driven Production and Processing of High Value Crops of Neh Region, pp, 84-87.

Heard, J., Cavers, C. and Adrian, G. 2006.Up in smoke - nutrient loss with straw burning. Better Crops, 90(3): 10-11.

Hepton, A. 2003. Cultural System. In: Bartholomew, D.P., Paull, R.E., Rohrbach, K.G. (eds) Pineapple: botany, production and uses. CABI Publication, Wallingford, pp, 109-142.

Hepton, A. and Bartholomew, D. P. 2003. Cultural system. In The pineapple: Botany, Production and Uses. (Eds D. P. Bartholomew, R. E. Paull and K. G. Rohrbach). Wallingford, UK: CABI Publishing, pp, 109-142.

Herman, D. C., Weinhold, P. S., Guskiewicz, K. M., Garrett, W. E., Yu, B., Padua, D. A. 2008. The effects of strength training on the lower extremity biomechanics of female recreational athletes during a stop-jump task.The American Journal of Sports Medicine, 36(4): 733-740.

IFA/FAO. 2000. Fertilizer and their use. FAO Rome Italy.

Isuwan, A. 2014. Agronomic traits and fruit quality of pineapple with different levels of chicken manure application. Silpakorn University Science & Technology Journal, 8(1): 67-73.

Kale, R. D. 2014. Earthworm biotechnology-source of sustainable agriculture. In: Chaudhuri P.S., Singh, S.M. (eds) Biology and ecology of tropical earthworms. Discovery Publishing House Pvt. Ltd., New Delhi, pp, 185-194.

Katyal, J. C. 2003. Soil fertility management – A key to prevent diversification. Journal of the Indian Society of Soil Science, 51: 378-387.

Kirkby E. A.1979. Maximizing calcium uptake by plants.Communication of Soil Science and Plant Analysis, 10: 89-113.

Kloepper, J. W., Lifshitz, R. and Zablotowicz, R. M. 1989. Free-living bacterial inocula for enhancing crop productivity. Trends in Biotechnology, 7: 39-43.

Kohli, R. R., Huchche, A. D., Lallan, R., Srivastava, A. K. and Dass, H. C.1996. Interaction effect of leaf nitrogen and potassium on growth, yield and quality of Nagpur mandarin. Journal of Potassium Research, 12: 70-74.

Krishan, H., Singh, R. K. D. and Langpoklakpam, B. 2017. Effect of variety and bio-fertilizer on growth and yield of pineapple [Ananascomosus (L.)Merr.]. Journal of Pharmacognosy and Phytochemistry, 6(6): 2568-2571.

Kumar, A., Tripathi, H. P. and Yadav, D. S. 2007. Correcting nutrient for sustainable crop production. Indian Journal of Fertilization, 2: 37-44.

Kumar, D., Pandey, V. and Anjaneyulu, K. 2008. Effect of planting density and nutrient management on growth, yield and quality of micro propagated banana Rasthali (AAB-Pathkapoora). Indian Journal of Horticulture, 65: 272-276.

Kumar, V., Ramjan, Md. and Das, T. 2019. Integrated Nutrient Management in Vegetable Crops. Biomolecule Reports- An International eNewsletter.

Kumari, U. and Deb, P. 2018. Effect of foliar application of zinc and boron on the physical parameters of pineapple cv. Mauritius. Internatinal Journal of Current Microbiology and Applied Science, 7: 397-402.

Lal, G. and Dayal, H. 2014. Effect of integrated nutrient management on yield and quality of acid lime (Citrus aurantifolia Swingle). African Journal of Agriculture Research, 9(40): 2985-2991.

Langenegger, W. and Smith, B. L. 1978. An evaluation of the DRIS system as applied to pineapple leaf analysis, In: A. R. Ferguson, R. L. Bieleski, and I. B. Ferguson (eds.). Proc. 8th Intl. Colloq. Plant Anal.&Fert. Problems, Auckland, New Zealand., pp, 263-273.

Li, J., Li, X. X., Peng, J. C. and Zhuang, G. Q. 1999. Effect of five mineral elements on nutritional status of Navel orange tree.Journal of Hunan Agriculture University, 25: 36-39.

Liu, C. H., Liu, Y., Fan, C. and Kuang, S. Z. 2013. The effects of composted pineapple residue return on soil properties and the growth and yield of pineapple. Journal of Soil Science and Plant Nutrition,13(2): 433-444.

Liu, C. H. and Liu, Y. 2012. Influences of organic manure addition on the maturity and quality of pineapple fruits ripened in winter. Journal of Soil Science and Plant Nutrition, 12(2): 211-220.

Ludders, P. 1980. Effects of time and amount of nutrient additives on nutrient status and distribution and on fruit quality, in: Atkinson D., Jackson J. E., Sharples R. O., Waller W.M. (Eds), Mineral nutrition of fruit trees, Butterworths, London, pp, 111-121.

Maeda, A. S., Buzetti, S. Boliani, A. C., Benett, C. G. S., Filho, M. C. M. T. and Andreotti, A. 2011. Foliar fertilization on pineapple quality and yield. Pesquisa Agropecuaria Tropical,Goiania, 41(2): 248-253.

Mahmud, M., Abdullah, R. and Yaacob, J. S. 2018. Effect of Vermicompost Amendment on Nutritional Status of Sandy Loam Soil, Growth Performance, and Yield of Pineapple (Ananas comosus var. MD2) under Field Conditions. Agronomy, 8: 183.

Malavolta, E., Oliveika, S. A. and Vitti, G. C. 1993. The use of diagnosis recommendation integrated system (DRIS) to evaluate the nutritional status of healthy and blight affected citrus trees. In: Proceedings of the Eight International. Colloquium on Optimization of Plant Nutrition, eds. M. A. C. Fragoso, and M. L. van Beusichem, pp, 157-159.

Malavolta, E., Vitti, G. C. and Oliveira, S. A. 1997. Avaliacao do estadonutricional das plantas: principios e aplicacoes. Piracicaba: Potafos.

Malezieux, E., Bartholomew, D. P. 2003. Plant nutrition. In the pineapple: botany production, and uses. Bartholomew, D.P., Paull, R.E., Rohrbach, K.G., Eds.; CAB International: Wallingford, UK, pp, 143-165.

Marzouka, H. A. and Kassem, H. A. 2011. Improving fruit quality, nutritional value and yield of Zaghloul dates by the application of organic and/or mineral fertilizers. ScientiaHorticulturae, 127: 249-254.

Mohamed, M. H. and Ahmad, H. 2004. Nutrient supply and dry-matter partitioning of pineapple cv. Josapine on sandy tin tailings.Technical paper, Department of Land Management, Faculty of Agriculture, University of Putra, Malaysia.

Molla, A. H., Shamsuddin, Z. H. and Saud, H. Md. 2001. Mechanism of root growth andpromotion of nodulation invegetable soybean by Azospirillum brasilense. Communications in Soil Science and Plant Analysis, 32(13-14): 2177-2187.

Nagavallemma, K. P., Wani, S. P., Lacroix, S., Padmaja, V. V., Vineela, C., BabuRao, M. and Sahrawat, K. L. 2004. Vermicomposting: recycling wastes into valuable organic fertilizer. Global Theme on Agroecosystems Report no 8. International Crops Research Institute for the Semi-Arid Tropics: Andhra Pradesh, India.

Ngampimol, H. and Kunathigan, V. 2008.The study of shelf life for liquid biofertilizer from vegetable waste. AU JT, 11(4): 204-208.

Ohno, T., Grunes, D. L. 1985. Potassium-magnesium interactions affecting nutrient uptake by wheat forage, Soil Science Society of American Journal,49: 685–690.

Omotoso, S.O. and Akinrinde, E.A. 2013.Effect of nitrogen fertilizer on some growth, yield and fruit quality parameters in pineapple (*Ananascomosus* L. Merr.) plant at Ado-EkitiSouthwestern, Nigeria. International Research Journal of Agricultural Science and Soil Science, 3(1): 11-16.

Orhan, E., Esitken, A., Ercisli, S., Turan, M. and Sahin, F. 2006. Effects of plant growth promoting rhizobacteria (PGPR) on yield, growth and nutrient contents in organically growing raspberry. ScientiaHorticulturae, 111: 38-43.

Orluchukwu, J. A. and Adedokun, O. M. 2015. Response of growth and yield of pineapple (Ananas comosus) on spent mushroom substrates and inorganic fertilizer in south-south, Nigeria. International Journal of Plant and Soil Science, 8: 1–5.

Owusu-Bennoah, E., Ahenkorah, Y. and Nutsukpo, D. X. 1997. Effect of different levels of N: K_2O ratios on the yield and quality of pineapple in the forest-savanna acetone of Ghana. In Second International Pineapple Symposium - Proceedings, International Society Horticultural Science, 393-402.

Patel, V. D., Patel, G. D., Desai, K. D., Patel, D. J. and Mangave, B. D. 2018. Effect of Integrated Nutrient Management on Growth and Flower Yield of African Marigold (Tagetes erecta L.).International Journal of Pure and Applied Bioscience, 6(1): 568-572.

Paula, M. B., Carvalho, V. D., Nogueira, F. D. and Souza, L. S. F. 1991. Efeito da calagem, potassio e nitrogenionaqualidade do fruto do abacaxizeiro. Pesquisa Agropecuaria Brasileiria, 26: 1337-1343.

Py, C., Lacoeuilhe, J. J. and Teisson, C. 1987. The Pineapple Cultivation and uses. Editions G. P. Maisonneuve, Paris, pp, 568.

Rajiv, K. S., Sunita, A., Krunal, C. and Dalsukh, V. 2010. The wonders of earthworms and its vermicompost in farm production: Charles Darwin's 'friends of farmers', with potential to replace destructive chemical fertilizers from agriculture. Agricultural Science, 1: 76-94.

Razzaquea, A. H. M. and Hanafi, M. M. 2001. Effect of potassium on growth, yield and quality of pineapple in tropical peat. Fruits, 56: 45–49.

Ribeiro, H. M., Romero, A. M., Pereira, H., Borges, P., Cabral, F. and Vasconcelos, E. 2007. Evaluation of a compost obtained from forestry wastes and solid phase of pig slurry as a substrate for seedlings production. Bioresource Technology, 98: 3294-3297.

Roy, R., Finck, A., Blair, G. and Tandon, H. 2006. Plant Nutrition for Food Security: A Guide for Integrated Nutrient Management. FAO Fertilizer and Plant Nutrition Bulletin; FAO: Rome, Italy, 16: 368.

Sakimin, Z., Samah, M. N. G. A., Juraimi, A. S., Alam1, M. A. and Aslani, F. 2017. Effects of different type of fertilizers on growth and physiology of MD2 pineapple siti. Bangladesh Journal of Botany, 46(1): 489-495.

Sankar, V., Vergagavathatham, D. and Kannan, M. 2009. Organic farming practices in white onion (Allium cepa L.). Journal of Ecofriendly Agriculture, 4: 17-21.

Sarkar, A. K., Pattanayak, S. K., Singh, S., Mahapatra, P., Kumar, A. and Ghosh, G. K. 2020. Integrated nutrient management strategies for acidic soil. Indian Journal of Fertilisers, 16 (5): 476-491.

Selamat, M. M. and Ramlah, M. 1993. The response of pineapple cv. Gandul to nitrogen, phosphorus and potassium on peat soil in Malaysia, in: Bartholomew D.P., Rohrbach K.G. (Eds), Proceedings of the first international pineapple symposium, Honolulu, Hawaii, pp, 247-254.

Shimbo, S., Zhang, Z. W., Watanabe, T., Nakatsuka, H. A, Matsuda-Inoguch, N., Higashikawa, K. and Ikeda, M. 2001. Cadmium and lead contents in rice and other cereal products in Japan in 1998-2000. Science of the Total Environment, 281: 165-175.

Sideris, C. P., Krauss, B. H. And Young, H. Y. 1938. Assimilation of ammonium and nitrate by pineapple plants grown in nutrient solutions and its effects on nitrogenous and carbohydrate constituents. Plant Physiology, 13: 489-527.

Sideris, C. P., Young, H. Y. And Chun, H. H. Q. 1947. Effects of nitrogen on the nitrogenous fractions of Ananas comosus (L.)Merr. Plant Physiology,22: 127-148.

Sideris, C. P. and Young, H. Y. 1946. Effects of different amounts of potassium on growth and ash constituents of Ananascomosus (L) Merr. Plant Physiology.20: 609–630.

Sideris, C. P. and Young, H. Y. 1950. Growth of Ananas comosus (L.) Merr. at different levels of mineral nutrition under greenhouse and field conditions. I. Plant and fruit weights and absorption of nitrate and potassium at different growth intervals' (with eight figures). Plant Physiology.20: 680–691.

Silva, A. L. P, Silva, A. P., Souza, A. P., Santos, D., Silva, S. M. and Silva, V. B. 2012. Resposta do abacaxizeiro 'Vitoria' a doses de nitrogenioem solos de tabuleiroscosteiros da Paraíba. Revista Brasileira Ciencia Solo, 36(2): 447-456.

Sinclair, E. 1993.Leaf and soil analysis.Queensland fruit and vegetable growers.Pineapple field day. pp, 64-66.

Singh, N. P., Awasthi, R. P. and Sud, A. 2000. Diagnosis and recommendation integrated system (DRIS) norms for apple (Malus x DomesticaBorkh. L. Cv. StarkingDelicius) in Himachal Pradesh. Indian Journal of Hoticulture, 53(3): 196-204.

Sinha, S., Aman, A. and Rajan, R. 2018.The Significance of 'D' Leaf in Pineapple. Biomolecule Reports, ISSN: 2456-8759.

Sossa, E. L., Agbangba, C. E., Accalogoun, S. G. G. S., Amadji, G. L., Agbossou, K. U. and Hounhouigan, D. J. 2017. Residues Management Practices and Nitrogen-Potassium Fertilization Influence on the Quality of Pineapple (*Ananascomosus* (L.) Merrill) Sugarloaf Fruit for Exportation and Local Consumption. Agronomy, 7(26).

Souza, L.F.d.S. and Reinhardt, D. H.. 2007. Pineapple: Chapter 3. Role of plant nutrient in small grain. pp,179-201.

Souza, L. F. D. S. and Reinhardt, D. H. 2007. Chapter 10: Pineapple. In Fertilizing for High Yield and Quality: Tropical Fruit of Brazil; IPI Bulletin No.18 ed.; Crisostomo, L.A., Naumov, A., Eds.; International Potash Institute: Horgen, Switzerland, pp, 179-201.

Spironello, A., Quaggio, J. A., Teixeira, L. A. J., Furlani, P. R. and Sigrist, J. M. M. 2004. Pineapple yield and fruit quality affected by NPK fertilization in a tropical soil. Revistra Brasileira de Fruticultura, 26(1): 155-159.

Srivastava, A. K. and Singh, S. 1999. Potassium Nutrition in Citrus, National Research Centre for Citrus, Nagpur, Maharashtra, pp, 1-113.

Srivastava, A. K. and Singh, S. 2003. Foliar fertilization in citrus: A review. Agricultural Review, 24(4): 250-264.

Su, N. R. 1975. Micronutrient problems in pineapples. ASPAC/Food ; Fertilizer Technology Center. Taipei. (Extension Bulletin, 51).

Sweet Kelly, D. E. 1993. Nutritional disorders. In: Broadley RH, Wassman RS, Sinclair RE ed. Pineapple pest and disorders, Brisbane, Department of Primary Industries. pp, 33-42.

Tagaliavini, M., Marangoni, B. 2002. Major and nutritional issues in deciduous fruit orchards of north Italy. Horticulture Technology, 12: 26-31.

Tandon, H. L. S. 1990. Integrated nutrient management for sustainable agriculture.In: Proceedings of the International Symposium on Natural Resources Management for a Sustainable Agriculture (Ed.) R.P. Singh, I: 203-222.

Tay, T. H. 1972. Comparative study of different types of fertilizer as sources of nitrogen, phosphorus and potassium in pineapple cultivation. Tropical Agriculture (Trinidad), 49: 51-59.

Teisson, C., Lacoeuilhe, J. J. and Combres, J. C. 1979. Le brunissement interne de l'ananas; V - Recharches des moyens de lutte. Fruits (Paris), 34(6): 399-415.

Teixeira, L. A. J., Quaggio, J. A., Cantarella, H., Mellis, E. V. 2011. Potassium fertilization for pineapple: effects on plant growth and fruit yield. Revista Brasileira de Fruticultura Jaboticabal-SP, 33: 618-626.

Teixeira, L. A. J., Quaggio, J. A., Cantarella, H. and Mellis, E. V. 2011. Potassium fertilization for pineapple: Effects on soil chemical properties and plant nutrition. Revista Brasileira de Fruticultura, 33: 627-636.

Teixera, L. A. J., Spironello, A., Furlani, P. R. and Sigrist, J. M. M. 2002. Parcelamento da adubacao NPK emabacaxizeiro. Revista Brasileira de Fruticultura, Jaboticabal, 24: 219-224.

Tejada, M. and Gonzalez, J. L. 2003. Effects of the application of a compost originating from crushed cotton gin residues on wheat yield under dry land conditions. European Journal of Agronomy. 19: 357-368.

Tejada, M. and Gonzalez, J. L. 2006.Crushed cotton gin compost on soil biological properties and rice yield. European Journal of Agronomy, 25: 22-29

Theunissen, J., Nhakidemi, P. and Laublsher, C.P. 2010. Potential of vermicompost produced from plant waste on the growth and nutrient status in vegetable production. International Journal of Physical Science, 5: 1964-1973.

Thuler, D. S., Floh, E. I. S., Handro, V. And Barbosa, H. R. 2003. Plant growth regulators and amino acids released by Azospirillum sp. in chemically defined media. Letters in Applied Microbiology, 37:174-178.

Tilak, K. V. B. R., Ranganayaki, N., Pal, K. K., De, R., Saxena, A. K., Nautiyal, C. S., Mittal, S., Tripathi, A. K. and Johri, B .N. Diversity of plant growth and soil health supporting bacteria. Current Science, 89(1): 136-150.

Vho, S. 1999. Pineapple production practices. National Agricultural Research Institute, Mon Repos, East Coast Demerara.

Viana, E. D. S., Reis, R. C., Rosa, R. C. C., Padua, T. R. P. D. and Matos, A. P. 2019. Quality and sensory acceptance of 'Perola' pineapple grown in soil with application of organic fertilizer. Ciencia Rural, Santa Maria, 49(7): e20170631.

Viegas, I. D. J. M., Silva, R. N. P., Silva, D. A. S., Neto, C. F. O., Conceicao, H. E. O., Mascarenhas, G. S., Okumura, R. S., Monfort, L. E. F. and Silva, R. T. L. 2014. Mineral composition and visual symptoms of nutrients deficiencies in Curaua plants (Ananas comosus var. erectifolius). Amity Journal of Computational Science, 8(5): 747-753.

Walia, S. S. and Kler, D. S. 2009. Organic vs chemical farming-A review. Journal of Research Punjab Agriculture University, 46(3and4): 114-26.

Walworth, J. L. and Sumner, M. E. 1987. The diagnosis and recommendation integrated system (DRIS). Advance in Soil Science, 6: 149-188.

Ward, M. H. 2009. Too much of a good thing? Nitrate from nitrogen fertilizers and cancer. Reviews on Environmental Health, 24(4): 357-363.

Wilkinson, S.R. 1983. Effect of soil application of magnesium and other minerals on plant magnesium content, in: Fontenot J. P. (Ed.), Role of magnesium in animal nutrition, Virginia Polytechnic Institute Press, Blacksburg, VA.

Wolfe, A. H. and Patz, J. A. 2002. Reactive nitrogen and human health: acute and long-term implications. Ambio: A Journal of the Human Environment, 31(2):120-5.

Zhang, Q. and Wang, G. H. 2005.Studies on nutrient uptake of rice and characteristics of soil microorganism in a long-term fertilization experiments for irrigated rice. Journal of Zhejang University of Science, 6: 147-154.

Zhao, L. M., Wu, L. H., Li, Y. S., Lu, X. H., Zhu, D. F. and Uphoff, N. 2009. Influence of the system of rice intensification on rice yields and nitrogen and water use efficiency with different application rates. Experimental Agriculture, 45: 275-286.

15

Status of Organic Pineapple Cultivation in India

[1]Mahabub Alam and [2]Agniva Halder

[1]Department of Pomology & Postharvest Technology, Faculty of Horticulture, Uttar Banga Krishi Viswavidyalaya, Pundibari, Cooch Behar West Bengal, India
[2]Precision Farming Development Centre, Department of Agriculture and Food Engineering, Indian Institute of Technology Kharagpur, West Bengal India

---

Organic products are grown under a system of agriculture without the use of chemical fertilizers and pesticides with an environmentally and socially responsible approach. This is a method of farming that works at grass root level preserving the reproductive and regenerative capacity of the soil, good plant nutrition, and sound soil management, produces nutritious food rich in vitality which has resistance to diseases. India is bestowed with lot of potential to produce all varieties of organic products due to its various agro climatic conditions. In several parts of the country, the inherited tradition of organic farming is an added advantage. This holds promise for the organic producers to tap the market which is growing steadily in the domestic and export sector.

As per the available statistics, India's rank 8th in terms of World's Organic Agricultural land and 1st in terms of total number of producers as per 2020 data (Source: FIBL & IFOAM Year Book, 2020). The APEDA, Ministry of Commerce & Industries, Government of India is implementing the National Programme for Organic Production (NPOP). The programme involves the accreditation of Certification Bodies, standards for organic production, promotion of organic farming and marketing etc. The NPOP standards for production and accreditation system have been recognized by European Commission and Switzerland for unprocessed plant products as equivalent to their country standards. Similarly, USDA has recognized NPOP conformity assessment procedures of accreditation as equivalent to that of NOP of US. With these recognitions, Indian organic products duly certified by the accredited certification bodies of India are accepted by the importing countries. APEDA is also in the process of bilateral equivalence with South Korea, Taiwan, Canada, Japan etc.

Area: As on 31st March 2020 total area under organic certification process (registered under National Programme for Organic Production) is 3.67 million Hectare (2019-20). This includes 2.299 million ha cultivable area and another 1.37 million Hectare for wild harvest collection. Among all the states, Madhya Pradesh has covered largest area under organic certification followed by Rajasthan, Maharashtra, Gujarat, Karnataka, Odhisa, Sikkim and Uttar Pradesh. During 2016, Sikkim has achieved a remarkable distinction of converting its entire cultivable land (more than 75000 ha) under organic certification.

Production: India produced around 2.75 million MT (2019-20) of certified organic products which includes all varieties of food products namely Oil Seeds, Sugar cane, Cereals & Millets, Cotton, Pulses, Aromatic & Medicinal Plants, Tea, Coffee, Fruits, Spices, Dry Fruits, Vegetables, Processed foods etc. The production is not limited to the edible sector but also produces organic cotton fiber, functional food products etc. Among different states Madhya Pradesh is the largest producer followed by Maharashtra, Karnataka, Uttar Pradesh and Rajasthan. In terms of commodities Oil seeds are the single largest category followed by Sugar crops, Cereals and Millets, Tea & Coffee, Fiber crops, fodder, Pulses, Medicinal/ Herbal and Aromatic plants and Spices & Condiments.

Pineapple is a popular fruit crop to be taken under organic cultivation in India. North-eastern India, comprising eight states namely Arunachal Pradesh, Assam, Manipur, Meghalaya, Mizoram, Nagaland, Tripura and Sikkim has a high potential for organic pineapple cultivation. The region has fertile soils (rich in organic matter), enough rainfall along with water resources and great climatic diversities. This region holds 40% of the total share of pineapple production in the country and notably 90 to 95% of the produce is organically produced. ‘Giant Kew’ and ‘Queen’ are the most common cultivars grown around here and the quality of the produce is different and are said to be among the “Best in the world as they are very sweet (high TSS) and have less fiber”. In the context of tremendous national and global market demand for organic pineapples as well as low volume of such products, this area of India is an ideal area to explore for pineapple cultivation. The region being an agrarian society with an average of 80% tribal population, this venture will result in a breakthrough of social empowerment of the tribal people of the North East states of India.

The total geographical area of the region is 2.62 million sq.km, which is about 8% of the country's total area (26.12 million ha) with a population of 3.9 million (i.e., 5.6% of the country's population). The physiographic of the region is divided into three, namely, Meghalaya Plateau (22%), North-Eastern hills and basin (65%), and Brahmaputra valley (13%). The region has an agrarian society with an average of 80% tribal population excepting the state of Assam and Sikkim. Agriculture and horticulture are the main sources of income for a majority of the population. The area and production of fruit crops in NER is 0.287 million ha and 2.6 MT respectively which accounts for 7.16% of the area and 0.06% of the fruits produced in the country (Horticulture data base, 2009-2010).

Traditionally, pineapples in the NER were cultivated in the backyard with no concept of commercialization. They were planted along the slope without any or only limited management practices and inputs. Unsystematic planting or single-row system with wide spacing was the most common practice adopted by pineapple growers, which accommodates only 8,000-10,000 plants/ha, thus resulting in low productivity. Planting is usually done between February and April and harvesting is done during two seasons, i.e., July to August (main season) and December to January (off season). The harvested produce was sold either on the road side of the national highway or at wholesale market with a small proportion sent to the processing unit. The Government is also implementing organic projects and helping the farmers by giving them various organic inputs and supporting a certification process through cluster approach. Many growers have now started adopting improved technologies like planting across the slope, double row system, high density planting, use of organic inputs, use of suckers and slips as planting materials, staggered planting, care of harvested produce and adoption of organic certification process.

Assam: Organic farming is catching up among the farmers in Assam, who have traditionally used manure with inorganic fertilizers. However, the use of inorganic chemical fertilizers has almost come to a standstill.

Nagaland: Molvom village, under Medziphema area, has 800 ha in pineapple, one of the largest pineapple plantations in the state, and almost all the households are engaged in pineapple cultivation. Marketing of the produce is a problem and, therefore, growers are distressed in sale of their produce. CIH in collaboration with National Horticulture Board (NHB) and International Competence Centre Organic Agriculture (ICCOA) have initiated a program whereby the organic pineapple growers' societies are facilitated to market their produce at Delhi under the central government scheme of NHB. The main objective of the venture is to showcase the premier quality of pineapple of the state and also expose the pineapple growers to different marketing channels so as to enable them to find an outlet for their produce in the future.

Pineapples from the region are chemical free (organic by tradition), which makes it even more acceptable to consumers.

Status of organic pineapple in North-East India.

State	Area (ha)	Production (t)	Productivity (t/ha)	Implementing or certifying agency
Arunachal Pradesh				Initiating process
Assam	500	4850	9.7	SGS
Manipur	700	5000	6-7	One cert.
Meghalaya	2100	27783	13.23	SGS
Nagaland	1120	17000	15-16	ICCOA/SGS
Tripura				Initiating process
Total	4420	54633		

Source: Directorate of Horticulture of North East States of India.

Odisha and Jharkhand: Area under pineapple cultivation for those two states is very low. Farmers are cultivating local pineapple cultivars and productions in these areas are mainly done by using organic inputs.

Besides this in various states of India people are producing pineapple in their domestic garden for home consumption. Generally they use organic inputs like FYM and other manures.

Organic Certification Standard for Pineapple in India

Organic certification system is a quality assurance initiative, intended to assure quality, prevent fraud and promote commerce, based on set of standards and ethics. It is a process certification for producers of organic food and other organic plant products.

Standards

- Standards defining production methods, not the product quality.
- Minimum requirements, not "best practice",
- Continuously developed and dynamic.
- Can be International, National or regional standards.

References

Akali, S., Jha, K. K. and Amod, S. 2009. Cultivation of pineapple for upliftment of rural economy in Nagaland, India. Acta Horticulturae, 822: 99-106.

Darnaudery, M., Fournier, P. and Lechaudel, M. 2018. Low-input pineapple crops with high quality fruit: promising impacts of locally integrated and organic fertilisation compared to chemical fertilisers. Experimental Agriculture, 54(2), 286-302.

Devadas, V. S. and Kuriakose, K. P. 2002. Evaluation of different organic manures on yield and quality of pineapple var. Mauritius. In IV International Pineapple Symposium, Acta Horticulturae, 666: 185-189.

Mathew, B., Pereira, L. S. and Suresh, C. P. 2010. Pineapple production in Meghalaya (India)-indigenous cultural practices and status. In VII International Pineapple Symposium, Acta Horticulturae, 902: 275-280.

Mitra, S. and Devi, H. 2016. Organic horticulture in India. Horticulturae, 2(4), 17.

Paul, U. K., Das, G., Mathur, T. and Debnath, A. 2017. Economic efficiency and its effect on cost: A case study of organic pineapple in India's Northeast. Organic Agriculture, 7(3): 281-291.

Sema, A., and Maiti, C. S. 2010. Pineapple cultivation in North East India - A prospective venture. In VII International Pineapple Symposium, Acta Horticulturae, 902: 69-78.

16

Cultural Strategies Adopted in Pineapple Cultivation

Arkendu Ghosh and Koyel Dey

Department of Fruit Science, Horticulture College, Khuntpani Birsa Agricultural University, Kanke, Ranchi, Jharkhand, India

Pineapple is one of the most economically important fruits in tropical and subtropical regions and gaining popularity to the consumers all over the world. The pineapple plant has many specific morphological, anatomical and physiological characteristics that determine crucial aspects of pineapple crop management, such as water use efficiency, nutrient management, weed management, flower induction etc.

Water Requirements and Irrigation

Determination of water requirements and plant moisture status: Pineapple plant responds well to as much as 5 cm of water per month from rain or irrigation though it is a xerophyte type and is capable of good production under relatively meager water supply. Maintenance of available soil moisture in the root zone areas requires less water for pineapple than for other crops due to lower transpiration rates. Plants achieve this type of adaptability through a process called assuberization, where suberin act as a physical barrier, preventing water loss from the tissues it surrounds.Stomata remains closed as an effect of drought during the entire day. But, whenever the plants are grown with sufficient moisture in well-aerated soils produced well-distributed roots with numerous white root tips. It helps the water use efficiency and carbon fixation that pineapple shows to drip irrigation. The pineapple leaf is a major water storage organ for the pineapple plant (Malézieux *et al.*, 1993). When the leaves are full of water the cells present in upper mesophyll layer can occupy almost half of its volume. Invery little water-supply, the water in the mesophyll layer is absorbed by the plant. Generally, representative 'D'-leaves are taken in a group of six to ten leaves and the total thickness is measured and compared with the thickness of fully turgid leaves. A shortcoming of this method is that it does not take into account the viability of the root system and its ability to extract moisture from the soil. Roots infested with pests, diseases and fungal pathogens may result in showing moisture deficit, while it is sufficient for a healthy plant. So, any type of defective leaves should be avoided.

Overhead irrigation: Overhead booms were found to be the most effective way toinitiate root growth and saturate dry soil after planting by using about 5 cm of water. The same overhead booms could be used to irrigate growing plants, usually applying 2.5-5 cm per month. The booms travelled directly above the planted areas and only applied water at a rate that could be absorbed by the soil with each pass. Sometimes, pineapple was planted through mulch in depressed beds, which helped direct the water to the roots. Other forms of overhead water application have been used with good effect but the irrigation efficiency (water applied/water available to the crop) is generally not as high as it is with the overhead boom (Bartholomew *et al.*, 2003).

Drip irrigation: Drip irrigation allows precise placement of water in the root zone in volumes that match the crop's consumptive demand. Drip irrigation is especially effective during the first few months after planting, when a consistent but low-volume supply of water is needed for the developing root system. Drip-irrigation systems involve the following major components: a water source, a filtration system, chlorination to prevent microbial growth in the drip tubes, a system of in-field distribution pipes and pressures to various parts of the fields, and the emitter tubes, which are usually buried just below the soil surface between two lines of plants. The emitter tubes are usually custommade for pineapple to allow for a slow release of relatively low volumes of water. Flow rates will depend on orifice spacing, orifice diameter and line pressure, and these variables will depend on the soil type and cost and availability of water. The length of run for emitter tubes will depend on the slope of the terrain. Drip irrigation may be used to safely deliver water-soluble fertilizers, nematocides and fungicides to the root zone. (Amaral *et al.*, 2014).

Weed Control: The requirements for effective weed control in pineapples vary considerably in various parts of the world. Weed-control practices are very difficult due to the thorny nature of the plant but weeds should be controlled to avoid their competition with crop growth and final yield. Special attention should be paid to the reduction and elimination of clump grasses and vines, as these can become extremely problematic. Mechanical weed control though earthing up is mainly done by men labours to cover shallow exposed pineapple plant around 2-3 months after planting (Singh, 2005). The pineapple plant is quite tolerant of many good herbicides, particularly in the few days immediately after planting and before root initiation. Pre-emergence herbicides.Bromacil (2-3kg/ha.), hexazinone, ametryne, diuron (2kg/ha.) and quizalofop-P-ethyl are generally use on pineapple and are used alone.

Plant Nutrition

Assessment of nutritional requirements: Pineapple responses to nutrient management are dramatically affected by the condition of the soil and the health of the developing root system. Pathogens, nematodes, waterlogging and impermeable soil easily inhibit the crop from assimilating nutrients and responding to fertilizers. When optimum growing conditions prevail, particularly for the roots, significant economical responses can be achieved with both soil-applied and foliar-applied fertilizers. A preliminary assessment of nutrient requirements can be made by soil testing. Soil samples should be taken prior to land preparation, such as just before knockdown, and should represent the rooting volume of the native soil, less any contributions made from the crop residue. The total nutrientmanagement programme for pineapple is determined for three important stages – broadcast amendments for soil improvement, localized preplant fertilizer to promote early rooting and early nutrient uptake, and postplant side-dressing or foliar fertilizers for the remaining nutrients which may still be limiting. Such a composite fertilizer programme has great flexibility and provides for maximum economic benefit. (Bartholomew *et al.*, 2003).

Broadcast amendments: Broadcast amendments are applied prior to planting to adjust soil pH or to increase the supply of Ca, K, N and P. Rock-phosphate, lime, animal manure and well-composted organic material have all been used successfully in pineapple culture systems under the right conditions. Amendments must be broadcast uniformly and incorporated to the desired depth to provide the most benefit for the developing root system. The serious problems is a localized high pH, which favours the infection of susceptible pineapple plants by *Phytophthora* and *Pythium* fungi (Rohrbach and Johnson, 2003). High soil pH resulting from the application of chicken manure or coral scraped from feedlot sites has led to spotty outbreaks of heart rot caused by *Phytophthora.*

Banded preplant fertilizers: The precision placement of preplant fertilizers can ensure both vigorous rooting and the early uptake of N, P and K prior to the development of the leaf canopy. Banded fertilizer should be applied in sufficient amounts to enhance rooting and carry the young plants for 3–4 months, until the canopy is sufficiently developed to make foliar fertilizer applications efficient and effective (Owureku-Asare *et al.*, 2015). Placement is usually just below or adjacent to the plant line to allow the earliest interception by the developing roots.

Postplant side-dressings and foliar fertilizers: Side-dressings or foliar fertilizers may be applied as a dry side-dressing to the soil, often close to the base of the plant or, in some cases, in the lower leaf axils of mature plants. Care must always be taken to avoid plant damage due to exposure to high osmotic concentrations from dissolving nutrients. The basal white tissue of young and expanding leaves is particularly sensitive to fertilizer burn. Foliar application is effective for pineapple plants as the leaves absorb nutrients through the cuticle and nutrients such as nitrogen, potassium, iron, zinc and boron are readily translocated throughout the plant (Arshad *et al.*, 2011). Spray volumes used range from about 225-2250 l/hadepends on the age of the crop (size of the canopy), the volume of spray required to cover the plant without injuring the basal white leaf tissue and the chemistry of the soluble fertilizers.

Plant Nutrient Sources and Method of Application by Element

Nitrogen: Broadcasting of organic compost or manure is mainly useful for pre planting of plants. Where N can be applied as band application in the soil as ammonium sulphate, ammonium phosphate, potassium nitrate, urea or urea ammonium nitrate at 25-100 kg/ha of N after planting. Repeated foliar applications of solutions of urea or urea ammonium nitrate, at 2, 3 or 4-week intervals, is effective for large scale production areas. The soilapplied nitrogen is converted to nitrates and these nitrates are accumulated in the fruit at the later developmental stages for early fruit ripening. High nitrates in the fruit causes damage during fruit canning.High nitrates reacts with the tin coating of the cans, leading to high levels of dissolved tin and hydrogen gas development results swollen cans (Heptonet *al.*, 1993). ***Potassium:*** Broadcast applications of manure, sulpomag (the natural mineral langbenite containing 22% sulphur, 22% potassium and 11% magnesium), potassium chloride (Muriate of potash, KCl) or potassium sulphate (K_2SO_4) are effective at the rate of 200 to 1000 kg/ha of K in potassium deficient soils. Side-dressings and foliar applications are also utilized. The requirement for potassium is high, as the plant and fruit remove significant quantities of this element. In sandy soils, potassium should be applied throughout the growing cycle to ensure adequate levels at the time of forcing. Potassium chloride is almost always the least expensive potassium fertilizer, while potassium nitrate and potassium sulphate are more expensive but provide other nutrients (Teixeira *et al.*, 2011). ***Phosphorus:*** Phosphorus is mainly important for early rooting and increasing fruit yield. Preplant applications of 25-150 kg/ha of P should be done. Where soil P levels are very low, mycorrhiza may also play an important role in the transfer of this nutrient from the soil to the pineapple plant (Aziz *et al.*, 1990). In acidic soils, finely ground rock-phosphate may be an effective method of supplying phosphorus. ***Calcium:*** Razzaque *et al.* (2000) found that lime, gypsum and manure are excellent sources of Ca and 200-2000 kg/ha Ca is broadcasted over the field based on the desired pH and Ca status of the soil. The judicious use of calcium results increase calcium content in soil and raise soil pH results in enhanced uptake of other nutrients. At the same time, the additional calcium produces stronger cell walls, which result in better plant growth and development of fruit that are more resistant to bacterial diseases. ***Iron:*** Sema *et al.* (2010) reported that repeated foliar applications of ferrous sulphate (1% solution) in the same foliar solution with N are very effective for controlling iron deficiency problem. Cumulative amounts from repeated foliar sprays may vary from 2-10 kg/ha of Fe in the plant crop and 1-5 kg/ha of Fe in the ratoon. Chelated sources of iron are effective, but can be prohibitively expensive and may deteriorate in storage.

Magnesium, zinc,boron and other micronutrients: Deficiencies of Mg, Zn and B may occur, but are not commonly found. Broadcasting of magnesite, dolomite or sulpomag (also called K-Mag) at 100-500 kg/ha of Mg can be used for this. For the immediate solution for Mg deficiency in pineapple, a series of foliar applications of magnesium sulphate (Epsom salts) are applied in combination with other sulphates, such as Fe or Zn. Since a deficiency of Zn or B can be especially damaging, they should be applied when marginally low levels are reached.. Zinc sulphate is applied in combination with other sulphates, such as Fe or Mg, to achieve a total application of no more than 0.1 kg/ha of Zn. Boron should be applied at about 1.0 kg/ha of B in one or two foliar applications. In Queensland, Australia, boron is commonly applied in the ethephon solution at 10 kg/acre of borax or solubor, at the time of forcing in midsummer (January–February) to raise the pH of the ethephon solution for enhancing forcing (Amorim *et al.*, 2013).

Flower Initiation and Fruit Development

Pineapple mostly starts flowering only after attaininga level of vegetative growth (formation of at least 36 leaves) and ripeness is attained 11-12 months after planting. It was also noticed that even after 18 months only half amount of total plantation commence flowering.

Natural differentiation: Natural flowering may be generally classified as winter or short-day-length,cold-weather-induced flowering and summer or dry-weather-induced flowering. Different stress factors, like root damage due to pathogens attack or waterlogged soils, may also induce flowering. The above factors usually slow down the vegetative growth, along with increase in starch accumulation in the leaves and stem. Summer flower-initiation period can be reduced or eliminated by the adequate supplies of nitrogen and water to promote active vegetative growth. The effect of low temperatures and short photoperiods during winter on natural flowering can be reduced in plants by maintaining high plant nitrogen status, but nitrogen alone cannot eliminate differentiation if air temperature goes below 15°C.

Chemically induced flowering: In the early 1700s, fumes from fires are used to induce flowering.After that Calcium carbide is also used for this purpose. A pea-sized amount of calcium carbide is dropped into the heart of the rosette of leaves of a vegetative plant of sufficient size. The carbide reacts with water to release acetylene, taken up by the leaves. Later research by Hossain *et al.* (2008) shows that Acetylene and ethyleneare the most effective forcing agent if applied at night at stomata opening duration. Kumari *et al.* (2020) found that synthetic auxin Naphthalene Acetic Acid (NAA) induce flowering by promoting the ethylene synthesis. The most widely used synthetic auxin was the sodium salt of naphthalene acetic acid (SNA) at the rate of 50 g/ha in 2800 l of water in two applications at weekly interval. The newest material used as a forcing agent is ethephon (2-chloroethylphosphonic acid) applied in combination with urea and in some areas, sodium borate or calcium or sodium carbonate (Hilton, 2008). These materials increase the solution pH, hastens the rate of breakdown of ethephon into ethylene.. More generally & commercially ethephon is used for knapsack and small and large boom applications for the uniformity of the harvest peak and the fruit yield.Application is done at 6–8 weeks after forcing and may be repeated at weekly intervals until 95% or more of the plants have differentiated.

Year round production and improvement of fruit quality: Singh (2005) reported some techniques for year round floweringand fruiting of pineapple

- Planting of pineapple should commence from September to March at monthly interval.
- Suckers of different weight (300g, 600g, 900g etc.) are used as planting material in different time of the year.
- Synchronies use of different planting materials like suckers, slips and crowns of different grade forprolonged harvesting period.
- 2-4 varieties with different maturity time should be planted for year round production.
- Etherel (0.25 ml/l) and Calcium carbide (20 g/l) should be applied approx. 1 year after planting.

- Chemical inducing flowering should be applied in the heart of the plant during evening time (5-6 p.m.) when the plant has almost 32 leaves to get the flower within 2 months.
- To maintain the fruit quality slips should be removed after emergence as it hampers the quality and sucker growth should be promoted as it improves the fruit quality.
- After 40 days of flower emergence innermost whorl of leaflets should be removed to enhance the fruit size 15-20%.

Plant weight–fruit weight relationships: To geta proper sized fruit both for table purpose and for canning, plants are grown to a prerequisite weight before forcing. The ideal plant weight depends on some factors, i.e. temperature, solar radiation, irrigation and planting density. When plants were grown under conditions that provided an identical rooting environment with proper nutrition, with only the climatic factors varying, a clear relationship was reported between dry matter accumulated in the plant stem, expressed as a fraction of total plant dry matter, and the subsequent fruit weight (Hepton *et al*., 1993). Differences in dry matter accumulation are due to photoperiodism and diurnal temperature variations.

Other Cultural Practices

Slip removal: Some of the pineapple varieties, like some clones of 'Smooth Cayenne' have a large number of slips. Though useful as planting material, if left on the plant at the time of fruit development, they can severely reduce fruit yield. Plants forced with SNA tend to have fewer slips than those forced with ethylene. This difference is attributed to the ability of ethylene to promote floral differentiation of lateral buds as well as the apical meristem. Each lateral bud, once differentiated, becomes a miniature fruit with a crown, i.e. a slip. To minimize the adverse effects of slips on fruit yield, they should be removed as soon as possible without damaging the peduncle of the developing fruit (Fassinou Hotegni *et al*., 2016).

Sucker pruning: Sucker development is a function of variety and plant vigour. Some varieties, e.g. 'Queen', produce numerous suckers from leaf axils at or above ground level. High numbers of suckers make ratooning very difficult, because ideally the plant should have one high quality sucker to replace the mother plant, thereby retaining the original planting density. Sucker number of 'Smooth Cayenne' is influenced by planting density, plant weight at the time of forcing and nutritional status. Large plants in a population will tend to develop one or more suckers early during fruit development, while sucker development on smaller plants may be significantly delayed. For this to provide uniformity in the ratoon crop, large suckers may be removed. Sucker pruning is mainly practiced to improve the uniformity of ratoon crops, but it will increase the time required to reach the predetermined average sucker weight at forcing. Pruning of suckers will also reduce yield per area per unit of time. Consequently, the decision to prune or not to prune suckers will depend on the purpose for which the crop is being grown, the projected harvest date and in some areas, the potential for using the pruned suckers as a source of planting material (Fassinou Hotegni *et al*., 2015). Because sucker pruning is labour intensive, it is generally practised in growing areas where the cost of removal is offset by the increased value of the larger and more uniform ratoon fruits and the value of suckers as planting material.

Crown gouging: When fruit develops under environmental conditions that suit vegetative growth, often a huge growth spurt to the crown in the last phase of fruit development. If the apical growing point of the crown is removed or damaged before this growth spurt, around 10-12 weeks before harvesting results an increase in fruit weight (6-10%). Crown gouging is not an option in areas where crowns are used as a primary source of planting material. As crowns are one of the least expensive forms of planting material; with modern equipment, crowns can be detached from the fruit, dipped in pesticides to control disease and insects and conveyed to separate bins for transport to areas where they will be planted. Crowns should also not be gouged in fields where high nitrates in the fruit could be a problem. The continued crown growth ensures nitrates movement through the fruit to the expanding crown leaves. If this process is stopped by gouging, nitrates will accumulate in the upper fruit part. As noted previously, high fruit nitrate levels are problematic in processing, and the situation is further aggravated if the fruit is not fully ripened at harvest.

Crown removal: Removal of crown is practiced to promote fruit expansion in multi crowned fruit. At the removal time crowns should be sufficiently well developed for easy removal without fruit damage. As this practice is labour intensive, it may not be practical in all areas, but, when it is performed by practiced workers, the scar tissue is rarely visible when the remaining crown is fully developed. If the crown is in large size to shade the fruit shoulders, retention of the crown is better to prevent fruit sunburn. This is practiced in areas of intense sunlight.

Leaf pruning: In some pineapple producing areas, leaves are cut for easy entry into the field mainly for harvesting. Cunha and Reinhardt (2004) reported that due to leaf pruning yields are reduced and the reduction is proportionate to the amount of leaf removed and the earlier the removal takes place. This pruning also has an adverse effect on the both slips and suckers growth. The reasons for yield reductions are due to leaf pruning a reduction in the supply of metabolites from the leaf canopy is occured and second is the loss of dry matter accumulated in the leaves, which would be available for fruit production.

References

Amaral, U., Maia, M.V., Pegoraro, F.R., Kondo, M.K. and Aspiazú, I. 2014. Water depths and macronutrients accumulationin 'Pérola' pineapple irrigated by drip. Revista Brassilia Frutiola, 36(3): 755-760.

Amorim, A. V., Garruti, D. S., Lacerda, C. F., Moura, C. F. H., Filho, E. G. 2013. Postharvest and sensory quality of pineapples grownunder micronutrients doses and two types of mulching. African Journal of Agricultural Research, 8(19): 2240-2248.

Arshad, A. M., Marzuki, A. A. and Aziz, A. 2011. Effect of foliar copper fertilizer on pineapple cv. n36 planted on bris soil at east coast of Peninsular Malaysia. Journal of Environmental Science and Engineering, 5(12): 1678-82.

Aziz, T., Yuen, J. E. and Habte, M. 1990. Response of pineapple to mycorrhizal inoculation and fosetyl-Altreatment.Communications in Soil Science and Plant Analysis, 21: 19–20.

Bartholomew, D. P., Paull, R. E. and Rohrbach, K. G. 2003. The PineappleBotany, Production and Uses. CABI Publishing. pp, 125-153.

Bartholomew, D. P. and Malézieux, E. 1994. Pineapple. In: Schaffer, B. and Anderson, P. (eds) Handbook of Environmental Physiology of Fruit Crops, Vol. II. CRC Press, Boca Raton, Florida. pp, 243–291.

Cunha, G. A. P. and Reinhardt, D. H. R. C. 2004. Propagação do abacaxi zeiro. Brasília, DF; EMBRAPA/CNPMF.P.70. (Coleção Plantar, 11).

Fassinou Hotegni, V. N., Lommen, W. J. M., Agbossou, E. K., and Struik, P. C. 2015. Selective pruning in pineapple plants as means to reduce heterogeneity in fruit quality. Springer Plus, 4(1): 129.

FassinouHotegni, V. N., Lommen, W. J. M., Agbossou, E. K., and Struik, P. C. 2016. Understanding the effects of slip pruning on pineapple fruit quality. ActaHorticulturae, 1111. DOI 10.17660/ActaHortic.2016.1111.34.

Hepton, A., Ingamells, L., Macion, E., Gonzales, J. and Sampongse, D. 1993. Pineapple plant and fruitgrowth and development in fertilized native soil and artificial root medium. ActaHorticulturae, 334: 131–139.

Hilton, B. 2008. Applying calcium carbide to induce flowering in pineapple.ECHO Development Notes (EDN) | EDN Issue #101.Pp.1-5.http://edn.link/yemkpk.

Hussain, M. J., Ulla, M. A., Salim, M. M. R., Rahman, M. M. and Sarker, M. M. R. 2008. Effect of different application time of Ethrel on off-season pineapple production at srimongal, Sylhet. International Journal of Sustainable Crop Production, 3(6):68-71.

Kumari, U., Jha, K. K., Sengupta, S., Mishra, S., Tiwary, A. K., Lal, H. C. and Kumar K. 2020. Studies on different growth regulators and stage of application on inflorescence induction and development of pineapple (Ananas comosus L. Merr.) Var. Queen. International Journal of Chemical Studies, 8(6): 1093-1095.

Malézieux, E. 1993. Dry matter accumulation and yield elaboration of pineapple in Côte D'Ivoire. Acta Horticulturae, 334: 149–157.

Owureku-Asare, M., Agyei-Amponsah, J., Agbemavor, S.W.K., Apatey, J., Sarfo, A.K., Okyere, A.A., Twum, L.A. and Dodobi, M.T. 2015. Effect of organic fertilizers on physical and chemical quality of sugar loaf pineapple (Ananas comosus L.) grown in two ecological sites in Ghana. 15(2): 9982-95.

Razzaque, A. H. M. and Hanafi, M. M. 2000. Effect of calcium on pineapple production in tropical peat soil. Indian Journal of Horticulture, 57(2): 110-113.

Rodriquez, A. R. 1932. Influence of smoke and ethylene on the fruiting of pineapple (Ananas sativus) Schult.University of Puerto Rico Journal of Agriculture, 26(1): 5-18.

Rohrbach, K. G. and Johnson, K. W. 2003. Pests, Diseases and Weeds. In: The Pineapple Botany, Production and Uses.CABI Publishing. pp, 214-262.

Sema, A., Maiti, C. S., Singh, A. K. and Bendangsengla, A. 2010. Dris nutrient norms for pineapple on alfisols of India. Journal of Plant Nutrition. 33(9): 1384-99.

Singh, N. P. 2005. Pineapple cultivation in Tripura. CSS Technology Mission. Publication No-20.ICAR Research Complex for NEH Region, Lembucherra, Tripura (West). pp, 4.

Teixeira, L. A. J., Quaggio, J. A., Cantarella, H. and Mellis, E. V. 2011. Potassium fertilization for pineapple: effects on plantgrowth and fruit yield. Revista Brassilia the Frutiola, 33(2): 618-626.

17

Uses of Growth Regulators and Flower Regulation In Pineapple

[1]Usha Kumari, [2]Prahlad Deb

[1]Department of Fruit Science, College of Agriculture, Birsa Agricultural University, Kanke, Ranchi, India
[2]Department of Horticulture & Postharvest Technology, Institute of Agriculture, Visva-Bharati, Sriniketan, West Bengal, India

---

The history of plant growth regulation was knownfor many years before the time of Christ, when it was common practice in the Middle East to place a drop of olive oil on figs to promote development. The use of plant growth regulators (PGRs) to modify growth and flowering of pineapple began in 1930s. In eighteenth century when a carpenter was working in a greenhouse in the Azores Islandaccidently set fire to a pile of shavings. To the surprise of the grower who had thought his plants were ruined, they burst into flower instead of being damaged (Collins, 1960). In Hawaii, Rodriguez (1932) noticed in the pineapple fields that smoke from brush fires adjacent to the field caused premature flowering. The ingredients of smoke were subsequently found to be ethylene and acetylene, the unsaturated hydrocarbons. During the 1940's, auxins were also shown to produce this effect,α-Naphthalene acetic acid (NAA), β-Naphthalene acetic acid (BNA), Fruitone and naphthaleneacetamide were the next forcing agent to become commercially used in pineapple. Since that time many works have been carried out with improving results like natural and synthetic compounds that alter function, shape and size of crop plants have been discovered. Today, specific PGRs are used to modify crop growth rate and growth pattern during the various stages of development from germination through harvest and post-harvest preservation. The use of regulating chemicals has become an important component of agro-technical procedures for most of the cultivated plants and especially for fruit plants.

Plant growth regulators or phytohormones are the organic substances (other than nutrients and vitamins) produced naturally in plants, synthesized in one part and usually translocated to other part where in every small quantity influence the growth and other physiological functions of the plant. A definition of plant hormones with still wider scope has been given by Van Overbeek *et al.* (1941). According to him, the plant hormones are defined as -organic compounds which regulate plant physiological process- regardless of whether these compounds are naturally occurring and/or synthetic; stimulating and/or inhibitory; local activators or substances which act at a distance from the place where they are formed. The Plant growth regulators generally include Auxins, Gibberellins, Cytokinins, ethylene, Growth retardants and growth inhibitors.

Flowering is a major developmental transition in plants, from the vegetative to the reproductive state, during which the plants structurally acquires reproductive competence by producing inflorescences. The pineapple is a perennial monocotyledonous plant belonging to the phylogenetic family Bromeliaceae with a terminal inflorescence which finally transformed to a single multiple fruit (sorosis). The pineapple inflorescence is formed by a group of 50-200 individual flowers that are connected with their subtending bracts and among each other around an axis – the core, which is an elongation of the peduncle, aligned in eight spiral rows, presenting a 8/21 phyllotaxy (Okimoto, 1948).The stage of inflorescence emergence is called 'red heart' due to the reddish peduncle bracts (usually five to seven) that are produced at its base and are shorter and narrower than the ordinary leaves (Bartholomew *et al.,* 2003).Flowers contain both the male and female reproductive organs (Purseglove, 1972). Pineapple flowers do not abscise; the petals, stamens and style wither and the entire flower develops parthenocarpically into a berry-like fruitlet. The fruit of pineapple is a seedless syncarp and polygonal in shape. A syncarp is derived from the fusion of many individual flowers/fruitlets into one fruit. The fruit consists of the fused ovaries, bases of sepals and bracts, and cortex of the central core. The initiation of pineapple flowering depends on the physiological state and nutritional reserve of the plant, day length and temperature (Bartholomew and Malezieux, 1994).

Natural flowering in Pineapple: Flowering in pineapple can occur naturally but it causes serious problems for pineapple growers in all producing areas around the world. Natural induction mainly occurs in the cooler winter months and the incidence is higher at subtropical latitudes where temperatures are cooler and winter day lengths are shorter (Bartholomew *et al.* 2003). Natural induction resulted from sudden drops in temperature associated with passing cold fronts. Natural flowering makes pineapple management difficult because assorted fruits are produced and harvest dates cannot be scheduled which increase harvesting costs or reduces yield, or both.That is the reason for minimizing or controlling its negative effects on the pineapple plant and fruit. Usually, intervarietal hybrids are more likely to flower naturally.

The Natural Flowering of Pineapple Can be Avoided

- By using optimum levels of nutrients and water to minimize stresses and promote rapid vegetative growth;
- If the vegetative growth of the plant is reduced in such a way that it will not be able to flower at the time when the weather conditions become favorable to natural flowering;
- By using floral inducers (synthetic or natural) at optimum concentrations;
- If the biosynthesis and action of ethylene in the plant are inhibited or blocked, with the use of some chemical substances;
- Genetic engineering method can control natural flowering. If antisense or trans-switch technology could be used to turn off the gene that codes for the ACC synthase associated with natural flower induction, then transformed plants would presumably not produce sufficient ethylene to initiate flowering.

Artificial flower induction: Artificial induction or forcing is a process where the buds of the fruit trees (originally vegetative) undergo metabolic changes that are subsequently prepared to transform into flower buds. This is done chemically by use of growth regulators which induces flowering and subsequent fruiting. Since pineapples flower erratically, artificial induction of flowers is a common practice. Forcing allows pineapple fruit to be harvested in most months of the year and it also induces almost 100% of the plants and produces a sharp ripening peak,which keeps harvesting costs low.

Advantages of Artificial Induction or Forcing in Pineapple

- It mainly homogenizes harvesting and flowering;
- Schedule the flowering of the plant and harvest at the desired time to synchronize with the season of high demand and price;
- Uniform size fruit production;
- Rational use of land;
- Regular supply of fruits for the industry and the fresh fruit market, and during more favorable commercial periods, without affecting fruit quality;
- Easier control of pests and diseases, since flowering can be induced during periods, when the inoculum potential is the least;

- Control of fruit weight and size, according to market requirements;
- Increase of the crop revenue, due to a larger number of harvested fruits per area;
- Better distribution of labor force and easier administration of the property;
- The products used to achieve this technology are not environmentally polluting, nor residual, on the contrary, they degrade rapidly under sunlight;
- Possibility of exploration of ratoon crop.

Generally we use chemical substances such as auxins, ethylene and ethylene releasing compounds because other chemicals may harm the plant. The reaction of carbide with water produced heat and usually causes damage to some heart leaves of the pineapple plants (Anonymous, 1974). A large quantity of water (about 3,000 litres/ha) is required to dissolve calcium carbide and promote subsequent acetylene production (Teisson, 1979). Another limitation of calcium carbide is that it causes plants to become more susceptible to *Phytophtora* sp. (Lacoeuilhe and Guyot, 1979). The use of calcium carbide is also labor and time consuming.

The susceptibility of pineapple plants to the forcing agent depends more or less on the size and/or age of the plant. Research conducted on Smooth Cayenne showed that the larger the plant, the more susceptible it was to a chemical forcing agent (Py and Guyot, 1970).The age of the plant at the period favorable for floral induction is also related to the process, in addition to environmental factors and cultural treatments that affect the vegetative growth of the plants (Friend and Lydon, 1979). Medium sized plants were easier to force than very large ones. A/c to Bondad (1973), one and two-month old plants did not respond to ethephon treatment (50 mg per plant) while older plants (5 to 8 months) flowered readily. The ease of floral initiation is also influenced by the nitrogen status of the plant at the time of forcing (Aldrich and Nakasone, 1975). If the nitrogen content of the leaves reaches a certain level, a greater quantity or multiple applications of the forcing agent is required to achieve a high flowering percentage.

Floral Inducers

Auxin: 1-Naphthaleneacetic acid, commonly abbreviated NAA and is also known as α-Naphthaleneacetic acid; Naphthalene-1-acetic acid.1-Naphthylacetic acid; α-Naphthyleneacetic acid. 1-Naphthaleneacetic acid (NAA) is a plant hormone in the auxin family. It is an organic compound with the formula $C_{10}H_7CH_2CO_2H$. NAA is used for control of pre harvest fruit drop, flower induction and fruit thinning in various crops such as apples, potatoes, olives and citrus fruits. It is a rooting agent and used for the vegetative propagation of plants from stem and leaf cutting. It is also used for plant tissue culture. NAA does not occur naturally. It is more stable in plant than IAA and for this reason has found wide practical applications and is found in the forms of white crystalline or amorphous powder' which is dissolved in alcohols, acetone, water, xylene, carbon tetrachloride, ether, chloroform, 2% sodium bicarbonate, 0.1 N ammonium hydroxide and is stable in 0.1N alkali, 3N HCl and in concentrated phosphoric acid. NAA induced flower formation in advance of normal period together with almost uniform flowering in pineapple has been reported from Hawaii (Clark and Kerns, 1942). Premature flowering occurred when NAA was applied in the upper leaf axils as a solution before normal floral differentiation (Collins, 1960).

According to van Overbeek (1946), Low concentration of NAA would promote flowering while high concentration inhibit it. Even very dilute concentration of NAA such as 0.05 mg/plant are capable of inducing flower in a limited number of plants while 0.25 mg/plant caused practically all the plants to flower (100%). The fact that promotion of induction of flowering in pineapple is caused by the applied auxin by stimulating the ethylene production and this ethylene helps the flowering in pineapple (Burg and Burg, 1966).The peak period of ethylene production after auxin treatment was about 5-7 days. From the work of Ao, (1995),it is clear that Chlorophyll and nitrogen content of 'D' leaf were also stimulated by NAA. Early flowering and delay in maturity together with increased quality of fruit resulted in NAA treated plants. NAA increased the TSS, sugar and vitamin-C content of the fruit.

The use of Sodium salt of NAA to induce flowering in the pineapple plants has been a commercial practicefor a number of years. Poignant (1969) applied this at the concentrations 110,150 and 200 ppm after 12-16 weeks of floral induction and found increased diameter of the fruit from 8 to 9 per cent with a tendency to increase the fruit length. Although naturally occurring auxins have their positive influences on the quality and quantity of the crops as that of synthetic auxins but the latter has found extensive use in agriculture. IAA is a naturally occurring auxin which has been synthesized and is extensively used in laboratory investigations. Due to its relative instability within the plant it is rarely used in practical agriculture. So, a search began for synthetic substance with hormonal activity.

Beta-hydroxy ethyl hydrazine (BOH) has also shown to be an effective forcing agent and is most reliable when applied to the heart of the plant at a rate of 0.1g/plant (Py and Guyot, 1970). The results are less consistent when BOH is sprayed over the whole plant. In commercial use, BOH was applied in the evening at a rate of 45 mg active ingredient per plant; however, some phytotoxic symptoms such as burning of the heart leaves may develop shortly after application of BOH. A/C to Bartholomew (2003), BOH is not registered for use in most areas because it is expensive and has been reported to be a potential carcinogen.

Ethylene: Ethylene, unlike the rest of the plant hormone compounds is a gaseous hormone.Of all the known plant growth substance, ethylene has the simplest structure. It is produced in all higher plants and has important roles in the ripening of many fruits, in the induction of senescence in leaves and flowers, and in the abscission of leaf petioles and flower peduncles. Ethylene has been used in practice since the ancient Egyptians, who would gas figs in order to stimulate ripening. In 1901, a Russian scientist named DimitryNeljubow showed that the active component was ethylene. In 1935, Crocker proposed that ethylene was the plant hormone responsible for fruit ripening as well as inhibition of vegetative tissues. The effects of the gas were noticed by Puerto Rican pineapple growers and Phiilipines mango growers when smoke from bonfire helped to initiate and synchronize flowering. Ethylene is now known to have many other functions as well. Ethylene in its gaseous form was very difficult to applyto plants. The practical application of ethylene started after the induction ofethephon (Cooke and Randall, 1968) which was patended as "Ethrel"and given the trivial name "Ethephon11 i.e. the compound, 2-chloroethyl phosphoricacid. Now a day, ethephon, which releases ethylene slowly to the plants is commerciallyavailable. The effectiveness of ethylene in stimulating fruit ripening first drew attention ofphysiologists and horticulturists. In addition to its use in hasteningthe fruit ripening ethylene has had some commercial applications to bring about flowering in pineapple.

Traub *et al.* (1949) studied histologically the effects ofethylene on initiation and development of inflorescence and observed that after 3-4 days of a 24 hour ethylene treatment the apices began to enlarge and produce flower primordia after 10-14 days on axils of bracts. Moderate pruning and spray of ethephon at 600 ppm produced the highest number of new shoots (Shaban *et al.,* 2009). Treating the plant with acetylene, NAA or ethephon transforms the vegetative apex to a reproductive one. As early as two days after treatment the apical height and width increase, attaining maximum dimensions in 8–14 days, after which they progressively decrease. During 4–6 weeks, when the full complement of florets has been initiated, the dimensions of the apex are again increased and leaves that form the fruit crown are initiated. A cambium like zone at the base of the corpus is noticed by 2–4 days after the application of acetylene and ethephon (Rao *et al.,* 1979). Acetylene and NAA result in a more pointed apex initially, while ethephon results in a more broadened apex. The largest increase in apical height and width is seen when NAA is used. Cell multiplication is the main cause of peduncle elongation and inflorescence head enlargement, although towards the later stages of growth, cell enlargement contributes more to the increase in size of the organs.

The increased efficacy of ethephon at a lower concentration for induction of flowering, with the addition of urea and calcium carbonate, is of great significance as this will lead to decreased costs and increased commercial use. Ethephon at a concentration as low as 25 ppm in combination with urea (2%) and calcium carbonate (0.04%) was able to induce more than 90% flowering after 50 days of treatment. Ethephon treatments were found to be markedly superior to planofix. Apart from earlier flowering, ethephon treatments resulted in uniform ripening, and harvests were complete within 10 days (Dass *et al.,*1975).

The number of slips produced per plant was decreased significantly with increasing rate of ethephon (Ouattara *et al.,* 1982). Ethephon-treated plots had a significantly lower incidence of 'fruit collapse' (a disease caused by *Erwinia chrysanthemi* entering *via* the open flowers) (Lim *et al.,* 1979). Application of ethephon increased the level of endogenous ethylene (C_2H_4), abscisic acid (ABA) and 2-isopentyl adenine (2-iP) while it decreased the concentration of indole-3-acetic acid (IAA), gibberellic acid (GA_3) and zeatin (ZT), and led to the transition of vegetative growth to inflorescence initiation (Liu *et al.,* 2011).

3 concentrations of BOH, calcium carbide and ethephon on 'Sugarloaf' pineapples increased leaf width, decreased leaf number and slip production, accelerated flowering and fruit maturity, caused uniform flowering and fruit ripening (Norman *et al.,* 2016). In the crown planting they decreased leaf length, plant height and peduncle size, increased total soluble solids (TSS), decreased acidity and TSS/acidity ratio, and produced cylindrical fruits.

Vences *et al.* (2018) worked on the flowering response of 'MD2' pineapple plants of different ages to rates of ethrel application. 'MD2' pineapple (10 and 11 months old) exhibited comparable growth characteristics except for D-leaf weight. Plants at 11 months after planting were relatively heavier compared to 10 MAP. Growth characteristics of the 9 MAP plantings were generally inferior compared with those of the 10 and 11 months-old. Rate of ethrel induced 95.56 to 100% apical differentiation of the plants. Apical differentiation of meristems among 'MD2' pineapple have significantly increased and become distinct with the increasing rates of ethrel. Percent red buds development at 41 days after flower induction increased from plants of 10 to 11 MAP and as the rates of ethrel application was increased from 800 to 1200 ppm. There was a significant interaction observed between the age of the plants and the rates of ethrel in red buds development of 'MD2' pineapple. Highest percentage of red buds development was attained in 11 month-old plants applied with 1200ppm ethrel. Eleven months old 'MD2' pineapple is the best age of plants being the most responsive to flower induction treatments. Application of ethrel at the rate of 1200 ppm induced the highest percentage of floral initiation as evidenced by the apical meristem differentiation, and development of red buds on 'MD2' pineapple plants. Application of ethrel at the rate of 1200 ppm on eleven months old 'MD2' pineapple plants is the best combination of treatments that will give the highest percentage flower initiation and red buds development.

Application of 50 ml solution of ethrel (100 ppm), calcium carbide (2 per cent) and NAA (20 ppm) at the core of plant of cv. Kew on 15 June, October and January, 1980–1981,has been done when the age of the plant was 335 days. Treatment with ethrel and calcium carbide in June caused appearance of inflorescence within 30 days; application in October delayed the emergence up to 58 and 60 days in case of calcium carbide and ethrel, respectively. Treatment in January resulted in flowering in about 50 days. The application of NAA at different seasons markedly affected the response to flowering and the time was reduced from 260 days to 62 days from June to January application. Treatment with calcium carbide, ethrel or NAA in February caused flowering in about 28 days. Treatment with chemicals in general showed a reduction in the percentage of flowering from June to January. Anatomical studies of shoot apices revealed that initiation to flowering started in 15 to 25 days after treatment by chemicals. Shoot apices from plants treated with ethrel and calcium carbide showed an increase in total nitrogen at the first (5 days) or second stage (15 days) after treatment. Application of ethrel and calcium carbide caused a greater release of ethylene irrespective of the time of treatment compared to control plants. The shoot apices treated with ethrel, calcium carbide and NAA mainly contained growth inhibitors at 5 days after treatment. An increase in growth promoting substances was found at 15 days while at 25 days after treatment the concentration of endogenous inhibitor again increased (Das Biswas *et al.,* 1983).

Gibberellin: Gibberellins commonly known as gibberellic acids was discovered in Japan by western scientists in 1950s.A fungal disease called *bakanae* disease of rice causes plants to grow taller and eliminated seed production. Plant pathologists found that these symptoms in rice plant were induced by a chemical secreted by a pathogenic fungus, *Gibberella fujikuroi.* Culturing this fungus in the laboratory and analyzing the culture in 1930s to obtain impure crystal of two fungal "compounds" possessing plant growth promoting activity. One of these, because it was isolated from the fungus *Gibberella,* was named gibberellin A. In 1950s scientists of Tokyo University separated and characterized 3 different gibberellins from gibberellin A sample, and named them gibberellin A_1, gibberellin A_2 and gibberellin A_3.

Gibberellins (GAs) consist of a family of diterpenoid acids, an important group of phytohormones that exercise different effects on growth and development of plants, such as germination, cell elongation, expansion of leaves, and development of flowers.The empirical formula of GAs is $C_{19}H_{22}0_6$ with molecular weight 346.37. GA_3 is found in the form of crystalline solid (melting point 221°C-223°C), crystalline potassium salt, compressed tablets of acid plus buffering products and as isopropyl alcohol solution. GA is soluble in methanol, ethanol, acetone and dilute alkali, moderately soluble in ethyl acetate, slightly soluble in water and ether.

Though much work has been carried out on the effect of different plant growth regulators like NAA and ethylene on pineapple plant, little information about the effect of GA_3 on pineapple is available. The most dramatic effect of GA_3 is the transformation of dwarf plants into tall ones by greatly increasing stemelongation (Brian and Hemming, 1955). Rapid stem elongation takes place when rosette plants bolt prior to flowering. GAs can overcome the effectsof unfavorable environmental conditions and caused rosette plants to elongate and flower.

GA_3 probably influences the distribution of assimilates between the stem apex and the flowers, and it acts as a substitute for some specific weather conditions, which control flower formation and improve fruit quality (Hamza *et al.,* 1983). A low GA_3 concentration did not delay flowering in the pineapple crop (Min *et al.,* 1993). However, it increases the peduncle length, reduced the pineapple inflorescence size, which can be an indication of its involvement in flower development.

Growth inhibitors/ Growth retardants: Some chemical substances play their role in inhibiting the vegetative growth of the plant to prevent, reduce or delay the natural flowering of pineapple.

ACP [2-(3-chlorophenoxy) propionic acid] and PBZ (paclobutrazole) were capable of inhibiting and delaying the floral initiation of the 'Perola' pineapple; the best product concentrations were between (90 and 240) mgL^{-1}, split into two and three equal applications, during April and May. ACP and PBZ significantly inhibited the pineapple flowering may be because there is no increase in the vegetative growth of the plant, as the products also caused a significant reduction in the 'D' leaf length.Flowering inhibition is related to a decrease in the plant mass, rather than a direct interference of ACP in the floral differentiation process (Scott, 1993).

The other products, Tebuconazole (TBZ) at (20 and 40) mgL^{-1} and Propaconazole(PPZ) at 40 mgL^{-1} also showed better inhibiting effects.The inhibition of natural flowering on 'Smooth Cayenne' pineapple, treated with emersion in (12.5 to 500) mgL^{-1}TBZ solutions, was recorded mainly at higher concentrations (Taniguchi *et al.,* 1999).

Uniconazole(a product of the same chemical group as TBZ), at (0.5 and 2.5) mgL^{-1}(25 mLplant^{-1}), not only delayed but also inhibited 'Smooth Cayenne' flowering. This product, when used in the highest concentration, kept the plants in the vegetative stage (Min, 1995). The induction of the floral axis is followed by a minimum level of growth promoters (GA_3) and a maximum level of inhibitors (ABA) (Maita *et al.,* 1998)

Method and Time of Application

Floral inducers differ in relation to the mode of application and efficiency: 2, 4-D are applied in solution into thecenter of the leaf rosette; ethylene and BNA are sprayed on the plants, while ethephon, NAA and BOHare applied either into the center of the rosette or onthe whole plant. NAA is more efficient when applied around to the period of natural differentiation (Gowing, 1961).

According to Cunha (1983), the recommended concentration is up to 4.000 mgL^{-1}, that is, one to fourliters of the commercial product for 1000 liters of water.The decrease of the acidity of the inducer solution toa pH 8 or 10 increases the efficiency of ethephon,making possible the use of reduced amounts of the product, once the release of ethylene is favored inalkaline solutions. Ethylene may also be applied directly toinduce pineapple flowering, being preferred in mechanized plantations, due to its known efficiency and beneficial effects on the inflorescence, fruit quality and planting material production (Py and Silvy, 1954).

Successful forcing is a key point for economical sustainability for pineapple farms. The time of plant growth regulators application is very important. It should be done preferably at night, from 8 pm to 5 am, or during cloudy days (Aldrich and Nakasone, 1975; Abutiate,1977). The greatest efficiency observed with night applications may be the result of a larger concentration of ethylene in the tissues of the plant under those conditions and/or toa better absorption of the applied product. Pineapple is a plant that exhibits, alternatively, the crassulacean acid metabolism (CAM), characterized by assimilation of CO_2 and stomata opening predominantly at night. It is important that the stomata stay open for a period of about four to six hours after the application of the inducer (Glennie, 1979).

Floral inducers sprays should be performed at high volumes, starting from the end ofthe afternoon, avoiding hot days in order to reduce failures in the artificial flowering (Turnbull *et al.,* 1993). The environmental temperature during the application of the products is also very important, being not recommended to besuperior to 26-28 °C. A/c to Ahmed *et al.* (1987), the increased synthesis of metabolites may be brought about by application of different flower inducing chemicals like ethrel, planofix and calcium carbide to the pineapple bud in right concentration at the right time.

Evening high-volume spraying and avoidance of hot days has significant advantage in reducing the frequency of flowering failure. The more effective hours were between 0:00 and 5:00 hours during the summer and being more flexible as the cooler months approach. In hot months, whole plant high pressure spraying was the most effective, followed by whole plant and low pressure and pouring in the plant's heart (Martinez *et al.,*1997).

Mwaule *et al.* (1982) worked on the effect of timing ethrel application (fifteen forcing dates) representing different plant ages between 44 and 72 weeks (10 and 17 months) and a control (without hormone) on plant growth, fruit production and quality of Smooth Cayenne variety of pineapples and found that vegetative growth in terms of leaf number and plant height (at flowering) slip and sucker production (at harvest) were significantly increased with plant age at forcing. Total soluble solids content of the juice increased while fruit acidity significantly decreased with plant age at forcing. The interval between forcing and flowering and between forcing and harvesting decreased while that between planting and harvesting significantly increased with late forcing. Flowering and harvesting were irregular and significantly prolonged when forcing was delayed. Forcing between 44 and 56 weeks (10 and 13 months) significantly reduced the interval between planting and harvesting.

Efficiency of ethephon was improved by the combination of 2% urea (pH 9.0) and applied at 4:00 p.m. Ethephon at 25 ppm + 2% urea (pH 9.0) applied at 4:00

p.m. had a similar effect on percentage of flowering.Bad weather conditions delayed flowering (Heenkenda, 1990). Ethrel applied in September on 300 days old plants induced flowering 146 days earlier in off-season than those in the control.September treated plants took minimum duration (520 days).Size of fruit was increased with increasing age of plants at forcing (Hussain *et al.*, 2008). Use of ethrel as growth regulator on 300 days old plants in September could be the effective and profitable for off season fruiting of pineapple.

Conclusion

An erratic flowering behavior is an obstacle for successful cultivation of pineapple. Even after 15-18 months of growth, under optimum nutritional and climatic conditions, less than 20% of the plants normally flower. Fruits produced by natural induction are costly to harvest because a relatively small percentage of these fruits mature at any given time in a field, and sometimes the fruits are too small to be marketable. These factors ultimately increase the cost of cultivation and reduce profits. Hence, flower regulation is very essential in pineapple.Artificial induction or forcing is a well-established commercial practice. Several identified growth regulators are efficient in initiating pineapple flowering. Among these products, the most commercially used are: NAA, BNA, IBA, 2, 4-D, succinic acid, ethephon, ethylene, acetylene, calcium carbide, hydroxietilhidrazine (HOH) and b-hydroxietilhidrazine (BOH). However, only a few are used because they are cheaper and easy to handle. Plant growth regulators play important role in synchronization of flowering in pineapple and produces uniform fruit throughout the year. The time and rate of plant growth regulators application is very important as it may reduce the efficiency of the artificial induction.Pineapple is a plant undergoes CAM (Crassulacean Acid Metabolism pathway) in which plant works at night when the stomata opens. It is important that the stomata stay open for a period of about four to six hours after the application of the inducer. Floral inducers sprays should be done at high volumes, starting from the end of the afternoon, avoiding hot days in order to reduce failures in the artificial flowering.

References

Abutiate, W. S. 1977. Effects of Ethrel on the flowering of Smooth Cayenne pineapple in 1978. Agris, 11(1): 205-210.

Ahmed, F. and Bora, P. C. 1987. Physico-chemical changes during flower bud differentiation in pineapple (Ananas comosus L. Merr). Indian Journal of Plant Physiology, 30(2): 189-193.

Aldrich, W. W. and Nakasone, H. Y. 1975. Day versus night application of calcium carbide for flower induction in pineapple. Journal of American Society for Horticultural Science, 100(4): 410-415.

Anonymous. 1974. Artificial flower induction. Queensland Fruits and Vegetable News, 263-266.

Ao, K. C. 1995. Studies on the effect of plant growth regulators on growth,flowering, fruit quality and yield of pineapple. Ph.D. Thesis Gauhati University.

Avila, M., Blanco, M. A., Nieves, N., Gonzalez, J., Marrero, P., Gonzalez, A., Quinones, J. and Martinez, T. 2005. Effect of Ethrel on Flowering Induction in Pineapple (Ananas comosus L. Merr.) Smooth Cayenne cv. Serrana. I. Changes in Levels of Polyamines, Proteins and Carbohydrates. ActaHorticulturae, 666: 175-182.

Bartholomew, D. P., Malezieux, E. 1994. Pineapple. In: Schaeffer, B.; Anderson, P. Environmental physiology of fruit crops: sub-tropical and tropical crops. Boca Raton. CRC Press. 2(1):243-291.

Bartholomew, D. P., Paull, R. E. and Rohrbach, K. G. 2003. The pineapple: botany, production and uses. CABI Publishing. Wallingford, UK. 1-301.

Bondad, N. D. 1973. Effect of ethephon on flowering, fruiting and slip production of 'Smooth Cayenne' pineapple. Philippines Geographic Journal, 17(1): 1-10.

Brian, P. W. and Hemming, H. G. 1955. The effect of GA_3 on shoot growth of pea seedlings. Physiologia Plantarum, 8: 669-681.

Burg, S. P. and Burg, E. A. 1966. The interaction between auxin and ethylene and its role in plant growth. Proceedings of the National Academy of Sciences, 55: 262-266.

Collins, J. L. 1960. The pineapple. Leonard Hill, London.

Cooke, A. R. and Randall, D. I. 1968. 2-Halo ethyl phosphonic acid as ethylene releasing agents for the induction of flowering in pineapples. Nature, 18: 974-975.

Crocker, W., Hitchcock, A. E. and Zimmerman, P. W. 1935. Similarities in the effects of ethlyene and the plant auxins. Boyce Thompson Institute, 7: 231–248.

Cunha, G. A. P., Costa, J. T. A. and Reinhardt, D. H. 2003. Natural flowering on pineapple: inhibition by growth regulators. Fruits, 58(1):27-37.

Cunha, G.A.P., Matos, A.P., Cabral, J.R.S., Souza, L. F. da S., Sanches, N.F., Reinhardt, D. H. R. C. 1993.Abacaxifor export: Technical aspects of production. Brasilia Embrapa.41.

Cunha, G. A. P. 2005. Applied aspects of pineapple flowering. Bragantia Campina, 64(4):499-516.

Das, B., Nath, V., Jana, B. R., Kumar, S. and Dey, P. 2007. Evaluation of different methods of crop regulation in guava grown under rainfed plateau conditions of eastern India. Indian Journal of Horticulture, 64(3): 294-299.

Das, N. 1965. Control of flower and fruit formation of pineapple by applications of auxin and anti-auxin alone and in mixtures. Journal of Indian Botanical Society, 43: 498-507.

Das Biswas, S., Dhua, R. S., Mitra, S. K. and Bose, T. K. 1983. Physiological studies on flowering of pineapple in response to chemicals and environment. ActaHorticulturae, 137: 231-242.

Dass, H. C., Randhawa, G. S. and Negi, S. P. 1975. Flowering in pineapple as influenced by ethephon and its combinations with urea and calcium carbonate. ScientiaHorticulturae, 3(3): 231-238.

Friend, D. J. C. and Lydon, J. 1979. Effects of day length on flowering, growth and CAM of pineapple (Ananas comosus (L) Merril.). Botanical Gazette, 140(3): 280-283.

Glennie, J. D. 1979. The effect of temperature on the flower induction of pineapples with ethephon. Australian Horticultural Research Newsletter, 50:49-52.

Gowing, D. P. 1961. Experiments on the photoperiodic response in pineapple. American Journal of Botany, 48:16-21.

Guyot, A. and Py, C. 1970. Controlled flowering of pineapple with Ethrel, a new growth regulator. Fruits, 25(5): 341-347.

Hamza, A. M. and Helaly, M. N. M. 1983. Interaction between chlormequat (CCC) and gibberellin (GA_3) on growth, flowering and mineral constituents of some ornamental plants. ActaHorticulturae, 137: 197–210.

Heenkenda, H. M. 1990. Improving the efficiency of flower induction in pineapple.Agris.112.

Hussain, M. J., Ulla, M. A., Salim, M. M. R., Rahman, M. M. and Sarker, M. M. R. 2008. Effect of different application time of Ethrel on off-season pineapple production at srimongal, Sylhet. International Journal of Sustainable Crop Production, 3(6):68-71.

Lacoeuilhe, J. J. and Guyot, A. 1979. The cultural techniques of the pineapple in open air.Fruits, 34:159-168.

Lim, W. H. and Lowings, P. H. 1979. Pineapple fruit collapse in Peninsular Malaysia: symptoms and varietal susceptibility. Plant Disease Reporter, 63:170-174.

Liu Sheng-hui, L., Xiao-ping, Z. and Guang-ming, S. 2011. Changes in endogenous hormone concentrations during inflorescence induction and development in pineapple (Ananas comosus cv. Smooth Cayenne) by ethephon. Academic Journals, 11: 124.

Lopez De Velez, A. M., Cunha, G. A. P. da.1983. Influence of pH and urea and 2 chloro ethyl phosphonic acid on floral induction of pineapple. Pesquisa Agropecuaria Brasileira, 18(11): 1199-1205.

Maita, A., Martnez, T., Perez, S. and Nogueira, J.1998. Study on the floral induction, growth and development of pineapple, in: Subradrabundhu, S., Chairidchai, P. Third Int. Pineapple Symposium. Thailand. pp, 20.

Min, X. J. 1995. Physiological effects of environmental factors and growth regulators on floral initiation and development of pineapple [Ananas comosus (L.) Merr.], University of Hawaii, Thesis, Honolulu, USA. pp, 111.

Min, X. J. and Bartholomew, D. P. 1993. Effects of growth regulators on ethylene production and floral initiation of pineapple, Acta Horticulturae, 334: 101–112.

Mwaule, Y. W. 1982. Effect of Ethrel on smooth cayenne pineapple slips at different stages of growth. Acta Horticulturae, 137: 243-250.

Neljubov, D. 1901. Uber die horizontale Nutation der Stengel von Pisum sativum und einigeran deren Pflanzen. Beih Bot Zentralbl, 10: 128–139.

Norman, K., Joseph, M., Paul, K., Karatininis, K. and Namuwooza, C. 2016. Tracing Uganda's global primary organic pineapple value chain. African Crop Science Journal, 24(2):15.

Okimoto, M. C. 1948. Anatomy and histology of the pineapple inflorescence and fruit. Botanical Gazette 110(2): 217-231.

Ouattara, A. 1982. Master's Thesis. University of Hawaii.

Poignant, A.1969. Effects of two hormones applied to pineapples during fruit formation. Fruits, 24: 353-364.

Purseglove, J. W. 1972. Tropical Crops. Monocotyledons. Purseglove, J.W. (eds.) Longman, London. pp, 75-91.

Py, C., Silvy, A. 1954. Hormone treatments on pineapple – practical methods for directing production. Fruits, 9(3): 101-123.

Rao, A. N. and Wee, Y. C. 1979. Development of the inflorescence and "crown" of Ananas comosus after treatment with acetylene, NAA and ethephon. American Journal of Botany, 66(4): 351-360.

Rebolledo-Martinez, A. and Aguirre-Gutierrez, D. 1997. Ethephon as pineapple flowering inductor in Mexico's weather conditions. Acta Horticulturae, 425:347-354.

Rodriguez, A. G. 1932. Influence of smoke and ethylene on the flowering of pineapples (Ananas comosus L.). Journal of the Department of Agriculture of Puerto Rico, 26:5-18.

Scott, C. H. 1993. The effect of two plant growth regulators on the inhibition of precocious fruiting in pineapple. ActaHorticulturae, 334: 77-82.

Shaban, A. E. A. and Haseeb, G. M. M. 2009. Effect of Pruning Severity and Spraying Some Chemical Substances on Growth and Fruiting of Guava Trees. American-Eurasian Journal of Agricultural & Environmental Sciences, 5(6): 825-831.

Sterry, J. R.1969. Ethrel, an ethylene evolving plant growth regulator. National Faculty of Agricultural sciences in Ghent. pp, 462-473.

Taniguchi, G. 1999. Effect of tebuconazole on natural flower induction. Pineapple News, 6:11.

Teisson, C. 1979. Internal browning of pineapples: Research on means of control. Fruits, 34:399-411.

Turnbull, C. G. N., Anderson, K. L., Shorter, A. J., Nissen, R. J. and Sinclair, E. R. 1993. Ethephon and causes of flowering failure in pineapple. Acta Horticulturae, 334: 83-92.

Van Overbeek, J., Conklin, M. E. and Blakeslee, A. F. 1941. Factors in coconut milk essential for growth and development of Datura embryos. Science, 94: 350.

Van Overbeek, J. 1946. Control of flower formation and fruit size in the pineapple. Botanical Gazette, 108:64-73.

Van Overbeek, J. 1951. Use of growth substances in tropical agriculture. In Plant growth substances. University of Wisconsin Press. Madison. pp, 224-225.

Vences, C. V. 2018. Plant Age and Rate of Flower Inducer Affects Flower Initiation of 'MD2' Pineapple (Ananas comosus L.). International Journal of Research & Review, 5(4):27.

18

Harvesting and Post-Harvest Operations of Pineapple

[1]Rocky Thokchom, [2]Nikita Khoisnam and [3]Goutam Mandal

[1]Department of Horticulture and Post-Harvest Technology
Institute of Agriculture, Visva-Bharati, Sriniketan, West Bengal, India
[2]Department of Agricultural Economics and Extension, School of Agriculture
Lovely Professional University, Jalandhar, Punjab, India
[3]Department of Horticulture and Post-Harvest Technology
Institute of Agriculture, Visva-Bharati, Sriniketan, West Bengal, India

Post-Harvest Losses

A post-harvest loss is defined as the measurable qualitative and quantitative food (including fruits and vegetables) loss along the supply chain, starting at the time of harvest till its consumption or other end uses (Hodges *et al*., 2011). Pineapple fruits are highly perishable in nature and continuously respired even after harvesting which limited its shelf-life leading to the storage problems (Hossain and Bepary, 2015) and decline in production due to post-harvest losses (Babalola *et al*., 2008). During the storage for distant marketing, pineapple fruits easily losses its moisture content through transpiration and become more susceptibility to microbial spoilage (Hossain and Bepary, 2015). In India the top five leading pineapple producing states are West Bengal, Assam, Karnataka, Meghalaya and Manipur (Anonymous, 2018). The measurable qualitative and quantitative losses of pineapple at post-harvest level are from the various stages of handling from the farm level up to the consumers (Table 1 and 2).

Table 1. Post-harvest losses of the pineapple from farm level up to the consumer in the Imphal-East district of Manipur.

SL. No.	Stages	Losses (%)
1.	Farm level	6.82
2.	Wholesale market including transportation	3.09
3.	Retailer	5.44
4.	Storage unit	5.64

5.	Processing unit	4.31
6.	Consumers	6.82
	Total post-harvest losses	**32.12**

Source: Ningombam *et al.* (2019)

Table 2: Postharvest loss of pineapple in Assam when the fruits are marketed within the states.

SL. No.	Stages		Losses (%)
1.	Farmer's level	During assembling	1.73
		Losses in market	1.22
		Others	1.27
2.	Middlemen's level	Local market	1.65
		Transportations (100-150 km)	1.24
		Others	2.14
Total post-harvest losses			**9.25**

Source: Deka (2015)

Factors Affecting Postharvest Life

Biotic factors: The biotic factors responsible for postharvest life of the pineapple fruits are species, cultivar, rate of respiration (non-climacteric fruit), transpiration, harvest maturity/ harvesting indexes, ethylene production, vulnerability to ethylene, changes in chemical composition of fruits, colour, flavour, texture, and physiological reaction towards low temperature, pathological infections and storage conditions.

Abiotic factors: The abiotic factors responsible for postharvest life of the pineapple fruits are temperature, relative humidity, gases of the storage environments, light, types of packaging materials, mode of transportation, pre and post-harvest treatments, etc.

Horticultural practices and facilities: Poor farm practices, improper harvesting, improper ripening practices, poor post-harvest handling, poor infrastructure and logistic facilities, ill equipped marketing system, etc.

Managing Postharvest Losses

Pre-harvest treatments on post-harvest shelf-life and quality: Pre-harvest treatment of 2% calcium chloride ($CaCl_2$), extend the shelf-life of pineapple cv. Kew during storage (Talukdar and Ahmed, 2002). The basal dressing of calcium oxide as a pre-harvest treatment reduce the peroxidase activity and reduces/eliminate the internal browning incidence in pineapple cv. Mauritius during cold storage. This pre-harvest treatment can also reduced the cold-induced injury to the cell membrane of the pineapple fruits during storage (Selvarajah *et al.*, 1998). Pre-harvest application of K_2SO_4 solution (50ml, 5%

w/v) along with alpha NAA (10ppm) or ethrel four weeks before harvesting in pineapple significantly resulted in decreases of the fruit core tissue deterioration and internal browning under cold storage conditions (Nanayakkara *et al.*, 1997). Pre-harvest application of high amount of potassium in pineapple cultivation resulted in to decrease of phenol biosynthetic enzymes (PAL: phenylalanine ammonia-lyase enzyme activities) which increase the membrane resistance making the flesh of the pineapple fruit firmer and also inhibited the polyphenoloxidase activities(PPO) which is the key enzyme for fruit browning during Post-harvest Conservation of fruits of pineapple cvs. Smooth Cayenne and MD2 (Souleymane *et al.*, 2019). Pre-harvest application of salicylic acid in winter pineapple significantly inhibited the polyphenoloxidase (PPO) activities which resulted in the incidence of internal browning during storage. This treatment also significantly inhibited the phenylalanine ammonia-lyase (PAL) enzyme activities which delayed the decline of ascorbic acid content and increase shelf-life of winter pineapple during storage (Lu *et al.*, 2011). Silicon in the form of partially burnt rice hull (2 or 4 kg/m) and sodium silicate (50 mg/l) when applied as pre-harvest treatments in pineapple cv. Mauritius resulted in physical strengthening of cells and thereby developed tolerance to cold stress and resulted in reduction of internal browning incidence (75%) of fruits during prolonged cold storage at 10°C (Weerahewa and Wicramasekara, 2020).

Maturity Indices/ Maturity Standards: Pineapple fruits should be harvested when they attained consumable stage (firm and mature). Since pineapple is climacteric fruits they will not improve in eating quality once they are harvested. Pineapple's fruit maturity and harvesting stage cannot be determined by a single factor, several external and internal fruit characteristics are usually considered in combination as indices to determine the harvest maturity (Joy and Rashida Rajuva, 2016). The maturity indices of fruits and vegetables can be determined in many aspects, including the assessment of duration of development, measurement of size, weight/density, physical (colour, firmness and moisture content) and chemical characteristics (starch content, sugar content and sugar/acid) (Shewfelt, 1993).

In general, harvesting of pineapple's fruits are carried out when the colour changes from green to yellow in ¼th to ½th portion of the fruit from its base, the eyes turn from green to yellow orange from the bottom, surfaces become smooth around the eyes, the tips of bract dry up, the number of rows of eyes that have changed colour (30-50% eyes turned yellow from the base), flatness of the eyes, when fruits attain TSS of 12-14°Brix and TSS: acid ratio of 14-20, the period from flower induction to ripening (141-150 days) etc. All the above indices may vary according to cultivars, seasons of growing, intended use, desired storage life, altitude, rainfall, micro-climate and field practices. Important maturity indices for Mauritius and Kew cultivars are illustrated in Table 4 and 5.

The harvesting of fruits for local market should be done at the full maturity stage and for distant markets it should be done at 75-80% maturity stage according to the references of Table 3.

Table 3. Maturation stages of Kew pineapples

Full ripening stage	**:**	**Approximately 165 days after flowering**
Pre-maturation stage	:	0-120 days after flowering
Early maturation stage	:	120-150 days after flowering
Late maturation stage	:	150-160 days after flowering

Source: Deka (2015)

Table 4. Maturity Indices of Mauritus Pineapple

Maturity Indices	**Descriptions**
Index 1	All Eyes Are Glossy Bluish Dark Green Immature fruit. All eyes are glossy bluish dark green with reddish bractea. Flesh very firm and sour.
Index 2	All Eyes Are Glossy Dark Green with Traces of Yellow Between Eyes at Base Breaker stage - Mature fruit. All eyes are glossy dark green with traces of yellow between eyes at base. Flesh firm and sourish. Not edible fresh.
Index 3	Eyes Are Dark Green with 1-2 Eyes Yellowish Green at Base Mature fruit. Eyes are dark green with 1-2 eyes yellowish green at base. Flesh crunchy and sourish. Edible fresh but not palatable.
Index 4	ABOUT 25% OF THE EYES FROM THE BASE ARE YELLOW Quarter ripe fruit. About 25% of the eyes from the base are yellow. Flesh firm, crunchy and sweet sour.
Index 5	About 50% of the Eyes are Orangy Yellow Half ripe fruit. About 50% of the eyes are orangy yellow. Flesh spongy, juicy, sweet and fragrant. Good for eating fresh.
Index 6	More than 75% of the Eyes are Orangy Yellow Three-quarter ripe fruit. More than 75% of the eyes are orangy yellow. Flesh spongy, very juicy, sweet and fragrant. Good for eating fresh.
Index 7	Full Orangy Yellow Fully ripe fruit. All eyes are orangy yellow. Flesh soft, very juicy and sweet sour

Source: Anonymous (2006)

Table 5. Maturity Indices of Kew Pineapple

Sl. No.	Maturity indices	Descriptions
1.	Days to harvest	146-150 days after flowering (1/2 color development stage)
2.	Specific gravity	0.93-0.98
3.	Respiration rate	7-10 mg/kg/hr
4.	TSS	18-19 °Brix
5.	Acidity	0.77-0.83%
6.	TSS: acid ratio	23-25

Source: Deka *et al.* (2005).

Harvesting: In India under natural conditions pineapple fruits are mainly harvested during May-August (Anonymous, 2018-19) and in the northeastern states during July-August (Deka 2015). Early harvesting of fruits may develop woody pulp on ripening whereas, the delay harvesting may reduces the shelf-life and summer season harvested fruits have more sensory score than the winter season harvested fruits (Deka 2015).

The harvesting of pineapple fruits are carried out manually by pickers (by hand) by twisting (with a downward motion) it from the stalk or by cutting the stalk with the help of sharp knife below the base of the fruit. The harvested fruits are usually placed in wooden or plastic field crates or in large canvas picking sack and immediately transferred to the central packing shed or packing facility. Fruits left in the direct sun in field will sunburn and turn soft and will have susceptible to postharvest decay. Harvested fruits should have intact crown and must be protected from damages because removal of the crown increases the risk of diseases and decay.

Pre-cooling: In India generally used pre-cooling methods includes room cooling under ambient conditions, forced air cooling, water cooling, vacuum cooling and package icing. Paull and Chen (2014) suggest room-cooling or forced-air cooling as the best pre-cooling methods of pineapple. Pre-cooled at a temperature below 7°C causes chilling injury of fruits (Deka 2015).

Grading: Pineapple fruits are graded on the basis of their weight, size and colour (Anonymous, 2018-19). Grade designation and sizing of pineapple as per AGMARK standard are mentioned in Table 6.

Post-harvest Treatments: After harvesting, the physico-chemical quality of the fruits can typically only be maintained but not improved (Prusky, 2011), so post-harvest treatments are treated prior to storage to prolong the storage period/shelf-life without deteriorating the quality. Post-harvest treatments of fresh horticulture produce (fruits and vegetables) with various physical (heat, irradiation, waxing/edible coatings etc.), chemical (antimicrobials, antioxidants and anti-browning agents, fungicide, growth regulators, nitric oxide, sulfur dioxide, calcium compound etc.) and gaseous treatments (ozone, ethylene, 1-Methylcyclopropene etc.) slow down physiological processes of senescence and maturation, reduce/inhibit development of physiological disorders and minimize the risk of microbial growth and contamination (Mahajan *et al.*, 2014; El-Ramady *et al.*, 2015).

In pineapple, post-harvest treatment of pineapple fruits with GA_3 (100 mg/l) showed delayed response of ripening, maintained quality and high shelf life during storage (Dhar *et al.*, 2008; Mandal *et al.*, 2015). Similar results of increasing shelf-life and maintaining physico-chemical quality of pineapple has been reported by Kamol *et al.* (2014) when treating fruits with NAA. Post-harvest treatment of salicylic acid (SA) and methyl salicylate (MeSA) increased the resistance to postharvest chilling injury as well as suppressed the polyphenoloxidase (PPO) activity in pineapple fruits during cold storage (Lu *et al.*, 2010). Further, post-harvest SA treatment delays the decline of ascorbic acid content and prevented its destruction, so high contents of ascorbic acid in treated pineapple improved the fruit quality (Lu *et al.*, 2011). Delay in the decline of ascorbic acid content of harvested pineapple fruits has also been observed when the fruits are post treated with 1- methylcyclopropene (1-MCP) before storage as 1-MCP partly inhibited the activity of oxidative enzymes (Selvarajah *et al.*, 2001).

Post-harvest waxing of pineapple fruits with Sta-Fresh 2952 is a potential effective method for commercial storage and marketing of pineapple fruits since it effectively alleviated chilling injury, maintained fruit's quality, effectively alleviated weight loss, retarding change of colour saturation, decreased respiration rate and hence extended the shelf-life of pineapple fruits under different storage conditions (Hu *et al.*, 2011; Lin *et al.*, 2013). Aloe Vera gel when applied as edible coating in pineapple fruits has beneficial effects in retarding the ripening process since it acts as a physical barrier and thus reduced the weight loss during post-harvest storage. In addition, it also delayed softening, ascorbic acid degradation and maintained the quality of pineapple fruit during storage (Adetunji *et al.*, 2012).

Table 6: Grade designation and sizing of pineapple as per AGMARK standard

Grade Designation	Grade Requirements	Minimum TSS (°Brix)	Grade Tolerances
Extra class	Pineapples must be of superior quality. They must be characteristic of variety and/or commercial type. They must be free of defects. Very slight superficial defects may be there, provided these do not affect general appearance of the produce, the quality, the keeping quality and presentation in the package. The crown, if present, shall be simple and straight with no sprouts and shall be between 50 and 150% of the length of the fruit with trimmed or untrimmed crowns.	12	5% by number or weight of Pineapples not satisfying the requirements of the grade, but meeting of those of Class I grade or, exceptionally coming within the tolerances of that grade.
Class I	Pineapples must be of good quality. They must be characteristics of the variety and/or commercial type. The following slight defects however, may be allowed, provided these do not affect the general appearance of the produce, the quality, the keeping quality and presentation in the package. • slight defects in shape. • slight defects in colouring; including sunspots; • slight skin defects (i.e. scratches, scars, scraps and blemishes) not exceeding 4% of the total surface area. The defects must not, in any case, affect the pulp of the fruit. The crown, if present shall be simple andstraight and slightly curved with no sprouts, and shall be between 50 and 150% of the length of the fruit for pineapples with trimmed or untrimmed crowns.	12	10% by number or weight of Pineapples not satisfying the requirements of the grade, but meeting those of class II grade or exceptionally, coming within the tolerance of that grade.

Grade Designation	Grade Requirements	Minimum TSS (°Brix)	Grade Tolerances
Class II	Pineapples which do not qualify for inclusion in the higher grades but satisfy the minimum requirements. The following defects may be allowed, provided the pineapples retain their essential characteristics as regards the general appearance, quality, the keeping quality and presentation • defects in shape, • defects in colouring; including sunspots; • skin defects(scratches, scars, bruises and blemishes) not exceeding 8% of the total surface area. The defects must not, in any case, affect the pulp of the fruit. The crown, if present, shall be simple or double and straight or slightly curved, with no sprouts.	12	10% by number or weight of Pineapples not satisfying the requirements of the grade but meeting the minimum requirements.

Source: Anonymous (2015).

Post-harvest heat shock treatment in the form of hot water dip immediately after harvest was found to induce fruit tolerance to internal browning in both Kew and Mauritius fruits (Weeraheva, 2002). Pineapple fruits when treated at 38°C for 60 minutes resulted in lesser browning development in the fresh (70%) and core (45%) region respectively during storage (Weeraheva and Adikaram, 2005). Three minutes hot water dip treatment at 54°C applied to the trimmed peduncle of pineapple was very effective in controlling the incidence of black rot (Wijeratnam *et al.*, 2005). The Mauritius pineapple fruit after harvest should be subjected to 53°C temperature for 5-7 minutes in hot water treatment to kill the mealy bug, scale insects, thrips, mites and prevent from storage rots (Joy, 2010).

Packaging: In many developing countries including India, pineapple fruits are not packed at all or packed in bamboo basket, wooden basket and gunny bags or is inadequate (Deka, 2015). Globally pineapples fruits are packed along with the crowns for prolonged shelf-life and to avoid infection at the crown attachment point. For local markets, fruits are packed in bamboo baskets lined with paddy-straw while for distant markets; fruit are wrapped individually with paddy straw and then packed having each basket weighing 20-25 kg. For export purposes or in the international trade the pineapples fruits are packed in a well ventilated full-telescopic two piece corrugated fiberboard carton. Packing in fiber boxes minimized the physiological loss in weight during transportation and storage (Singh, 2009). Generally fruits are usually packed into cartons of two different sizes (on the basis of color and size); in a large telescoping fiberboard carton holding 18 kg (containing 8 to 10 fruit in two layers) in flat or upright position, for surface and air shipment and in a smaller container of 9 kg (5-6 fruit in a single layer) laid flat for air shipping. Pineapples will only fit into boxes in small numbers (4 or 6) when packed vertically or horizontally with a crown of leaves cut to fit snugly into the box. (Thompson, 2003; Joy, 2010).

Important Consideration for Packaging

1. The choice of proper packaging materials and containers minimizes damage and losses.
2. Packing should be in such a way as to protect the fruits properly.
3. The materials used in packaging must be new, clean, and of a quality such as to avoid causing any external or internal damage to the fruits.
4. All fruit packed in the same carton or container should be uniform in size, shape, and extenal color.
5. Immature or over mature fruit should not be packed (especially for export purpose).
6. Packing boxes must be well ventilated and be of sufficient strength to withstand the rigors of distribution without collapsing.

Transportation: In the post-harvest chain of pineapple, transportation is an important factor which takes the fresh fruits from the grower to the consumer via marketing. The cost of transport contributes significantly to the price fixed by the seller. In India, for domestic transportation it is usually done either by railways or roadways (trucks) which may be refrigerated or non-refrigerated. For export in other countries both sea and air ways are used. Cool chain is essential during the transport of export quality commodity all the way from the farm to the customer.

Important Considerations for Packaging

1. Fully ripe fruits are not suitable for transportation to distant markets. 1/4th colour development stage from the base of the fruits is suitable for long distant transportation, while for local market, the fruits may be harvested at 1/2-2/3rd colour development stage. Immature fruits are not

shipped since they do not develop good flavour, have low TSS and are more prone to chilling injury.

2. Fruit should be transported carefully to market in order to minimize bruise damage and subsequent post-harvest decay.
3. Use of strong protective well ventilated packaging will help to minimize bruise damage. Proper loading and stacking of the cartons in the bed of the transport vehicle is essential to protect the product.
4. The pineapple cartons should be covered and never exposed to direct sunlight during transport.
5. In the absence of refrigeration, the fruit should be transported during the cooler part of the day.
6. For domestic transportation, the vehicles should offers substantial advantages of convenience, availability, flexibility permitting door-to-door delivery, and reasonable cost of transport.
7. Where sea shipment is to be done in distant countries, the fruit should be harvested on the day prior to shipment.
8. Where exports are made to be by airways, fruits should be harvested at more advanced stages of maturity.

Storage: In India the middlemen only practices the storage of pineapple fruits and in North east India the fruits are mainly stored in open floor, gunny bags, bamboo basket or in the thatch (Deka, 2015). For pineapple maturity at harvest is the most important factor to determine the shelf-life and quality of the fruits. Generally overripe fruits become soft and mealy with insipid flavour after harvest and cannot be stored for long duration.

Pineapple fruits can be best stored at 7-8°C and above and 80-90% relative humidity under adequate air circulation for 4-6 weeks but cold storage at a temperature of 4.44°C and lower causes chilling injury and breakdown in fruits (Morton, 1987). Temperatures of 7 to 12°C and relative humidity of 85 to 95% are recommended for storage of pineapples for 14 to 20 days, provided fruit are at the color break stage (Paull, 1993). Quarter-yellow fruit at harvest gain about one additional week of storage for every 6°C decrease in storage temperature (Dull 1971). The maximum storage-life at 7°C is about 4 weeks (Paull and Rohrbach, 1985). In tropical areas, partially ripe, healthy and unbruised pineapple could be stored for almost 20 days when refrigerated at 10-13°C with RH 85-90%. Fruits harvested in early stage of ripening are stored at 7-10°C (Anonymous, 2018-19).

Table 7: Storage conditions of pineapple fruits as per APEDA

Temperature (°C)		Relative humidity (%)	Storage Period (Weeks)	Freezing point (°C)
Half ripe or fully ripe fruits	Fully green fruits			
7 - 10	10 - 13	85 - 95	3-5 weeks*	-1

Source: Anonymous (2015); *depending on cultivar and ripening stage

Conclusion: The issues related to post-harvest losses of pineapple fruit in India begin right from the standardization of harvest maturity up to the storage levels before reaching the consumer. Since pineapple has huge demand for consumption, processing and has high export potential in India, therefore the role of proper post-harvesting operations and management is very crucial for quality assurance and market expansion.

References

Adetunji, C. O., Fawole, O. B., Arowora, K. A., Nwaubani, S. I., Ajayi, E. S., Oloke, J. K., Majolagbe, O. M., Ogundele, B. A., Aina, J. A. and Adetunji, J. B. 2012. Effects of edible coatings from aloe vera gel on quality and postharvest physiology of Ananas comosus (L.) fruit during ambient storage. Global Journal of Science Frontier Research, 12(5): 38-43.

Anonymous. 2006. Menuju Ke Arah Kualiti Malaysia Best. Federal Agriculture Marketing Agency (FAMA), Ministry of Agriculture and Food Industry, Selangor, Malaysia. Retrieved from: http://www.fama.gov.my/spesifikasi-gred-dan-piawaian1.

Anonymous. 2018. Horticultural Statistics at a Glance 2018. Horticulture Statistics Division, Department of Agriculture, Cooperation & Farmers Welfare, Ministry of Agriculture & Farmers Welfare, Government of India. Retrieved from http://www.agricoop.nic.in.

Anonymous. 2018. Pineapple, In: Fruit Crop High Tech Cultivation in Open Field. Indian Horticulture Board (NHB), Ministry of Agriculture, Government of India. Retrieved from: http://www.nhb.gov.in

Anonymous 2015. Pineapple. In: Agri Exchange. The Agricultural and Processed Food Products Export Development Authority (APEDA), Ministry of Commerce and Industry, Government of India.

Babalola, D. A., Megbope, T. A. and Agbola, P. O. 2008. Post harvest losses in pineapple production: case study of Ado-odo Otta local government area of Ogun State. Bowen Journal of Agriculture, 5(1&2): 55-62.

Deka, B. C. 2015. Pineapple. In: Chadha, K.L. and Pal, R.K. (Ed.), Managing postharvest quality and losses in Horticultural crops Daya Publishing House, New Delhi, India, 2: 497-515.

Deka, B. C. Saikia, J. and Sharma, S. 2005. Standardization of maturity indices of 'Kew' pineapple. Acta Horticulturae, 682: 2215-2220.

Dhar, M., Rahman, S. M. and Sayem, S. M. 2008. Maturity and post harvest study of pineapple with quality and shelf life under red soil. International Journal of Sustainable Crop Production, 3(2): 69-75.

Dull, G. G. 1971. The pineapple: General. In: Hulme, A.C. (Ed.), The biochemistry of fruits and their products. Academic Press, London, 2: 303-331.

El-Ramady, H .R., Domokos-Szabolcsy, E., Abdalla, N. A., Taha, H. S. and Fari, M. 2015. Postharvest Management of Fruits and Vegetables Storage. In: Lichtfouse, E. (Ed.), Sustainable Agriculture Reviews. Springer International Publishing, Switzerland. 15: 65-152.

Hodges, R. J., Buzby, J. C., and Bennett, B. 2011. Postharvest losses and waste in developed and less developed countries: opportunities to improve resource use. Journal of Agricultural Science, 149: 37-45.

Hossain, M., and Bepary, R. H. 2015. Post-Harvest handling of pineapples: A key role to minimize the post-harvest loss. International Journal of Recent Scientific Research, 6(9): 6069-6075.

http://apeda.in/agriexchange/Market%20Profile/one/PINEAPPLE.aspx

Hu, H., Li, X., Dong, C. and Chen, W. 2011. Effects of wax treatment on quality and postharvest physiology of pineapple fruit in cold storage. African Journal of Biotechnology, 10(39): 7592-7603.

Joy, P. P. 2010. Protocol for pineapple "Kerala Mauritius" variety for export to shorter duration (15-18 days). In: Pineapple Technology. Pineapple Research Station (PRS), Kerala Agricultural University, Kerala, India.

Joy, P. P. and Rashida Rajuva, T. A. (2016). Harvesting and post-harvest handling of pineapple. Pineapple Research Station (Kerala Agricultural University) Vazhakulam-686670 Muvattupuzha, Ernakulam, Kerala. Retrieved from: Web: http://prsvkm.kau.in

Kamol, S. I., Howlader, J., Sutra Dhar, G. C. and Aklimuzzaman, M. 2014. Effect of different stages of maturity and postharvest treatments on quality and storability of pineapple. Journal of the Bangladesh Agricultural University, 12(2): 251-260.

Lin, X., Li, X. and Chen, W. 2013. Effect of wax treatment on the quality and postharvest physiology of pineapple fruits. Acta Horticulturae, 975: 519-526.

Lu, X., Sun, D., Li, Y., Shi, W. and Sun, G. 2011. Pre- and Post- harvest salicylic acid treatments alleviate internal browning and maintain quality of winter pineapple fruit. Scientia Horticulturae, 130: 97-101.

Lu, X. H., Sun, D. Q., Mo, Y. W., Xi, J. G. and Sun, G. M. 2010. Effects of post-harvest salicylic acid treatment on fruit quality and anti-oxidant metabolism in pineapple during cold storage. Journal of Horticultural Science and Biotechnology, 85: 454-458.

Mahajan, P. V., Caleb, O. J., Singh, Z., Watkins, C. B. and Geyer, M. 2014. Postharvest treatments of fresh produce. Philosophical Transactions of the Royal Society A, 372: 1-19.

Mandal, D., Lalremruata, Hazarika, T. K. and Nautiyal, B. P. 2015. Effect of Post-harvest Treatments on Quality and Shelf Life of Pineapple (Ananas comosus [L.] Merr. 'Giant Kew') Fruits at Ambient Storage Condition. International Journal of Bio-resource and Stress Management, 6(4): 490-496.

Morton, J. 1987. Pineapple. In: Fruits of warm climates. Julia F. Morton, Miami, Florida. pp, 18-28.

Nanayakkara, K. P. G. A., Herath, H. M. W. and Senanayake, Y. D. A. 1997. Effects of some pre-harvest treatments of potassium and hormones on fruit core tissue deterioration and internal browning of pineapple under cold storage. Tropical Agricultural Research, 9: 90-95.

Ningombam, S., Noel, A. S. and Singh, J. 2019. Post-harvest losses of pineapple at various stages of handling from the farm level up to the consumer in Manipur. International Journal of Agriculture Sciences, 11(22): 9235-9237.

Paull, R. E. 1993. Pineapple and papaya. In: Seymour, G., Taylor, J. and Tucker, G. (Ed.), Biochemistry of Fruit Ripening. Chapman & Hall, London, UK. pp, 291–323.

Paull, R. E. and Chen, C. C. 2014. Pineapple: Postharvest Quality-Maintenance Guidelines, In: Fruit, Nut, and Beverage Crops. The CTAHR publications database. College of Tropical Agriculture and Human Resources, University of Hawai'i in Manoa. Retrieved from: https://www.ctahr.hawaii.edu/oc/freepubs/pdf/F_N-32.pdf

Paull, R. E. and Rohrbach, K. G. 1985. Symptom development of chilling injury in pineapple fruit. Journal of the American Society for Horticultural Science, 110: 100-105.

Prusky, D. 2011. Reduction of the incidence of postharvest quality losses, and future prospects. Food Security, 3(4): 463-474.

Selvarajah, S., Bauchot, A. D. and John, P. 2001. Internal browning in cold-stored pineapples is suppressed by a postharvest application of 1-methylcyclopropene. Postharvest Biology and Technology, 23(2): 167-170.

Selvarajah, S., Herath, H. M. W., Bandara, D. C. and Abeysinghe Banda, D. M. G. 1998. Effect of pre-harvest calcium treatment on post-harvest quality of pineapple. Tropical Agricultural Research, 10: 214-224.

Shewfelt, R. L. 1993. Measuring quality and maturity. In: Shewfelt, R.L. and Prussia, S.E. (Eds.), Postharvest Handling: A Systems Approach. Academic Press, San Diego, California. pp, 99-124.

Singh, I. S. 2009. Post-Harvest Handling and Processing of Fruits and Vegetables. Westville Publishing House, New Delhi, India.

Souleymane, C., Edwige-Salome, Y., Laurent, N., Samuel, K. and Kouakou, T. 2019. Effects of potassium fertilization for pineapple on internal browning of fruit in post-harvest conservation. Journal of Agriculture and Crops, 5(6): 100-108.

Talukdar, S. and Ahmed, F. 2002. Effect of pre-harvest chemical treatments on shelf life of pineapple fruit. Journal of Food Science and Technology, 39: 53-54.

Thompson, A. K. 2003. Fruit and vegetables: harvesting, handling, and storage (2nd Ed.). Wiley-Blackwell Publishing Ltd., Hoboken, New Jersey, United States.

Weerahewa, D. and Adhikaram, N. K. B. 2005. Some biochemical factors underlying the differential susceptibility of two pineapple cultivars to internal browning disorder. Journal of Biosciences, 34: 7-20.

Weerahewa, H. L. D. 2002. A study of internal browning of two cultivars of pineapple with special reference to heat shock treatment as a control measure. (Unpublished Ph.D. Thesis), University of Peradeniya, Sri Lanka. pp, 274.

Weerahewa, H. L. D. and Wicramasekara, I. 2020. Preharvest application of silicon reduces internal browning development of pineapple (Ananas comosus ‹Mauritius›) during cold storage: A novel approach. Acta Horticulturae, 1278: 39-44.

Wijeratnam, W. R. S., Hewajulige, I. G. N. and Abeyratne, N. 2005. Postharvest hot water treatment for the control of Thielaviopsis Black rot of pineapple. Postharvest Biology and Technology, 36: 323-327.

19

Physiological Disorders of Pineapple

Sananda Mondal and Debasish Panda

Department of Crop Physiology, Institute of Agriculture, Visva-Bharati Sriniketan, West Bengal

Pineapple, are well adapted crops for the farmers towards water economy due to its morphological, anatomical, and physiological characteristics. However, the plant contains rosette leaf structure can able to channels the light rainfalls and dew through stem and finally into the soil. It also have the ability to absorb water and nutrients through their waxy leaves and aerial roots and can capable to store water in specialized aquiferous leaf tissue; multicellular trichomes that cover the leaves and protect the plant from excessive transpiration by reflecting radiation; a thick cuticle covering the leaf epidermis that restricts water loss by evaporation, location of the stomata in furrows at the lower leaf surface where they also are covered by trichomes, limiting evapotranspiration; and its nocturnal carbon dioxide fixation via the crassulacean acid metabolism (CAM) carbon assimilation pathway, which results in stomatal opening primarily at night when evaporative demand is lowest (Reinhardt *et al.*, 2018). The most common propagating material of pineapple is suckers or slips. Suckers are used to arise from the underground parts of the plant are used for propagation. Slips arise from the fruiting stem and the crown region on top of the fruit.

In this context, it is important to know the nutritional aspects of pineapples which are as attractive as their unique anatomy. "Pineapples contain high amounts of vitamin C and manganese," said San Diego-based nutritionist Laura Flores. These tropical fruits are also important for the dietary fiber and bromelain (an enzyme). "As well as having high amounts of manganese, which is important for antioxidant defenses, pineapples also contain high amounts of thiamin, A, B vitamin that is involved in energy production," Flores said. Pineapples are fat-free, cholesterol-free and low in sodium (Arpaia *et al.*, 2004; De Costa, 2010). Side by side they contain sugars of about 14 grams per cup (www.livescience.com, 2019). In continuation with this, the chapter was focused on physiological disorder of pineapple caused by various reasons.

Basically, non-pathological conditions are the reasons of physiological disorders of any kind of crop plants; it includes poor light, adverse weather, water-logging, phytotoxic compounds or a lack of nutrients, which affects the normal functioning of the plant system. Sometimes physiological disorders may be initiated before harvest, continue their development after harvest, or appear after harvest. In was observed several times that physiological disorders are causative agent for the deterioration of horticultural commodities which leads to economic losses. An important commercialized fruit i.e., pineapple is also facing the same problem of different kind of physiological disorders; it includes multiple crowns, fruit and crown fasciation, collar of slips, dry fruit and bottle neck, sun-scald/ sunburn, blackheart etc.

Multiple Crowns: Multiple crowns are a common physiological disorder that can be of genetic or environmental origin. Multiple crowns increase the size of the fruit core and result in flattening of the upper portion of the fruit, which reduces the value of the fruit for the fresh market and canning. Generally, fruit bears a single crown but in some cases, a fruit bears more than 1 or even up to 25 crowns. With this context, Collins (1960) states that fasciation - an abnormal growth resulting in two to numerous crowns - is relatively uncommon in 'Smooth Cayenne' and 'Queen,' but common in 'Singapore Spanish' and 'Pernambuco'. In addition, Collins (1960) also reported the existence of mutant clones of 'Smooth Cayenne' which produced 50% or more fruit with multiple crowns. The taste of such fruits is insipid and corkier. It is supposed to be a heritable character, found mostly in the Cayenne group to which var. Kew belongs.

Environmental conditions are the promoting factors of multiple-crown disorder, in which high fertility with rapid growth habit followed by a period of prolonged drought occurs at the time of inflorescence initiation (Collins, 1960; Py *et al.*, 1987). An increased incidence of multiple crowns is correlated with periods of high irradiance and high temperature that occur during early inflorescence development, although the disorder is generally thought to be due to high-temperature injury. Increased planting density reduced the incidence of multiple crowns (Linford and Mehrlich, 1934; Norman, 1977; Scott, 1992), presumably because mutual shading in the more dense plantings reduced the temperature of the reproductive apex. The significance of this shading effect is confirmed by the observation that there was a higher incidence of deformed crowns on outside rows than in the interior of a field. Optimum irrigation during the time of inflorescence development reduced the incidence of multiple crowns disorder of pineapple (Py *et al.*, 1987).

Fruit and Crown Fasciation: Fasciated fruits are deformed to such an extent that they are useless. In some cases, proliferation is so extreme that fruit is highly flattened and twisted with innumerable crowns. Fruits and crowns fasciation is associated with high vigour of plants, which take a longer time to flower than the normal ones. High fertile conditions of soil and warm weather is highly congenial for vigorous vegetative growth, may favour fasciation in fruits. The incidence of fasciation was found to increase with the advancing ratoons. There may be some cases, where deficiency of calcium or zinc may cause fasciation.

Collar of Slips: A slip is a rudimentary fruit with an exaggerated crown. Slips develop from buds in the axils of leaves borne on the peduncle (fruit stalk). Because they must grow outward, then upward from under the fruit, slips are curved at the base. On slip-producing clones, the number of slips can vary from none (common in hot, equatorial zones) to as many as 10 or more, as is observed in the undesirable mutation known as "collar of slips" (Bartholomew *et al.*, 2002). The excessive slip growth is at the expense of the fruit, resulting in small, tapered fruits, often with knobs at the base. High nitrogen fertilization and rainfall along with relatively low temperatures are supposed to be congenial for such an abnormality (Ganapathy *et al.*, 1977). Some pineapple varieties are not able to produce slips and hapas in any significant number, for example, some clones, e.g., Champaka F153, of cv. Smooth Cayenne and there are others, for example, cv. Pérola and clones of Smooth Cayenne having the "collar of slips" condition, that bear them in large numbers, reducing average fruit weight and yield (Collins, 1960; Lima *et al.*, 2001, 2002). Temperature also influences the slip number. The cv. MD-2 produces relatively large numbers of slips in subtropical environments (Hawaii, at high elevation in Kenya) but produces few or no slips in warm tropical environments such as Ghana. In Brazil, slips are the most important planting material for Pérola, the main traditional cultivar, as well as the more recent ones, such as 'Imperial' and 'Vitória', and these cultivars all produce large average numbers of them (Cabral and Matos, 2003).

Dry Fruit and Bottle Neck: Another disorder i.e., dry fruit and bottleneck fruit types are very similar and may be derived from the same parent. In the dry fruit type, the fruit is small, flowers are absent and fruitlets do not develop. In bottleneck, lower fruitlets develop normally and upper ones do not develop and give the same appearance as dry fruits. Suckers are produced freely from both the types.

Sun-Scald/ Sunburn: Pineapples are susceptible to heat stress and sun damage, which can significantly reduce marketable yield. Extreme sun intensity can limit overall plant production and directly damage the fruit. In many cases, the damage is not realized until after the fruit is harvested. Early symptoms of sunburn exhibit yellowing, "bleached" skin that turns pale grey/ brown as the tissue deteriorates. This damaged tissue is susceptible to disease and infestation. Internal damage occurs when the fruit overheats and ends up "cooking" on the inside (Lal and Sahu, 2017).

Literature suggests that, when the plant leans or falls over to one side, thus exposing one side of the fruit to direct sunlight. The cells of the exposed surface get damaged. Later shell surface assumes a brownish to black colour and cracks may appear between fruitlets. Affected fruits soon rot and become infested with pests. They must be cut as soon as noticed and safely disposed of where they will not contaminate other fruits. This disorder can be controlled by high-density planting system; the intensity of sun-scald is minimized. Under favourable climates where leaf growth is luxuriant, leaves can be tied around the fruits to protect them from sunscald. The other controlling method is to cover the sun-exposed portion of the fruit with dry straw or grass or with any other locally available materials. Along with this Joy (2010) observed that during the months of summer it is necessary to protect the fruits from the scorching sun by putting dried grasses, coconut, or areca nut leaves to prevent sunburn. Inter row spaces should be utilized by growing intercrops as per crop-specific. Maize or pigeon pea is most useful in pineapple orchard to protect from sunburn/sunscald (Lal and Sahu, 2017).

Black heart (BH): Blackheart (BH), a low-temperature-related physiological disorder of is a worldwide problem. But induction of BH can be takes place at a wider temperature range of up to 21°C. BH may be occur at a pre-harvest stage and is associated with low field temperatures during the time of growth. After harvesting, BH occurs in fruits subjected to exposure to low temperatures during storage, handling, and transportation. Factors influence the development of BH which includes temperature, length of exposure, varieties, stage of maturity, and growing climatic conditions. The disorder is also known as "endogenous brown spots" and "internal browning". The characteristics symptom of the disorder is the development of dark spots in the flesh area close to the core. In its early stage of development, the spots appear watery but they then enlarge and turn brown as the severity of the disorder increases. In severe cases, the entire flesh and core tissue of fruit may be visibly affected. Fruits affected by BH appear normal externally and its presence is only known after the fruit has been cut open for processing or consumption. The lack of detectable external symptoms creates substantial problems in the marketing of fresh fruit. BH has been associated with a reduction in ascorbic acid and an increase in polyphenol oxidase enzyme (PPO) activity. However, the initial ascorbic acid concentration and PPO activity do not indicate fruit susceptibility to BH. Partial control of BH was reported to be achieved by pre-harvest applications of chemicals such as parachlorophenoxyacetic acid (PCPA), α-naphthaleneacetic acid (ANA), potassium, and calcium. Several postharvest methods have also been reported to provide some control of BH, including heat treatment, controlled and modified atmosphere, surface coatings, and 1-methyl cyclopropane (1-MCP) treatment. Plant breeding to develop BH-resistant cultivars provides an attractive alternative to postharvest treatments to control the disorder. New hybrids resistant to BH have been produced from conventional breeding while a genetic engineering approach has also shown promising development (Hassan *et al.*, 2010).

Physiological breakdown: Physiological breakdown of the fruit of the pineapple is characterized by a darkening of the flesh of the fruitlets, beginning near their attachment to the core in a region an inch or two above the base of the fruit. The disorder progresses upward and outward, subsequently involving all of the fruitlet except the ovary and a layer approximately one-quarter inch deep below the skin. Only in advanced stages do the surface of the fruit and the core become involved. Hence in most fruits, there may be no external symptoms of the disorder. Severe cases of physiological breakdown are frequently discovered in pineapples that appear normal and attractive from the exterior. The disorder is not a form of over-ripeness as is true of the "internal breakdown" of apples (Miller 1951).

Conclusion

Although, pineapple is a globally accepted fruit crops but facing problems due to different kinds of physiological disorders which leads to below marketable quality of fruits and finally economic loss. There are different kinds of problems associated with cultivation package and practices of pineapple among them physiological abnormalities are one of them. In these particular aspects very less information is available in the scientific database. So in a nutshell, it can be concluded that much work/ experiment is needed to meet up the gap and protect the pineapple fruits/ plants from such physiological disorders, make them marketable and improve the economic conditions of the pineapple growers.

References

Arpaia, M. L., Rooyen, Z. V., Bower, J. P., Hofman, P. J. and Woolf, A. B. 2004. Growers practice will influence postharvest fruit quality. 2nd Seminario International Depaltos. 29 Septiembre – 1 Octubre, 2004. Sociedad ardiazabaly Magdahl Ltda.Quillota, Chile.

Bartholomew, D. P., Rohrbach, K. G. and Evans, D. O. 2002. Pineapple cultivation in Hawaii. Fruits and Nuts, Co-operative Extension Service, College of Tropical Agriculture and Human Resources University of Hawaii at Manoa.

Cabral, J. R. S., Matos, A. P. de, and Jughans, D. T. 2003. Desenvolvimento de híbridos de abacaxi resistentes à fusariose. Cruz das Almas: Embrapa-CNPMF, 4p. (Comunicado Técnico, 88).

Collins, J. L. 1960. The Pineapple: Botany, Cultivation and Utilization, Leonard Hilll Publisher, New York.

De Costa, D. M. 2010. Quality retention of dessert banana and other tropical fruits by controlling postharvest diseases using non-chemical methods. In: Sivakumar D (Ed) New Trends in Postharvest Management of Fresh Produce II. Fresh Produce 4 (Special Issue 1), pp, 1-14.

Hassan, A., Wills, R., Atan, R. Md, Othman, Z., Fatt, L. P. and Smith, M. 2010. Blackheart disorder in fresh pineapple. Fresh Produce 4 (Special Issue 1), 29-35. Global Science Books.

https://horticult.wordpress.com/pineapple/html.

https://www.livescience.com/45487-pineapple-nutrition.html.

Joy, P. P. 2010. Production technology for pineapple variety 'Kew'. Pineapple Research Station (KAU), Kerala, India.

JSTOR, www.jstor.org/stable/4258386. Accessed 30 Sept. 2020.

Lal, N. and Sahu, N. 2017. Management Strategies of Sun Burn in Fruit Crops-A Review. Int. Journal of Current Microbiology and Applied Science, 6(6):1126-1138. https://doi.org/10.20546/ijcmas.2017.606.131.

Lima, V. P. de, Reinhardt, D. H., and Costa, J. A. 2002. Desbaste de mudas tipo filhote do abacaxi cv. Pérola. Análises de crescimento e correlações. Revista Brasileira de Fruticultura, 24(1):101-107.

Lima, V. P de, Reinhardt, D. H. and Costa, J. A. 2001. Desbaste de mudas tipo filhote do abacaxi cv. Pérola. Produção e qualidade do fruto. Revista Brasileira de Fruticultura, 23(3):634-638.

Linford, M. B. 1934b. Fruit quality studies VII. Variations in fruit and crown characters associated with amounts of fertilizer and with time of ripening in C.P.C. experiment D-2, Kahuku, 1932. Pineapple Quarterly, 4:144-151.

Medicinal Plant Archive, 2020. https://www.medicinalplantsarchive.us/pineapple-fruit/multiplecrowns.html.

Miller Erston, V. 1951. Physiological Studies of the Fruits of the Pineapple [Ananas Comosus (L.) Merr.] with Special Reference to Physiological Breakdown. Plant Physiology, 26(1): 66–75.

Norman, J. C. 1977. Chemical regulation of growth, flowering and fruiting in 'Sugarloaf' pineapple. Scientia Horticulture, 7(2):143-151.

Py, C., Lacoville, J. J. and Teisson, C. 1987. The pineapple, cultivation and uses, maisonneuve et Larose G.P. (eds) Paris, France.

Reinhardt, D. H. R. C., Bartholomew, D. P., Souza, F. V. D., Carvalho, A. C. P. P de, Padua, T. R. P. de, Junghans, D. T. and Matos, A. P. de 2018. Advances in pineapple plant propagation. Revista Brasileira de Fruticultura, 40(6):(e-302). DOI: http://dx.doi.org/10.1590/0100-29452018302.

Scott, C. H. 1992. Effect of two plant growth regulators on the precocious fruiting in pineapple. I International Pineapple Symposium, Acta Horticulturae, 334.

20

Important Pests of Pineapple and Management

[1]Panchal Bhattacharjee, [2]Sayan Sau, [3]Prahlad Deb and [4]Pallab Datta

[1]Department of Horticulture, Anand Agricultural University, Anand, Gujarat, India
[2]Purba Medinipur Krishi Vigyan Kendra, BCKV, Mulakhop, Dayaldasi Purba Medinipur, West Bengal, India
[3]Department of Horticulture & Postharvest Technology Institute of Agriculture, Visva-Bharati, Sriniketan, West Bengal, India
[4]Department of Fruit Science, Faculty of Horticulture, Bidhan Chandra Krishi Viswavidyalaya, Mohanpur, Nadia, West Bengal, India

Pineapple i.e. *Ananas comosus* (L.) Mer. belongs to the Bromeliaceae family, which has around 46 genera and 1700 species, mainly cultivated in the tropical and sub- tropical regions. It is ranked third in production of tropical fruit after banana and citrus. The pineapple market has been growing extensively due to the attractive aroma compounds and nutritional values as well as huge demand and competitive retail prices (Ali *et al.*, 2020). But, several sanitary problems have been reported in pineapple and they are considered as the main constraints for low yield of that crop in several producing areas. Significant yield losses have been caused by infestation of different pests like mealy bug, fruit borer, mites, scale insects, as well as with different nematode species (Pires de Matos, 2017). The crop lifecycle phases, soil, climatic variables and crop management practices influence the pest infestation intensity. In order to prevent severe crop losses, an integrated approach comprising prophylactic, biological, cultural, mechanical and need-based chemical steps is most appropriate for controlling pineapple pests and maintaining their population below the threshold level. This chapter deals with globally important pests of pineapple with their nature of damages and effective control measures in depth.

Pineapple Mealybug

Dysmicoccus brevipes (Cockerell) (Order-Hemiptera; Family-Pseudococcidae)

Pineapple mealybug is ovoviviparous, *i.e.* the eggs hatch and give birth to larvae within the female. Considering worldwide significance different species of mealybug are pink pineapple mealybug [*Dysmicoccus brevipes* (Cockerell)], grey pineapple mealybug [*Dysmicoccus neobrevipes* Beardsley] and long-tailed mealybug [*Pseudococcus adonidum* (L.) syn. *Pseudococcus longispinus* (Targioni-Tozzetti)] (Satyagopal *et al.,* 2014).

Brief about pest and Nature of damages: They appear as cottony, tiny, circular, sucking insects with a soft body. Though they are in a number of shapes, usually pink-colored ones referred to as pineapple mealybug. They live predominantly on seed matter. Such bugs only have versatility at a younger stage. These insects are normally static in position although being competent of movement, congregate together in groups of 20 odd in numbers. The first-stage (or first-instar) crawlers (0.6–0.7 mm) are typically the most active stage, and they move around the plant host seeking a place to settle down to feed.

If their population grows, mealy bugs become deadly as the bugs suck the sap from leaves. Spread of mealy bug further speed up by ants. The mealy bug's sugar secretion attracts ants and they bear them, making them wide spread. Preventing the establishment of new ant colonies in new plantation is crucial for preventing mealybug wilt (Rohrbach and Schmitt, 1994). The excretion of honey like sticky substance often impedes the capacity of plants to grow food, form chlorosis of the plant tissues and induce fruit rotting and leakage.

Effective control measures: Pineapple mealybug's role as vector is often the biggest problem for pineapple farmers. The pineapple mealybug wilt usually referred to as pineapple wilt, mealybug wilt, or edge-wilt (Sether *et al.,* 1998). Viruses of three distinct species (*Pineapple mealybug wilt-associated virus-1, virus-2, and virus-3*) of the genus Ampelovirus (family Closteroviridae) have been reported in many pineapple-growing regions worldwide (Nyarko and Asare-Bediako 2019), such as Hawaii (Hu *et al.,* 1993), Australia (Wakman *et al.* 1995), Central and South America (Borroto *et al.,* 1998), Ecuador (Alvarez *et al.* 2015), and West Africa (Querry *et al.,* 2016). Development of peculiar red pigmentation in leaves is the primary symptom, subsequently the infected plant losses stiffness and looks wilted. In wilting regions, the reddish-yellow hue manifests. Under severe infestation, it causes quick wilt (result in yellow or red leaves) or slow wilt (no definite colour change) in mild attack due to root damage. In some instances, plants may recover and continue to expand, but weight, leaf size and root length are limited (Mau and Kessing, 2007). If this wilt continued in fruit development stage, then fruit become completely layered with waxy white coating which make them unfit for marketing.

Pre-planting sucker treatments are successful in managing the mealy bugs of pineapple. Basal portions of the planting materials have to be dipped in 0.02-0.04% Methyl parathion as a prophylactic measure. It is inevitable to pick good suckers (avoid choosing 5-6 year old sucker), curing them in upside down position also proven helpful. Mealybug infestation can be reduced by destroying ant colonies, using Hydramethylnon or sodium octaborate. Removal of alternate hosts such as hibiscus, custard apple, guava from plant's vicinity and use sterilized machinery for cultural operations in an uninfected field reduces the spread of mealybug.

Certain natural enemies are well known such as *Anagyrus ananatis* Gahan, a parasitoid of the *D. brevipes*. A new species of *R. leucophenga*, a potential predator of pineapple mealy bugs (Culik, 2009) was introduced in Brazil. Biological protection can only be successful if ants are controlled because ants defend this pest against natural enemies. Under laboratory conditions, if ants are present *A. ananatis* does not attack *D. brevipes*. Other biocontrol agents of the pineapple mealybug are *Euryrhopalus propinguus* Kernich, *Hambletonia pseudococcina* Compere, *Lobodiplosis pseudococci* Felt, *Nephus bilucenarius* Mulsant and *Scymnus uncinatus* Sicard.

Successful chemical control measures are use of Imidacloprid 200SL @ 0.3 ml/L or Buprofezin @ 2ml/L or Quinalphos @ 2 ml/L @ 500-600 L/ha and regulation of ant colonies by drenching soil with Chlorpyriphos 20 EC at 2.5 ml/L (Tanwar *et al.*, 2007).

Pineapple Scales

Diaspis bromeliae (Kerner) (Order-Hemiptera: Family-Diaspididae)

Brief about pest and Nature of damages: Scales are omnipresent on pineapple leaves and fruit worldwide, the crown build-up can be noticed in case of high infestation. Other shrubs known to flourish in vicinity of pineapple fields such as agave, billbergia, and bromelia can serve as an alternate host to this pest (Petty, 1978).The insect community displays high degrees of sexual dimorphism; female insects maintain immature bodies even when sexually mature. Many adults have wings but do not usually feed, and die within a day or two.

Effective control measures: For scale insects, a light, superior oil/insecticide blend may be added only as plants begin to develop to combat emerging crawlers. Typically, this time span lasts only a few days, so scheduling is important. Before the leaves open, the oil must be added, but late enough to kill the crawlers with this combination. When crawlers are around, the safest time to add any insecticide is as this stage does not have a defensive coating, and is thus vulnerable to almost any chemical registered for this purpose.

Many natural enemies can biologically regulate this insect. The scales are parasitized by tiny wasps, including *Aphytis chrysomphali* (Mercet), *Aphytis diaspidis* (Howard) and *Aspidiotiphagus citrinus* (Craw) (Hymenoptera: Aphelinidae), contributing to mortality on the scale. The scales also preyed by lady birds such as *Rhyzobius lophanthae* Blasid and *Telsimia nitida* (Coleoptera: Coccinellidae) (Carter, 1967; Waite, 1993).

Thrips

Holopothrips ananasi Costa Lima (Order-Thysanoptera: Family-Phlaeothripidae)

Brief about pest and Nature of damages: Pineapple thrips (*Holopothrips ananasi* Costa Lima) is a major pineapple pest that causes significant damage to pineapples. Worldwide, as many as 39 species of thrips have been verified in the cultivation of pineapple crops (Sakimura, 1937; Carter, 1939; Petty, 1978). Thrips tend to be infective at dry hours. Irrigation helps to minimize thrips infestation and their displacement by wind. *Thrips tabaci* and *Frankliniella schultzei* are also known as significant pineapple pests. Thrips mainly thrives on flowers and led to "dead-eye" development in the fruit. Feeding on fruit crowns contribute to the development of concentric ring patterns on its leaves.

Thrips feed on the plant mostly during day hours, lesions are formed around the region of piercing. The damaged leaves will normally have silvery flecks and turn brown, leaves a tiny blackish spot as key symptom of infestation, ultimately it hinders in photosynthetic activity. Thrips are active in the spread of diseases in plants, as it acts as a vector of tomato spotted wilt virus which causes Pineapple yellow spot.

Effective control measures: Mulching eliminates infestation of thrips, intercropping among plant that repels thrips such as citronella, garlic and pyrethrum beside use of windbreaks to suppress thrips population are proved as effective cultural measures.

Different plant extract as biological control measures for thrips also advocated are spray consisting of garlic and pepper (filtered water extract of 2 bulbs of garlic and hot peppers mixed up with upto 5liters of water for spray); extract from *Andrographis paniculata* (2 kg of fresh plant material of *A. paniculata* and 250 ml of water is ground to get the spray material); dried chilli (21 litres of cow urine mixed with 10 g of smashed dried chilli added to 10 L of water for getting the final spray matter) and root extract of *Derris eliptica* (extract of small dried cut root pieces added to 4 parts of soap and 225 times water for developing the end spray matter) (Satyagopal *et al.*, 2014).

Pineapple Fruit Borer

Strymon megarus Godart (Order-Lepidoptera, Family-Lycaenidae)

Brief about pest and Nature of damages: *Strymon megarus* Godart is a pest of pineapple, affecting in fields from Mexico to Argentina and considered a principal pest of pineapple in Brazil. However, it is present in all pineapple growing regions and damage intensity varies with pest population and environmental conditions. During flowering and fruit forming, fruit borer attacks occur in most instances, although this borer can target slips but rarely act as a leaf miner. Infestation rate of borer is recorded maximum in drier climate. The larvae of the fruit borer open galleries in the pulp, creating an oozing called, "Gummosis" on the fruit's outer side.

Adults activity found both during the day or night, they fly in a rapid fashion and in haphazard manner. Eggs are mainly laid on flowers and larval stage mostly completed within the fruit. The reddish coloured caterpillar penetrates the inflorescence and stays in the tissue for 15 days, made tunnel in the tissues and destroying it. It travels to the base of the peduncle after this point and transforms into a pupa and emerges as a butterfly 7 to 10 days later (Satyagopal *et al.*, 2014).

Effective control measures: *Bacillus thuringiensis* appears to be an acceptable biological alternative to carbaryl for control of *S. megarus* on pineapple. Mirid bug (*Dicyphus hesperus*), reduviid bug, big eyed bugs (*Geocoris* sp) rove beetle, dragonfly, damselfly, spider, robber fly, praying mantis, red ants, lacewings etc act as predetors for this pest. Chlorpyriphos (Hilban 20EC, 2.5 ml/l) at 500l/ha acts to control the borer.

Pineapple Fruit Fly

Melanoloma canopilosum Hendel, *M. viatrix* Hendel (Order-Diptera: Family-Richardiidae)

Brief about pest and Nature of damages: *Melanoloma canopilosum,* and *M. viatrix* (Diptera: Richardiidae) are found commonly in fields of Paraguay and Peru, mostly neotropical in nature. Maggots feed inside the fruits (Bello, *et al.,* 1997).

Female fruit flies puncture fruit with their long and sharp ovipositors. Through breached skin, bacterial penetration occurs within fruits, which causes decay. The maggot hatching from laid eggs, feeds (saprophagous) on rotting fruit. Within the fruit, all the developmental stages of larvae take place. Larval infestation causes irregular ripening, discoloration and decay of fruit through fermentation (Bello, *et al.*, 1997; Martínez *et al.*, 2000).

Effective control measures: A major control of fruit flies is plucking off infested fruits. Fermented fruits must be cleared up or exposed to the sun for larval death and fruit decomposition. The removal of leftover fruits or residues is very efficient as it decreases the possibility of using the field as a breeding ground. Traps can be placed near its breeding sites with fermented fruit in an inverted funnel. In order to save fruit damage, bagging and netting of fruits is also advisable. Fruit harvesting without much ripening can limit its assault.

Pineapple Red Mite

Dolichotetranychus floridanus (Banks) (Order-Trombidiformes: Family-Tetranychidae)

Brief about pest and Nature of damages: The bright orange pineapple mite is the largest mite present on pineapple. Adult mites are approximately 0.3-0.4 mm long and 0.1 mm wide. The mite is often found on the white basal portion of the leaves where it feeds especially on the crown (Petty, 1978). This cause leaves and fruits to be harmed. Extreme attacks can lead to large and dark brown lesions which can cause death of the leaves. It's a major pest in newly established orchard. Infested plants remain in the initial stages, and the production of fruit is either curtailed or non-existent. Heavily infested plants can die without producing fruit (Jeppson *et al.*, 1975).

Effective control measures: Control programmes are focused on crop surveillance to look for colonies of leaves and/or leaf indentations and "rusty" feeding sites on the basal white tissue. Planting, flower initiation and close to fruit harvest are three stages in the crop cycle that should be observed. For effective monitoring, the timing of treatments is crucial.

The *Hirustella thompsoni* fungus has been used successfully to manage this mite. Recurrent spray application of Dicofol@4 ml/L at 500 L/ha based on need is suggested. Four sprays can be useful at fortnightly intervals at flower initiation stage.

Some other pests harbor on pineapple with limited or regional significance is listed in following table

Name of the pest	Scientific name	Major damage symptoms	References
Root-knot nematode	*Meloidogyne javanica*	A high incidence of terminal galls and branching are an indication of an early, but severe root-knot nematode infection. This results in a stunted root system with poor anchorage and plants are then more susceptible to nutrient and moisture stress.	Joy *et al.*, 2012; Sipes *et al.* 2005
Root lesion nematode	*Pratylenchus brachyurus*	Black, necrotic lesions may be visible where the nematodes penetrate but can extend across the entire root surface. In severe stage, plants can be devoid of secondary roots and consist only of poorly developed primary roots	Sipes *et al.* 2005
Reniform nematode	*Rotylenchulus reniformis* Linford and Oliveira.	They attacks lateral and feeder roots but remaining roots are elongate normally. Thus, it also develops same symptoms except lodging.	Pires de Matos, 2017
Symphylids (very small arthropods)	*Hanseniella sp.*	Symphylids feed on pineapple roots and root hairs inciting symptoms similar to those caused by nematode feeding in and on the roots. Roots of affected plant develop witches-broom symptoms and in severe case the plant ultimately dies.	Pires de Matos, 2017
Sugarcane midget	*Elaphira nucicolora* (Guenée)	Larvae infest both inflorescence and fruit. Galleries in fruit peel and gum like exudates observed for its infestation. They also lead to deformation of fruit.	Py *et al.*, 1987; Sanches, 1999
Bud moth	*Opogona sacchari* (Bojer)	Larvae feed on decaying plant and also attacks of leaves. Further, it bores into fruit peel and damages extensively, as a result gum like substances exudes. Ultimately at severe stage, plant becomes perished.	Davis *et al*, 1990; Vorsino, 2005
White grubs	*Phyllophaga* spp.	Grubs are mostly feed on the root and thus plants shows wilting, stunting, yellowing and ultimately become dry.	Joy *et al.*, 2012

Name of the pest	Scientific name	Major damage symptoms	References
Fig beetle	*Cotinis mutabilis* (Gory and Percheron)	Larvae feed on root parts but adults prefer flower parts like pollen, nectar, petal as their feed matter. Due to severe infestation of larvae, plants become wilted.	Camino-Lavín *et al.*, 1996
Pineapple weevil	*Diastethus bromeliarum* Champion	Grubs result in major infection; they mainly feed on inner tissues of the flower stalk. Due to its infection crown initiation hampered and gelatinous substances found in infected areas.	Barbra and Howard, 2010

References

Ali, M. M., Hashim, N., Abd Aziz, S. and Lasekan, O. 2020. Pineapple (Ananas comosus): A comprehensive review of nutritional values, volatile compounds, health benefits, and potential food products. Food Research International, 109675.

Alvarez, R. A., Martin, R. R. and Quito-Avila, D. F. 2015. First report of Pineapple mealybug wilt associated virus-1 in Ecuador. New Disease Reports, 31(1), 15-15.

Barbra, L. and Frank, J. H. 2010. Eeny-161 (IN318), one of a series of Feat ured Creatures from the Entomology and Nematology Department, Florida Cooperative Extension Service, Institute of Food and Agricultural Sciences, University of Florida. (http://entomology.ifas.ufl.edu/creatures)

Bello, A., Otiniano, J. A., and León, V. H. 1997. Mancha de la fruta de piña tipo galerias asociada a Melanoloma canopilosum Hendel. In: Martin-Prével, P. and Hugon, R. (eds). Proceedings of the 2nd International Pineapple Symposium, Trois- Ilets, Martinique, 20–24 February 1995. Acta Horticulturae, 425: 493–500 (Spanish).

Borroto, E. G., Cintra, M., González, J., Borroto, C. and Oramas, P. 1992. First report of a closterovirus-like particle associated with pineapple plants (Ananas comosus cv. Smooth Cayenne) affected with pineapple mealybug wilt in Cuba. Plant Disease, 82(2), 263-263.

Camino-Lavín, M., Jiminez-Perez, A., Castrejon-Gomez, V., Castrejon-Ayala, F. and Figueroa-Brito, R. 1996. Performance of anew trap for melolonthine scarabs, root pests. South Western Entomologist.

Carter, W. 1968. Insects and Related Pests of Pineapple in Hawaii. Pineapple Research Institute of Hawaii, Honolulu. pp, 105.

Caswell, E. P., Sarah, J.L. and Apt, W. J. 1990. Nematode parasites of pineapple. In: Luc, M., Sikora, R.A. and Bridge, J. (eds) Plant Parasitic Nematodes in Subtropical and Tropical Agriculture. CAB International, Wallingford, UK, pp. 519–537.

Culik, M. P., and Ventura, J. A. 2009. New species of Rhinoleucophenga, a potential predator of pineapple mealybugs. Pesquisa Agropecuária Brasileira, 44(4): 417- 420

Davis, D. R., and Peña, J. E. 1990. Biology and morphology of the banana moth, Opogona sacchari (Bojer) and its introduction into Florida (Lepidoptera: Tineidae). Proceedings of Entomological Society Wash., 92: 593–618.

Hu, J. S., Gonsalves, A., Sether, D., and Ullman, D. E. 1992. Detection of pineapple closterovirus, a possible cause of mealybug wilt of pineapple. In I International Pineapple Symposium, Acta Horticulturae, 334: 411-416.

Jeppson, L. R., Keifer, H. H., and Baker, E. W. 1975. Mites Injurious to Economic Plants. University of California Press, Berkeley, California.

Joy, P. P., Anjan, R., and Soumya, K. K. 2012. Insect Pests of Pineapple and Their Management. Pineapple Research Station, Vazkhulam, Muvattupuzha, India.

Martínez, N. B. de, Rosales, C. J. and Godoy, F. 2000. La mosca del fruto la piña Melanoloma viatrix (Diptera: Richardiidae) nuevo insecto plaga en Venezuela. Agronomia Tropical, 50: 135–140.

Mau, R. F. L. and Kessing, J. L. M. 2007. Dysmicoccus brevipes. (Cockerell). The Crop Master Program, extension entomology and UH-CTAHR Integrated Pest Management Program.

Nyarko, J. and Asare-Bediako, E. 2019. First report of Pineapple mealybug wilt-associated virus-1 and-3 in Ghanaian pineapple. New Disease Reports, 40(18), 2044-0588.

Petty, G. J. 1978. H.14 Pineapple Pests: Pineapple Scale. Pineapple Series H: Diseases and Pests, Government Printer, Pretoria, Republic of South Africa, pp, 3-8.

Pires de Matos, A. 2017. Main pests affecting pineapple plantations and their impact on crop development. In IX International Pineapple Symposium, Acta Horticulturae, 1239: 137-146).

Py, C., Lacoeuilhe, J. J. and Teisson, C. 1987. The Pineapple: Cultivation and Uses. G.P. Maisonneuve et Larose, Paris.

Sakimura, K. 1937. A survey of host ranges of thrips in and around Hawaiian pineapple fields. Proceedings of the Hawaiian Entomological Society, 9: 415–427.

Sanches, N. F. 1999. Pragas e seu Controle. In: Cunha, G.A.P.d., Cabral, J.R.S. and Souza, L.F.S.d. (eds) Oabacaxizeiro: Cultivo, agroindústria e economia. EMBRAPA-SCT, Brasilia, Brazil, pp, 307–341.

Sanewski, G. and Scott, C. 2000. The Australian pineapple industry. In: Subhadrabandhu, S. and Chairidchai, P. (eds) Proceedings of the Third International Pineapple Symposium. International Society for Horticultural Science, Pattaya, Thailand, pp, 53–55.

Satyagopal, K., S. N. Sushil, P. Jeyakumar, G. Shankar, O. P. Sharma, D. R. Boina, S. K. Sain, N. S. Rao, B. S. Sunanda, Ram Asre, R. Murali, Sanjay Arya, Subhash Kumar, S. Lingaraju, V. K. Koshta, A. K. Awasthi, R. N. Ganguli, M. P. Thakur, A. S. Kotasthane, H. S. Yadava, B. Gangadhar Naik, S. Deshmukh, S. Gangopadhyay and B. R. Patel. 2014. AESA based IPM package for Pineapple. pp, 47.

Sether, D. M., Ullman, D. E., Hu, J. S. 1998. Transmission of Pineapple mealybug wilt associated virus by two species of mealybug (Dysmicoccus spp.). Phytopathology, 88(11): 1224–1230.

Sipes, S. B. 2000. Crop profile for pineapples in Hawaii, NSF Centre for integrated pest management, North Carolina State University.

Tanwar R. K., Jeyakumar P. and Monga D. Technical Bulletin. 2007. Mealy bugs and their management, National Centre for Integrated Pest Management, New Delhi. pp, 4-13.

Vorsino, A. E., Taniguchi, G. Y., and Wright, M. G. 2005. Proceedings of Hawaiian Entomlogical Society.

Waite, G. R. 1993. Pests. In: Broadly, R. H., Wassman, R. C. and Sinclair, E. R. (eds) Pineapple Pests and Disorders. Department of Primary Industries, Brisbane, Queensland, Australia

Wakman, W., Teakle, D. S., Thomas, J. E., and Dietzgen, R. G. 1995. Presence of a clostero-like virus and a bacilliform virus in pineapple plants in Australia. Australian Journal of Agricultural Research, 46(5), 947-958.

21

Ecofriendly Management of Insect Pests Infesting Pineapple

Manasij Das and Swarnali Bhattacharya

Department of Agricultural Entomology, Palli Siksha Bhavana
Visva-Bharati, Sriniketan, West Bengal

Pineapple is an important tropical fruit crop of India introduced by Portuguese from Southern Brazil early in the 16th century. It is a versatile fruit acting as a rich source of nutrients in many favourite recipes. In addition, it is an excellent source of vitamins, minerals, and antioxidants,highlighting its role as anti-inflammatory, digestion, and immunity support.It resembles epiphytes where roots are sensitive to poor soil aeration (Joy and Sindhu, 2012). In the present day, the demand for pineapple is increasing gradually in different parts of the country. However, insect-pests infestation turns out to be one of the major impediments in its quantum of production.

Pineapple is infested by several insect pests like mealy bug, scale insect, thrips, fruit borer, white grubs, bud moths, midges, fruit flies, termites etc. during its entire crop duration. However, Mealy bugs appeared as the key pest of pineapple while otherslike scale insects, thrips and fruit borer could cause economic damage to the crop only under favourable environment. Therefore, emphasis must be given for development of suitableecofriendly management techniques by proper prophylactic (cultural, mechanical, biological etc.) and protective measures to reduce the insect-pest population below the economic damage through proper monitoring and regular survey of the crop field.Thus, this chapter emphasizes on the important insect-pests infesting Pineapple along with their appropriate management options.

Insect Pests of Pineapple and Their Eco-Friendly Management

Pineapple Mealy Bug: *Dysmicoccus brevipes*, Family: Pseudococcidae, Order: Hemiptera

Description: This is a pink coloured, cottony, small, oval, soft-bodied sucking insect mainly remaining on the seed material (Bartholomew *et. al.*, 2003). Adult mealy bugs are elliptical in shape with 17 pairs of wax filaments (Tanwar *et. al.*, 2007). They are soft-skinned with waxy secretions giving the body surfaces chalky appearance. They also have white, waxy filaments of various lengths extending from the pleural margins of their bodies. In a single colony, their number of individuals can exceed 20. The first-instar crawlers are the most active and motilestage having body length of 0.6-0.7mm and move around the host plant seeking a place to settle down and feed from that place. Further honey dew secretion by them create attraction to the ants and thereby transport them to nearby areas. This honey dew invites secondary infection by *Capnodium sp.* causing sooty mould which indirectly hampers with the photosynthetic and respiratory activity of the plant. The adult females produce cottony sac holding golden-coloured eggs parthenogenetically.

Fig. 1: Pineapple mealy bug (*Source*: www.bugwood.org, retrievedon 14 Oct 2013)

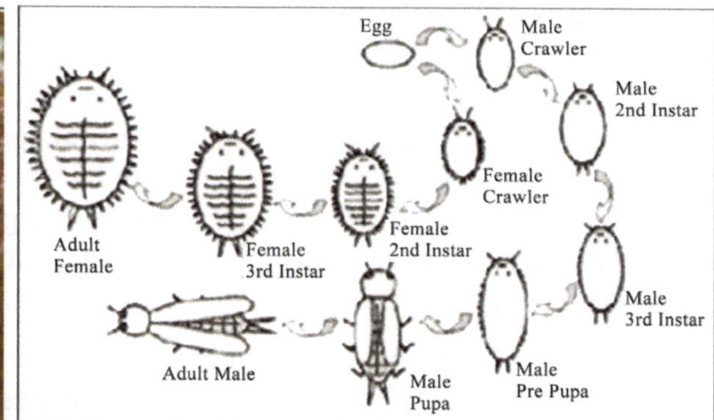

Fig. 2: Life cycle of mealy bug (Annecke and Moran, 1982)

Mealy bugs first appear on roots and infestation at this time is difficult to detect and manage. They feed from the root tissues thereby reducing the growth of the plant. Secondly, they are also capable of infesting the above ground part of the plant by sucking the plant sap and colonizing in the leaf axils and on developing fruit.

Biology: The mealy bugs of Pineapple are ovoviviparous i.e. the eggs hatch within the female and give birth to the nymph. The life cycle of the mealy bug includes three stages viz; egg, nymph and adult stage having life span upto 95 days depending upon the weather conditions (Ito, 1938).

Eggs: These are minute, and pink in colour.

Larvae: There are three larval/nymphal stages. The first instar nymph is known as crawlers and it is the dispersal stage of the insect since their body is extensively covered with hairs. The total nymphalduration ranges from 40-50 days depending upon the prevalent temperature. Larvae or nymph feed as a first instar and in the early part of second instar. The nymphs resemble the adults having white flattened bodies with waxy coating and long filaments. The nymphs at the later stage are carried by wind to newer locations.

Adults: Adults are represented by females only and they have soft, convex, pinkish body surrounded by 17 pairs of wax filaments. Lateral wax filaments are usually less than one fourth as long as the breadth of the body, and those towards the back of the insect are one-half as long as the body. The pre-larval oviposition period for adult females lasts around 27 days. The larvi position period lasts for an average of 25 days. They give birth to about 234 progenies but may produce up to 1000 crawlers. The adult females may then live for another 5 days before dying. Duration of adult female varies from 31- 80 days with an average of 56 days (Balachowsky, 1957). There may be as many as 15 generations. Adult males are having a distinguishable number of eight antennal segments. The species survives in cold as eggs or other stages, both on the host plant and in the soil. In warm climates, the insects stay active and reproduce round the year (Tanwar *et al.*, 2007).

Fig 3: (a) Mealy bugs infested pineapple sucker (b) Ant interaction with mealybugs (c) Mealy bug Infested Pineapple fruit (d) Pineapple Mealy bug Wilt Associated virus attacked pineapple field.
(*Source*: a. www.forestryimages.org, b. ucce.ucdavis.edu).

Nature of Damage: The pineapple mealybug damages pineapple in two ways viz; an indirect damage and a direct damage all of which lower the market value of the fruit. Direct feeding occurs when both the nymphs and adults suck sap from the plants and the plant shows wilting symptoms. Infestation due to mealy bug shows Quick wilt symptoms resulting in yellow or red leaves under heavy infestation, whereas slow wilt symptom with no prominent colour change under mild attack restricted to roots and rootlets. When fruits are infested, they become entirely covered with white, waxy coating making them unfit for human consumption. Leaves appear pale green to yellow streaks with brown tips. Feeding by the pineapple mealybug can weaken the plant, increasing susceptibility to other pests and diseases. Indirect damage by mealy bugs appears when due to their act of feeding they secrete copious amount of honey dew which results in the development of Black sooty mold and other molds. This mold ultimately hampers with the photosynthetic and respiratory activity of the plant. But another indirect damage by pineapple mealy bug posing a great threat for the pineapple growers is that they act as vectors of *Pineapple mealybug wilt-associated virus* commonly referred to as pineapple wilt, mealybug wilt, or edge-wilt (Sether *et al.*, 1998).

Management: Mealy bugs infest commercial pineapple plantings having impact on production in several ways (Beardsley, 1993). Their greatest impact is associated with the disease mealy bug wilt of pineapple (MW) (Rohrbach *et al.*, 1988). They may infest developing plants and fruit, thereby reducing fruit quality and quantity by influencing plant development (Carter, 1933). Mealy bug stripes appear in plants because of their feeding (Carter, 1967). Sometimes mealy bugs are found as contaminants in canned fruit and their presence on fresh market fruit may interfere with quarantine restrictions at port entrances. However, a sound management strategy is required against the pest to maintain good crop yield. The different strategies adopted are;

Prophylactic measures: Pre-planting treatments are effective in controlling pineapple mealy bugs. Pineapple cultivation using five to six years old suckers should be avoided. Selection of healthy suckers is advisable. Before planting, suckers can be dipped in botanical extracts (1200 ml/ 600 l of water/ha) or insecticidal solution for about 5 minutes (Carillo, 2011). Besides, destruction of all alternate host plants, sowing of ecologically engineered crops surrounding the field, sowing of barrier crop like maize or sorghum all around the field and ploughing deep will be effective in checking the quick spread of the pest in the field.

Biological control: There are several natural enemies associated with mealy bug under natural conditions which can be introduced in mealy bug infested fields for timely control.

a. *Cryptolaemus montrouzieri*: It is commonly known as the red headed ladybird beetle or the mealy bug destroyer. The predator was introduced in Karnataka to reduce large populations of pink mealy bug. The adult female ladybird beetle lays egg among mealy bug egg masses. The larvae of the beetle grow up to 1.3 cm in length and have woolly appendages of wax making them look like mealy bugs. The beetle larvae feed on mealy bug eggs and young crawlers. The life span of the *C. montrouzieri* is two months within which they lay up to 400 eggs. It is capable of eating upto 5,000 mealy bugs in its larval stage.

b. *Anagyruskamali and Anagyrusanatis*: This parasitoid was introduced in Karnataka from China for the control of pink mealy bugs. It feeds on mealy bug in two ways: (i) the female wasp punctures the bug and sucks the sap and (ii) it lays egg within the bug. When the egg matures it comes out of the bug's body. The entire process takes only half the entire life span of mealy bugs. A. anatis is also highly effective against the pink mealy bug (Gonzalez- Hernandez, 1995).

c. *Chrysoperlacarnea*: Release eggs or first instar grubs of Golden eyed insect @ 2-3 grubs per plant to predate upon all the stages of mealy bug.

d. Other predators prevalent in the field capable of reducing the mealy bug population in pineapple are; *Scymnus* sp., *Dicyphusheperus*, Dragonfly, Damselfly, spider, Robberfly, Preying mantid, *Chilomenessexmaculata* etc.

e. Follow conservation of natural enemies through Ecological engineering.

f. *Verticillium lecanii*/ *Beauveria bassiana*: Foliar spray of *Verticillium lecanii* (White Halo fungus) or *Beauveria bassiana* (White Muscardine fungus) @ ($2x10^8$cfu/ ml) at 5 g/ml per litre of water is effective during high humid months in reducing the populations of mealy bugs (Tanwar *et al*., 2007).

Cultural Control: Mealy bug spread can be minimized by the destruction of ant population. The mealy bug infested fields must be prepared by removing all the plant residues and burning them. Weeds present in the field should also be destroyed before sowing pineapple suckers to destroy the food source on which mealy bug population thrives. Removal of alternate hosts like *Hibiscus*, custard apple, guava in and nearby crop area is also essential. Use of sterile equipment is recommended for use in an uninfested field.

Chemical control: Need based chemical control of mealy bugs as a final option of management are to be followed.

- Application of Chlorpyriphos 20 EC (2.5ml/l) or Imidacloprid 17.8% SL (0.3 ml/l) at 500-600 l spray solution/ha.
- Ant colonies can be removed by soil drenching with Chlorpyriphos 20 EC @ 2ml/l of water.
- Indirect control of mealy bugs can be achieved by treating the soil either with 2.75 kg/ha of chlordane or heptachlor (2.25 kg/ha) to kill the ants.
- Basal portions of the planting materials have to be dipped in 0.02-0.04% methyl parathion as a prophylactic measure (megapib.nic.in/pppineapple.htm.)

Management of Pineapple wilt disease

- Use planting materials from wilt free areas or from fields with a low level of wilt disease
- If 10% plants show wilt symptoms do not use the field as a source of planting material
- Eradicate badly affected areas immediately after harvest
- Keep main fields and field boundaries free of weeds and trash which may act as reservoirs for ants and mealybugs
- Use of suitable insecticides to manage the insect vector.

Scale Insects: *Diaspis bromeliae* (Kerner), Family: Coccoideae, Order: Hemiptera

The pineapple scale insects are verysmall measuring 1–2 mm and found on upper surfaces of pineapple leaves and fruit worldwide (Waite, 1993). It has varying size and shape like oystershells, musselshells and to shiny pearl-like objects (about 5 mm). Adult female scales area lmost always immobile and permanently attached to the plant secreting a waxy coating for defence.which resemble reptilian scales or fish scales. Yellow spots may develop on leaves when scale densities are low (Waite, 1993). Other scales have been reported infesting pineapple but they do not cause severe problem.

Fig. 4: Left: *Melanaspis bromeliae*, Right: *Diaspis bromeliae.* (*Source*: www.sel.barc.usda.gov).

The brown (or red) pineapple scale *Melanaspis bromeliae*, is similar in appearance to *Diaspis bromeliae*, but it is a chocolate-brown colour with an elevated centre (Carter, 1967). Nigra scale, also known as black coffee scale, *Parasaissetia nigra*, may be found on pineapple (Zimmerman, 1948). The group shows high degrees of sexual dimorphism *i.e.* female scale insects retain the immature external morphology even when sexually mature, which is known as neoteny. Adultmales usually have wings (depending on their species) but never feed, and die within a day ortwo.

Biology: Scale insects being hemimetabolous have three distinct life stages (egg, nymph and adult) and maycomplete severalgenerations in a year. The first instar nymphs have functional legs and are informally called "crawlers" because immediately after emergence they crawl around insearch of a suitable spot to settle down and feeding untilmaturity. Sometimes, the young scale insects are associated with species of antsthat act as herdersand carry theyoung ones to protected sites tofeed. In many species ofscale insects the crawlers, after their first moult, become sedentary by losing their legs and start feeding on the same plant part. Only the malesretain their legs and use them in seeking females for mating. The reproductive systems of scale insects vary considerably within the group, including hermaphroditism and at least seven forms of parthenogenesis. Adult females produce eggs beneath the scale covering or in acottony material, and in many cases spend the cold winter months in this stage. Tiny six legged crawlers emerge from the eggs, move to newer growth on the plant, insert their proboscis and begin tofeed. Ascale-like covering produced from waxy filaments is formed over each individual scale. Adult scalespecies are identified by the colour and shape of the covering. The covering protects scales from predation by other insects and from insecticides.

The male scaleis often a slightly different shape than the female and passes through a tiny, winged stage which is seldom noticed. Females are wingless and usually remain in one place after inserting their mouth parts into plant tissues.

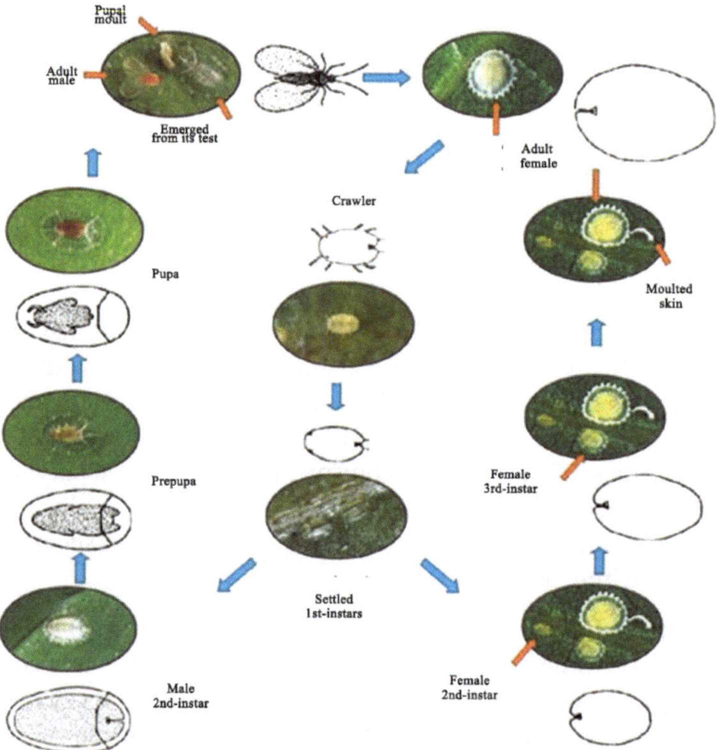

Fig. 5: Life Cycle of Scale Insects. (*Source*: www.landcareresearch.co.nz).

Nature of Damage: Scale insects are sap sucking phytophagous insects injecting toxic substances at the time of feeding.Development of rust coloured spots are noticed as an initial symptom. The insect is found beneath the secretion, which serves as a shield. The scale or protective armour is made up partly of a waxy secretion of the insect and partly of moulted skins. The insect itself in the adult stage is quite well buried beneath the epidermis of the plant and hence there is necessity of combating the pest in its early stages. During periods of large populations, some scales become so abundant that an infestedplant tissue is totally encrusted with insects. They are phloem feeders and secrete huge amount of honey dew rich in sugar at the time of feeding helping in development of mutualistic association with ants. Indirectly the honey dew favours development of sooty mold interfering with the photosynthetic and respiratory activity of the plant.

Management: This pest may be biologically controlled by natural enemies (Waite, 1993). Tinywasps, including *Aphytis chrysomphali* (Mercet), *Aphytis diaspidis* (Howard) and *Aspidiotiphagus citrinus* (Craw) (Hymenoptera: Aphelinidae) and different species of *Encarsia* sp., parasitize the scales, resulting in scale death (Zimmerman, 1948). Ladybirds, such as *Rhyzobius lophanthae* are active predators of scale insects (Carter, 1967; Waite,1993).

A light, superior oil/ entomopathogenic fungi like *Verticillium lecanii*/ insecticide mixture may be applied justas plants begin to grow to control emerging crawlers. The oil must be applied before leaves open, yet late enough so that this mixture will kill the crawlers. The best time to apply any insecticide is the crawler stage, as this stage does not havea protective covering, and is therefore vulnerable to almost most of the chemicals registered. Several conventional insecticides and insecticidal soaps are registered for crawler management. The plant must be thoroughly covered to kill the crawlers with one application.

Systemic insecticides, applied as a foliar spray, can help control adult scale insects during the growing season. The management of scale insects has to be under taken by knowing the biology, host preference, preferred sites of feeding, area of origin and natural enemies of different species of scale insects prevalent in the pineapple field.

Pineapple Thrips: *Holopothrips ananasi*, Family: Phaleothripidae, Order: Thysanoptera

Pineapple thrips is a major pest of pineapple which can cause serious damage to pineapple (Cavalleri and Kaminski, 2007). As many as 39 species of thrips have been reported worldwide in and around pineapple fields (Sakimura, 1937; Carter, 1939;Petty, 1978d). They are small (1.5 mm long), slender, brown insects with pale yellow hind wings that appear as a yellow line down the back of the body when the insect isat rest. Adult thrips have transparent fringed wings. *Thripstabaci, Frankliniella schultzei* are also considered as important pests of pineapple. They feed mainly on flowers and its feeding results in the development of“dead-eye” in the fruit. The crown of pineapple fruits develops concentric ring patterns.Thrips are mobile and can travel great distances with wind. They remain active all throughout the year and high moisture favours its activity.

Fig. 6: Left: *Thrips tabaci*, Right: *Frankliniella occidentalis*, (*Source*: ecoport.org, retrieved on 23 Oct 2013)

Biology: The life cycle of thrips is unique and fast.Females have a saw-like ovipositor that helps to make an incision in plant tissue particularly in the epidermis of the leaves and stem of young plants for egg laying. Eggs are elliptical, white, and placed singly. Theyare whitish at deposition and change to anorange tint as development continues. It hatches within 4 to 10 days. The newly hatched out nymphs immediately begin to suck sap from plant. Thereare two larval/nymphal stages which complete their development in about 9 days. Thereare twonon-feeding stages called the prepupa and pupal stage dwelling in the soil.The prepupal and pupal development completes within 4-7 days. Adults are 0.02 cm long, pale yellow to dark brown coloured body and wings are unbanded dirty grey incolour. Each female can produce up to 80 eggs thereby large populations can be generated within a short period. Mating for reproduction is not necessary. Females produce only female off spring without mating. The entire life cycle is estimated as three weeks producing 5-10 generations in a year.

Nature of Damage: Thrips are primarily active during the daylight hours and feed on the plant sap by damaging the leaves. They feed by piercing individual cells and lacerating thecontents due to which cells lose their normal colour, and when many adjacent cellsare damaged, the tissue appears as whitish spots or silvery spots. The typical appearance of the damage is asilvery-flecked leaf surface which in severe cases turns brown and their photosynthetic activity is reduced. Small black spots on the leaves along with insect excreta are the external identifying character of thrips infestation. Thrips are also responsible for the transmission of fungal and viral diseases in pineapple.

Management: Thrips are highly active during the dry season. Several cultural practices are followed to reduce the thrips population among which removal of weeds following crop rotation, mulching of crops reduces and application of timely & proper irrigation can be adopted. Establishment of wind breaks reduces thrips population because thrips movement is facilitated by wind. Recommended economic threshold for small scale farmers is when 20 per cent of the plants are infested with thrips.

Many botanicals can act as natural repellence to thrips like citronella, garlic and pyrethrum. A spray made of garlic and pepper reduces the population of thrips wherein, two bulbs of garlic and some hot chilli peppers are blended in water. The solution is then filtered out and 5 litres of water are added to this solution.This solution can be directly applied on plants to get better results. In another set of botanical preparation we can mix 2 kg of fresh plant material of *Andrographis paniculata* with 250 ml of water and grind it well. Then we can add 2 litres of cow urine and 10 g of crusheddried chilli fruits. Add 10 litres of water and leave the solution for some time. Filter the solution and it is ready for spraying. Wash fresh roots of *Derris eliptica* and cut them into short pieces of 5 cm length. Add small amount of water and pound the roots until they are finely shredded. Filter thesolution and dilute with soap and water at the ratio of 1 part soap: 4 parts root solution: 225 parts waters. The prepared solution should be applied immediately. A soap spray also kills thrips which needs to be repeated twice a week.

Under conditions of heavy infestations by thrips, chemical insecticides can be used maintaining the proper rules for pesticide application.

Pineapple Fruit Borer: *Strymon megarus* Godart Family: Lycaenidae, Order: Lepidoptera

Pineapple fruit borer was considered as one of the major pest of Mexico, Braziland then invading Argentina. However, it is present in all pineapple growing regions causing yield losses in almost all pineapple variety according to pest population and environmental conditions. Fruit affected from fruit borer show alteration in their shape and presence of galleries in the pulp. Recent evidence, however, has indicated that the insect is able to attack fruits even after the dry petal stage and to affect developing slips as well.

The adult is greyish coloured butterfly which can be found during the day ornight,flying in a rapid and haphazard fashion. After mating moths lay eggs which are white, circular and slightly flat and approximately 0.8 mm in diameter on the flower. Larvae complete their development inside the fruit. Burrowing and feeding activities produce visible damage in the form of frass production and a sticky, gummy exudate. The reddish coloured caterpillar remains in the tissue for 15 days then undergo pupation at the base of the peduncle. The pupa is about 12 mm long and5 mm wide with a brown colour and a few dark spots. Adult emergences take place within 7 to 10 days.

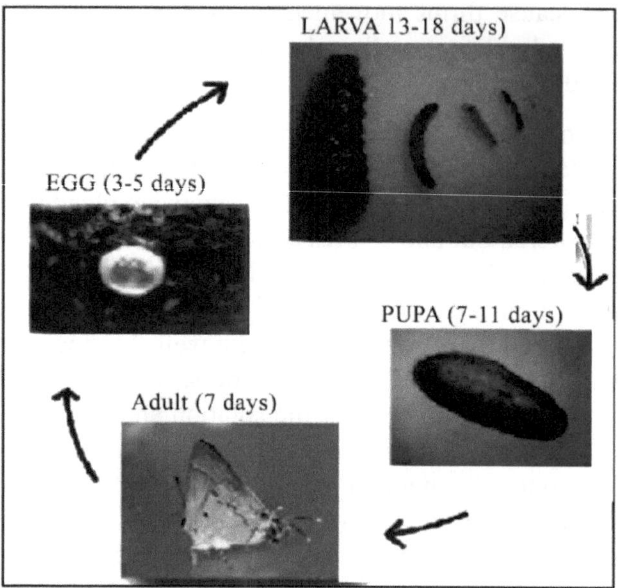

Fig 7: Life Cycle of Fruit Borer (*Source*: www.cep.unep.org, retrieved on 23 Oct 2013)

Damage: The larvae bore into the fruit causing holes and uneven fruit development. It destroys the tissues of the inflorescence due to which a resin-coloured liquid gum isexuded from the fruitlets. After coming in contact withair, the exuded gum becomes reddish in colour and later it solidifies in dark brown coloured flake. Damage from this pest varies greatly that can reach up to 90 per cent and dry climates seem tofavour borer attack. In most cases fruit borer attacks occur during flowering and fruit formation stage.

Management: Among the microbial control agent, *Bacillus thuringiensis* appears to be an acceptable biological alternative to carbaryl for control of *S. megarus* on pineapple. However, application of Chlorpyriphos 20EC @ 2.5 ml per litter of water has been proved to be very effective for the management of this pest.

Other Eco-friendly Management Practices that Can be Adopted for PestManagement in Pineapple

- Dilute one litre of cow urine in ten litres of water and wet the whole plant at the rate of 200-300 l/ha at regular intervals.
- 12.5 kg of fresh cow dung and 12.5 litres of cow urine are collected in a near then pot and mixed thoroughly with 12.5 litres of water. The pot is covered and the mixture is allowed to ferment for a week. Occasionally

it is stirred with a stick. After a week of fermentation, the mixture is filtered and 100 g of lime is added. The concentration is diluted with water in a 1:10 ratio and sprayed on the crop at 200–250 l/ha.

- Neem Oil Spray: Neem oil spray: 2 per cent neem oil is mixed with any detergent powder at 40–50 g/100 l and used as a spray solution.
- Neem Seed Kernel Extract (NSKE) can be used as a prophylactic measure before the on set of pests.
- Spraying of chili garlic spray is an effective insect repellent.
- Herbal Mixture Spray: About 500 g of tobacco leaves, 1 kg of neem kernel, 500 g limepowder, 500 g *Datura* leaves and 500 g pods and seeds of oleander (*Nerium oleander*) arepowdered and mixed together, then soaked in 15 litres of water for 15 days. On alternate days, the mixture needs to be stirred witha stick. After 15 days, one litre of filtrate is mixed in 15 litres of water and sprayed on the crop. It is enough for 2.5 ha and is a multi-pest repellent.
- Tobacco tea - Tobacco tea is effective against most pests.

References

Annecke, D. P. and Moran, V. C. 1982. Insects and mites of cultivated plants in SouthAfrica. Butterworths, Durban/Pretoria.

Arevalo, E. A., and Osorio, M. 1995. General Considerations viatrix Melanoloma Hendela new pest of pineapple. Journal of Entomology, 21 (1): 1-8.

Barbra, L. and Frank, J. H. 2010. EENY-161 (IN318), one of a series of Featured Creaturesfrom the Entomology and Nematology Department, Florida CooperativeExtension Service, Institute of Food and Agricultural Sciences, University of Florida. (http: //entomology. ifas.ufl.edu/creatures).

Bartholomew, D. P., Paull, R. E. and Rhrbach, K. G. 2003. The pineapple botany,production and uses. CABI Publishing. New York, United States. pp, 301

Bearsdsley, J. W. 1993. The pineapple mealy bug complex; taxonomy, distribution,and host relationships. In: First International Pineapple Symposium, D. P.Bartholomew and K. G. Rohrbach, (eds), Acta Horticulture, 334: 383–386.

Bello, A., Otiniano, J. A., and León, V. H. 1997. Mancha de la fruta de piña tipogaleria sasociada a Melanoloma canopilosum Hendel. In: Martin-Prével, P. andHugon, R. (eds). Proceedings of the 2nd International Pineapple Symposium, Trois-Ilets, Martinique, 20–24 February 1995. Acta Horticulturae, 425: 493–500 (Spanish).

Billen, W. 1987. Information on the banana shoot borer Opogona sacchari. Gesunde Pflanzen, 39: 458-465.

Broughton, S. and Lima, De F. 2002. Monitoring and control of Thrips, Farmnote Department of Agriculture Western Australia.

Camino-Lavín, M., Jiminez-Perez, A., Castrejon-Gomez, V., Castrejon-Ayala, F. And Figueroa-Brito, R. 1996. Performance of a new trap for melolonthine scarabs,root pests. South Western Entomologist.

Carillo, E. V. 2011. Guide for integrated pest identification and management inpineapple, pp, 6-8.

Carter, W. 1933. The pineapple mealy bug, Pseudococcus brevipes (Ckl.) and wilt ofpineapple. Phytopathology, 23: 207–242.

Carter, W. 1948. The effect of mealy bug feeding on pineapple plants grown in finelyatomized nutrient solution. Phytopathology, 38: 645–657.

Carter, W. 1962. Mealy bug wilt of pineapple; a reappraisal. Ann. N.Y. Academia Science, 105:741–764.

Carter, W. 1967. Insects and related pests of pineapple in Hawaii. A manual for fieldmen. Pineapple Research Institute (restricted publication).

Cavalleri, A. and Kaminski, A. L. 2007. A new Holopothrips species (Thysanoptera: Phlaeothripidae) damaging Mollinedia (Monimiaceae) leaves in Southern BrazilZootaxa. 1625: 61–68.

Cave, R.D.1997. Admontia sp., a potential biological control agent of Metamasius callizona. Journal of Bromiliade Society, 47: 244-249.

Culik, M. P., and Ventura, J. A. 2009. New species of Rhinoleucophenga, a potential predator of pineapple mealybugs. Pesquisa Agropecuária Brasileira, 44(4): 417-420.

Frank, J. H., Cave, R. D. 2005. Metamasius callizona is destroying Florida's native bromeliads. In: Hoddle MS. USDA Forest Service Publication FHTET-2005-08.Vol1. Second International Symposium on Biological Control of Arthropods;2005 Sep 1216; Davos, Switzerland. Washington D.C.: USDA Forest Service. pp, 91-101.

Gowdey, C. C. 1922. The principal Agricultural Pests in Jamaica. Dept. of Agriculture Jamaica, Entomological Bulletin, 2: 68.

Heppner, J.B., Pena, J.E., Glenn, H. 1987. The banana moth Opogonasacchariin Florida. Entomology Circular No. 293. Division of Plant Industry, Florida Department of Agriculture and Consumer Services, Gainesville, USA.ipm.ait.asia, retrieved on 5Sep 2013.

Ito, K. 1938. Studies on the Life Histories of the Pineapple Mealy bug, Pseudococcus brevipes (Ckll.). Journal of Economic Entomology, 31(2): 291-198.

Jeppson, L. R., Keifer, H. H., and Baker, E. W. 1975. Mites Injurious to Economic Plants. University of California Press, Berkeley, California.

Joy P. P. and Sindhu G. 2012. Diseases of Pineapple (*Ananaus comosus*): Pathogen, symptoms, infection, spread and management. Pineapple Research Station,Vazhakulam, Muvattupuzha, India.

Lacey, L. A., Frutos, R., Kaya, H. K. and Vail, P. 2001. Insect pathogens as biological control agents: do they have a future? Biological Control, 21: 230-248.

Le Roux, W. 1992. Pineapple Production in the Eastern Cape: a Technical Manual.Gem Instant Print, Cape Town.megapib.nic.in/pppineapple.htm., retrieved on 11 Oct 2013.

O'Brien, C. W. 1994. Two new species in the Cholusspinipes group (Cholini, Curculioninae, Curculionidae). Transactions of the American Entomological Society, 120: 412–421.

Petty, G. J. 1977a. Beetle pests of pineapples: some biological aspects. Citrus, Subtropical Fruit Journal, 529: 4–7.

Petty, G. J. 1977b. H.2 Pineapple Pests: Leathery Pocket in Pineapples. Pineapple Series H: Diseases and Pests, Government Printer, Pretoria, Republic of South Africa, pp.3.

Petty, G. J. 1978b. H.14 Pineapple Pests: Pineapple Scale. Pineapple Series H: Diseases and Pests, Government Printer, Pretoria, Republic of South Africa, pp, 3.

Petty, G. J. 1978d. H.17 Pineapple Pests: Thrips. Pineapple Series H: Diseases and Pests, Government Printer, Pretoria, Republic of South Africa, pp, 8.

Petty, G. J. 1990. Effect of endosulfan on leathery pocket disease in pineapples, and evidence for the association of the mite, Steneotarsonemus ananas (Acarina: Tarsonemidae) with leathery pocket. In: Robinson, J.C. (ed.) International Symposium on the Culture of Subtropical and Tropical Fruits and Crops. International Society of Horticultural Science, Wageningen, pp. 663–670.

Py, C., Lacoeuilhe, J. J. and Teisson, C. 1987. The Pineapple: Cultivation and uses. G.P. Maisonneuve et Larose, Paris.

Sakimura, K. 1937. On the bionomics of Thripoctenus brivuillet, a paradisite of Thripstabaci Lind. in Japan. Kontyu Vol. XI.5, pp, 370-390.

Salas, J. and O'Brien, C. W. 1997. Cholusvaurieae O'Brien (Coleoptera: Curculionidae), nuevaplaga de la piña enelestado Lara, Venezuela. Boletin de Entomologia Venezolana, 12: 157–158.

Sether, D. M., Ullman, D. E, Hu, J. S. 1998. Transmission of Pineapple mealybug wilt associated virus by two species of mealy bug (*Dysmicoccus* spp.). Phytopathology, 88 (11): 1224-1230.

Sherwood, M. 2010. The pineapple weevil (Metamasius ritchiei Marshall; Coleoptera: Curculionidae), Entomology Circular Ministry of Agriculture and FisheriesResearch and Development, Jamaica.

Singh, S. J. and Sastry, S. M. 1974. Wilt of pineapple, a new virus disease in India. Indian Phytopathology. 27: 298–303.

Sipes, S. B. 2000. Crop profile for pineapples in Hawaii, NSF Centre for integrated pest management, North Carolina State University.

Stone, M. W. 1982. The peach beetle, Cotinis mutabilis (Gory and Percheron) in California (Coleoptera: Scarabaeidae). Pan-Pacific Entomologist, 58: 159–161.

Tanwar R. K., Jeyakumar P. and Monga D. Technical Bulletin.Sep. 2007. Mealy bugsand their management, National Centre for Integrated Pest Management, NewDelhi. 19: 4-13.

Veenenbos, J. A. 1981. Opogona sacchari, a pest risk from imports of ornamental plantsof tropical origin. Bulletin OEPP/EPPO, 11: 235-238.

Vorsino, A. E., Taniguchi, G. Y., and Wright, M. G. 2005. Proceedings Hawaiian Entomlogy Society, Waite, G.R. 1993. Pests. In: Broadly, R.H., Wassman, R.C. and Sinclair, E.R. (eds) Pineapple Pests and Disorders. Department of Primary Industries, Brisbane, Queensland, Australia.

Wood, D. M. and Cave, R. D. 2006. Description of a new genus and species of weevil parasitoid from Honduras (Diptera: Tachinidae). Florida Entomology, 89(2): 239-24.

Zimmerman, E. C. 1948. Insects of Hawaii. A Manual of the Insects of the Hawaiian Islands Including Enumeration of the Species and Notes on the Origin, Distribution, Hosts, Parasites, etc. University Press of Hawaii, Honolulu, Hawaii.

22

Diseases of Pineapple and Management

Pradipto Kumar Mukherjee

Department of Horticulture and Postharvest Technology, Institute of Agriculture, Visva-Bharati, Sriniketan, West Bengal, India

Pineapple (*Ananas comosus*) is an edible fruit bearing plant native to South America belonging to family Bromeliaceae. Pineapple is grown over an area of 1.06 million hectares with an annual production of 29.36 million metric tonnes (FAO, 2022). Pineapples are rich in vitamin C, minerals, fiber and low in calories. The enzyme bromelain present in pineapple is of considerable importance to the pharmaceutical and cosmetic industry (Varilla *et al.*, 2021). The market for pineapples has been expanding significantly because of its appealing aroma compounds and nutritional benefits. The climate and distribution of rainfall in tropical and sub-tropical regions make these areas ideal for pineapple cultivation. Worldwide, Brazil is the leading producer of Pineapple followed by Philippines, Thailand, Costa Rica and India (FAO, 2022). In India the leading pineapple producing states are West Bengal, Northeastern region, Bihar, Kerala, Karnataka, Goa, and Maharashtra.This 'Kew' of Cayenne group and 'Mauritius' of the rough leaf Queen group are the two varieties dominating in India (Devi *et al.*, 2013; Joy and Sindhu, 2012).

Botanically, Pineapple is a herbaceous perennial having short stem and tough waxy leaves. It produces a fleshy, edible, yellow fleshed which is a fusion of 100-200 berry like fruitlets. Most of the members of Bromeliaceae family are epiphytes *i.e.* grown on rocks and living trees though pineapples grow on the soil bearing sensitive adventitious root that cannot regenerate once damage and intolerant to poor soil aeration. The cultivated varieties of pineapple can be categorized into three groups namely, Cayenne, Queen, and Spanish. The Spanish and Cayenne groups can be cultivated for dual purposes *viz.* fresh fruit and canning. However, the Queen group are only cultivated for fresh fruit markets due to the presence of deep eyes making it unsuitable for canning (Devi *et al.*, 2013).

Pineapple is affected by several abiotic and biotic stresses causing considerable decrease in yield and quality. Amongst biotic stresses, diseases caused by fungi, bacteria and viruses cause stresses, considerable economic impact to pineapple production. *Phytophthora* heart and root rot, base rot, and mealy bug wilt are important disease affecting pineapple. Diseases like black rot, fruitlet core rot, yeasty fermentation and pink disease are the major postharvest losses in pineapple (Baiswar *et al.*, 2021). The details of the diseases affecting pineapple are given below.

Diseases affecting pineapple

Fungal Diseases		
S.No.	**Disease**	**Causal organism**
1	Phytophthora heart rot	*Phytophthora cinnamon* and *Phytophthora nicotianae*
2	Phytophthora root rot	*Phytophthora cinnamoni*
3	Base rot	*Ceratocystis paradoxa*
4	Black rot	*Ceratocystis paradoxa*
5	Fruitlet core rot	*Fusarium annatumand Talaromyces funiculosus*
6	Fusarosis	*Fusarium guttiforme*
7	Green fruit rot	*Phytophthora cinnamoni*
8	Interfruitlet corking	*Penicillium funiculosum*
9	Leathery pocket	*Penicillium funiculosum*
10	White leaf spot	*Ceratocystisparadoxa*
11	Yeasty disease	*Hanseniaspora valbyensis, Saccharomyces* sp. and *Candida* sp.
12	Leaf Spot Disease	*Exserohilium rostratum, Curvularia clavata*
13	Stem end rot	*Neoscytalidium dimidiatum*
Bacterial diseases		
1	Bacterial heart rot	*Dickeyazeae*
2	Fruit collapse	*Dickeyazeae*
3	Pink disease	*Pantoea citrea*
4	Marbling	*Pantoea ananatis* and *Acetobacter* sp
Viral Diseases		
1	Mealybug wilt	*Pineapple Mealybug Wilt-associated Virus* (PMWaV)
2	Yellow Spot	*Capsicum Chlorosis Virus, Tomato Spotted Wilt Virus*

Fungal Diseases

Phytophthora heart (top) rot and root rot

Heart rot causal organism: *Phytophthora nicotianae* and *Phytophthora cinnamoni.*

Root rot causal organism: *Phytophthora cinnamoni*

Etiology: It belongs to Phylum Oomycota commonly known as water molds. It produces non-septate mycelium and asexual fruiting body called sporangia which is ellipsoid, ovoid, pyriform, obpyriform, or spherical in shape (Hardham and Blackman, 2018). The sporangia contain zoospores which are kidney shaped motile spores having anterior tinsel and posterior whiplash flagella. The fungi are mostly heterothallic and require two mating types for formation of oospore.*P. nicotianae* causes heart (top) rot only, whereas *P.cinnamoni* can cause both heart rot as well as root rot symptoms.

Heart rot symptoms: The typical symptoms are characterized by yellow-red leaves that detach easily from the plant. Thereafter the leaves show wilting and browning symptoms and eventually die. The white leaf bases too display a distinct water-soaked grey-brown lesion which gradually extends from the leaf's base to its tip. Additionally, the affected tissue emits a foul odor due to the invasion of secondary microorganisms (Joy and Sindhu, 2012; Sapak *et al.*, 2021).

Root rot symptoms: The plants can be easily pulled from the ground as the roots die due to necrosis and rotting. The above ground symptoms are not distinctive and are manifested by the change of leaf color from green to various shades of yellow and red. Reduction in plant growth and browning of leaf margins is also seen in this disease. The fruits from the infected plants are small and have less marketable value (Joy and Sindhu, 2012).

Disease cycle: This disease is soil-borne, and the pathogen can survive for several years as chlamydospore. The primary source of inoculums is mostly chlamydospore and to a lesser extentoospore which remains dormant in soil for several years (Rohrbach and Johnson, 2003). The chlamydospore germinates under favorable condition forming a multinucleate sporangium which then undergoes cytoplasmic cleavage giving rise to 20-30 biflagellate motile zoospore. On encountering the host tissue, the zoospore develops into a cyst by shedding its flagella and gets firmly attached to the surface of the host plant. The cyst produces germ tube which penetrates the plant and forms extensive mycelium in the process of acquiring nutrients. Once inside the host the asexual cycle continues wherein numerous sporangia are produced eventually giving rise to zoospores. These zoospores act as secondary source of inoculum and can be easily spread by rain splashes and irrigation water. Sexual reproduction occurs when both the mating types A1 and A2 are present (Panabières *et al.*, 2016).The oogonia and antheridia are produced on separate thallus which on fusing forms the oospore. The oospore overwinters in soil and on encountering appropriate host germinates giving rise to zoospores.

Epidemiology: This disease is mostly seen in soils having soils having pH above 4.0. *Phytophthora* being a soil inhabiting fungus requires water for spore germination and infection. Heavy rainfall after planting, poor drainage and low temperature aggravate the disease situation. Zoospore production occurs at a temperature of 10-24^{0}C. Generally, pineapple plants of 1–2-year age are more susceptible to the disease. *P. cinnamomi*is prevalent in the cool, wet soil at upper elevations, whereas *P. nicotianae* is found in a wider range of soil conditions (Rohrbach and Apt, 1986). The root rot symptoms are severe in high rainfall areas and in ratoon crop. The rough leaved and low acid varieties are generally more susceptible to *Phytophthora* than the Smooth Cayenne variety.

Management: Appropriate site of planting having pH less than 4.0 must be chosen for pineapple plantation. Proper sanitary practices must be followed to prevent the occurrence of the disease. Use of clean planting material, sanitized farming tools and ensuring good drainage are essential for managing this disease. Deep planting of pineapple suckers must be avoided as it increases the risk of infection. Also, the infected plants must be destroyed and burnt to prevent the spread of the pathogen. Systemic insecticides like metalaxyl and Fosetly AL @ 0.1% are the most effective in combating this pathogen (Rohrbach and Schenck, 1985). Chemicals may be applied as pre-plant dips, foliar applications, and soil drenches etc. for reducing the disease.

Base (But) rot and Black rot

Causal organism: *Ceratocystis paradoxa* (Telomorph-Sexual stage)

Etiology: The anamorph (asexual stage) of this fungus is denoted as *Thielaviopsis paradoxa* or *Chalara paradoxa*. It belongs to Kingdom fungi, Phylum ascomycota and produces well branched septate hyphae. This fungus produces thick-walled restingspores called chlamydospores.It produces two types of asexual spores, microspores, and macrospores. The microspores are cylindrical, hyaline and are smaller in size. The macrospores are brown, oval shaped and larger in size. The sexual spore is represented by ellipsoidal ascospores which are produced on long necked perithecia.

Base (But) rot symptoms: Rotting of the butt tissue leads to development of cavity due to dissolution of the soft tissue leaving behind only the stringy fibers. The symptoms can be observed on suckers, crown or slip before or immediately after planting and the infected plants are easily broken off from the ground level. Rotting of the butt tissue leads to reduced plant growth and less yield. Severe cases lead to rapid wilting, death of leaf tissue and development of black lesions in the fruit. The characteristic symptom differentiating base rot from *Phytophthora* rot is that the young leaves stay intact firmly at the top of the stem. This pathogen attacks the soft butt tissue causing a grey to black rot to the area (Joy and Sindhu, 2012).

Black rot symptoms: It also called as water blister of pineapple which is a post-harvest disease. The pathogen enters the fruit through broken peduncle and forms circular and water-soaked spots. Due to development the symptoms on the fruit flesh, the overlying skin become glossy, water-soaked and develops black patches.

Disease cycle: The pathogen overwinters as chlamydospore in the infected planting material and acts as primary source of inoculum. It enters the plant through fresh wounds and develops extensive mycelium in the host tissue. The conidia are formed which can then act as secondary source of inoculum and spreads by wind or rain (Sapak *et al.*, 2021).

Epidemiology: The occurrence of the disease is higher in warm and wet season. High humidity favors the formation of conidia which then disseminates to surrounding areas by wind. Improper storage of the fruit or planting material in damp conditions makes it vulnerable to the disease. Prolonged wet period during planting season increases the chances of the disease.

Management: The planting material must be cured in sun for 2-3 days before planting in the field to reduce the incidence of the disease. Proper sanitation practices must be followed in the plantation site. Systemic fungicides like Propiconazole or Hexaconazole @ 0.1% can be sprayed in the field. The fruits can be dipped in benomyl (0.2%) or thiabendazole (0.1%) solution after harvest for three minutes. The fruits must be handled carefully, and bruising must be avoided.

Fruitlet Core Rot (Green eye)

Causal organism: *Fusarium annatum* and *Talaromyces funiculosus* (Previous name: *Penicillium funiculosum*)

Etiology: Two genera, namely *Fusarium* and *Talaromyces* belonging to phylum ascomycota are believed to cause fruitlet core rot. Earlier *Fusarium guttiforme* was thought to be the causal agent of fruitlet core rot, however molecular analysis revealed *F. annatum* as the main causal agent (Jacobs *et al.*, 2010; Petty *et al.*, 2006).*F. oxysporum*, *F. proliferatum* and *F. verticilloides* have also been reported in this disease complex. *Fusarium sp.* produces chlamydospore which serves as perennating structure. Sporodochium is produced which is an asexual fruiting body consisting of conidiophore and two kinds of conidia i.e. macro conidia (sickle shaped) and micro conidia (nearly oval). These conidia are responsible for perpetuation and spread of the disease.

Penicillium funiculosum was earlier reported in this disease. However, molecular analysis led to the transfer of this fungus to genus *Talaromyces* (Samson *et al.*, 2011). *Talaromyces funiculosus* produces enormous conidia in a long chain on a branched conidiophore. *Talaro mycesstollii* has also been recently found associated with this disease (Barral *et al.*, 2020).

Symptoms: The fruitlets show brownish discolouration and become sunken during ripening. Internal browning of the centre of the fruit takes place which starts just below the floral cavity and sometimes may extending up to the core (Rohrbach and Johnson, 2003). Browning discoloration remains quite firm inside the fruitlets but varies in size from a speck to complete discolouration. The browning is due to the oxidation of phenolic compounds into quinones by the enzyme's polyphenol oxidase and laccase. Smooth Cayenne fruits do not usually show any external symptoms. The fruit of the rough-leaf (Mauritius) variety may produce fruitlets that fail to colour (green eye).

Disease Cycle: The chlamydospores of *Fusarium* are present in soil and in infected crop debris serving as primary source of inoculum. The pathogen takes entry through the open flowers or injury sites. The risk of disease establishment is high when the crop encounters warm condition specifically from flower initiation to fruit maturity. The hyphae enters the tissue forms extensive mycelium and finally the sporodochia bearing the two kinds of conidia. These conidia can be spread by rain splashes to nearby plants. Air borne conidia of *Talaromyces* also serve as the source of infection for fruitlet core rot (Vignassa *et al.*, 2021). The mycelium proliferates in the tissue and produces numerous conidia which can be spread by air, water, and rain splashes.

Epidemiology: Cool temperature of 16–20 °C during the five weeks after flower initiation favours infection. Mite damage causes fungal build up in the leaf hairs. Infection occurs between flower initiation and open flower stage.

Management: Proper agronomic practices should be followed as recommended. Being an internal fruit disease, external application of fungicides after establishment of the disease is not effective. Fungicides should be applied as protective spray on the opening of the terminal leaves at the time of emerging-inflorescence.

Fusarosis

Causal organism: *Fusarium guttiforme*

Etiology: *Fusarium guttiforme* has been reported as the causal agent of fusarosis and is a part of *Fusarium fujikuroi* species complex (FFSC). It produces chlamydospores and two types of conidia micro conidia and macro conidia.

Symptoms: This pathogen can affect all parts of the plant and exudation of gum from infected tissues can be seen. Due to girdling and killing of the stem tissue rosetting and curvature of the plant occur. Soft rot of the flesh, gum accumulation in the locules of the ovary and gum exudation are the characteristic symptoms seen in the fruit (Sapak *et al.*, 2021).

Disease cycle: Infected planting materials like suckers, slips and crown act as a primary source of inoculum and when get introduced in the new sites and establishes infection. Soils can also harbour the chlamydospores of the pathogen for several months. The entry of the pathogen takes place during the flower initiation stage thereafter forming conidia which then serves as secondary source of inoculum.Abandoned plantings also serve as an important source of inoculums.

Epidemiology: Dry soil favours the survival of the pathogen. Optimum temperatures recorded for growth are 25°C, Injuries made by insects like pineapple fruit caterpillar aid in infection (*Thecla basilides*).

Management: Destruction of plant debris must be done. The rough leaf pineapple cultivars are more susceptible than smooth-leaf varieties. Hot-water treatment at 54°C for 90 min reduces pathogen infection. Fungicide applications at flower induction and three weeks after fruiting can reduce disease (Joy and Sidhu, 2012).

Green Fruit Rot

Causal organism: *Phytophthora cinnamoni*

Etiology: It is an oomycete pathogen forming zoospores as asexual spore and oospore as sexual spore. Details are discussed previously.

Symptoms: The initiation of the disease symptoms takes place in the green fruit which are in contact with the soil. Development of water-soaked spots in the interior takes place with no visible external symptoms. With the progression of the disease, water-soaked rot occupy greater area of green fruit with distinct brown margin (Joy and Sidhu, 2012).

Disease Cycle and Epidemiology: A temperature range of 10-24°C associated with wet or humid weather is congenial. The primary source of inoculum is present in soil as chlamydospores. Details discussed above.

Management: Following proper sanitation practices and destruction of crop debris minimizes the incidence of the disease. Clean planting material must be chosen for plantation. Use of systemic fungicides like metalaxyl can be done.

Interfruitlet Corking and Leathery Pocket

Causal organism: *Penicillium funiculosum*

Etiology: It belongs to phylum ascomycota and produces well branched septate mycelia. Numerous conidia are produced on conidiophores termed as phliades.

Inter fruitlet corking symptoms: Corky tissue develops between the fruitlets and shiny patches appear on the fruit during early development. In severe conditions the corkyness develops extensively in the fruit leading to malformed and underdeveloped fruit. The entry of the pathogen occurs through trichomes (hair) which is aided by mite feeding. Narrow, transverse cracks may develop on the sepals and bracts (Joy and Sidhu, 2012).

Leathery pocket Symptoms: In this case too no external symptoms are produced. Formation of corky tissue on the internal walls of the fruitlets makes them leathery and brown.

Disease cycle and epidemiology: Cool temperature of 16-20°C is congenial for the development of this disease. Sporadic occurrence of the disease takes place. The entry of the pathogen takes place between flower initiation and open flower stage. Thereafter the fungus proliferates in the host tissue and produces conidia.

Management: Fungicides are generally not effective as the disease is confined to the interior of the fruit. Inter fruitlet corking is often confused with boron deficiency. As mite infestation aggravates the disease application of miticides is beneficial.

White Leaf Spot

Causal agent: *Ceratocystis paradoxa* (Telomorph: Sexual stage)

Etiology: See Base (but) rot

White leaf spot symptoms: A small brown spot appears on the leaf at the site of where the leaves have been rubbed by strong winds. During wet weather the spots lengthen up till the tip and in dry weather these spots become cream coloured and paper(Joy and Sidhu, 2012).

Disease cycle and epidemiology: See Base (but) rot

Management: See Base (but) rot

Yeasty Disease

Causal organism: *Hanseniaspora valbyensis, Saccharomyces* spp. and *Candida spp.*

Etiology: These fungi belong to the phylum ascomycota. True mycelium is not formed and the thallus is represented by single cells. Asexual reproduction takes place by budding and sexual stage is represented by formation of ascospore.

Symptoms: It is a post-harvest disease affecting ripe fruits. The characteristic symptom of this disease is the formation of gas in the fruit tissue where the sugars are converted into alcohol and carbon dioxide. The fruits become hollow and with the formation of air pocket inside make the fruit spongy (Rohrbach and Johnson, 2003). Decaying flesh turns to bright yellow and develops large gas cavities. The shell then turns brown and leathery.

Epidemiology: Rapid changes in fruit growth occur during spring due to shift from cold and dry to warm and wet weather. This fluctuations in temperature results in skin cracking between fruit lets of pineapple. Fruit may also get affected by even minor frost damage (prone to cracking) as they ripen in spring. Yeast invades through the wounds and destroys the ripened fruits. The disease is severe just before or after harvest.

Management: Appropriate sanitary conditions must be maintained to avoid injury to the fruit. Over ripening of the fruits under field conditions during spring should be avoided. Fruit that even slightly exhibits inter fruitlet cracking shouldn't be sold fresh. To reduce losses, fruit with fractures between fruitlets should be picked as soon as possible after it reaches maturity.

Bacterial Diseases

Bacterial heart rot and fruit collapse

Causal organism: *Dickeyazeae* (Previously, *Erwinia chrysanthemi*)

Etiology: It is a gram-negative bacterium belonging to Phylum Proteobacteria, Class Gamma proteobacteria. It is a motile and facultative anaerobic bacterium which multiplies by binary fission.

Heart rot Symptoms: Water-soaked lesion on the white basal portion of the central whorl is the initial symptom indicative of bacterial heart rot. Blister like lesions filled with gas appear in the lamina and mesophyll tissues which is a typical symptom of this disease. Infected leaves show olive green and puffy appearance. With the progression of infection a light brown exudate comes out of the blisterturning leaves to light brown to dark brown in colour. 1–2 weeks after initial symptoms the pineapple heart and stem can be easily detached from the lower portion of the plant around. About 21 days after the initial infection, the whole infected area on the leaf eventually rots. Fungal heart rot canbe distinguished from bacterial heart rot, based on the progression of infection beyond the basal region. In the case of fungal heart rot, the infection does not extend beyond the white basal portion into the mature green areas

Fruit collapse symptoms: The symptoms are visible on the mature fruits 2-3 days before ripening. The fruit collapse disease is characterized by exudation of juice and foul-smelling gas because of fermentation. With the progression of disease the pineapple fruit shell changes colorto olive green.

Disease cycle: Juice from infected fruits and heart rot tissues serves as main source of inoculum of the pathogen. Bacteria, being passive invaders, enter the plant through stomata and mechanical wounds which are aided by insects like ants, flies, and mites (Sapak *et al.*, 2021). In the plants the bacteria multiply in the tissue and are disseminated by wind and rain splash. Rainy and wet weather favors the disease which subsides with the onset of dry weather.

Epidemiology: The plant is most vulnerable to disease during the early growth stage following planting and during rapid vegetative development between the ages of 5-13 months. Young plants' leaf wounds, open flowers, and cracks in immature fruit are all potential entry points for infection. The bacterium invades the vascular system during its systemic phase. Young plants may experience an arrest in leaf symptoms during dry weather (latent stage). However, the symptoms frequently recur during early flowering or the fruiting stage.

Management: Choosing a planting site free from the history of bacterial rot disease is essential. Proper sanitation practices including decontamination the farming tools and destruction of host debris reduces the chances of the disease. Healthy planting material must be chosen for planting. Care should be taken during fertilization like reducing nitrogen application during the vegetative stage and use of fertilizers low in ammonium nitrogen to reduce salt injury, are useful. Biocontrol agents like *Bacillus cereus* have been reported to be effective against bacterial heart rot (Husin and Sapak, 2022). No complete chemical controls measures are available, although bactericides may provide limited control. Decreases in disease incidence have been observed by managing the vectors with miticides and insecticides.

Pink Disease

Causal Organism: *Pantoea citrea*

Etiology: It belongs to Phylum Proteobacteria, Class Gamma proteobacteria and family Enterobacteriaceae. It is a gram-negative, non-spore forming and facultative anaerobic bacteria. Earlier, a bunch of bacteria were found to be associated with this disease however with recent molecular analysis *Pantoea citrea* was found to be the causal agent.

Symptoms: The symptoms of pink disease are not recognizable in the field due to no external symptoms. It causes significant impact in the canning fruit industry. During the process of canning, the pineapple develops a rusty red color which is due to the enzymatic secretions of the bacteria (Kado, 2003). *Pantoea citrea* produces gluconate dehydrogenase enzymes which converts glucose into 2,5-diketogluconic acid (Cha *et al.*, 1997). Further due to heating a dimer of 2,5-diketogluconate is formed which gives the characteristic rusty red colour. Under severe invasion of the pathogen, a translucent appearance may be observed in the sub-dermal fruit tissue.

Disease cycle and epidemiology: Sporadic occurrence of the disease is seen when fruits develop under cool and wet climate. The bacteria enter the fruit through numerous florets growing out of the fruit. Insects are presumed to transfer bacteria from one flower to another. Once inside the fruit the bacteria multiply and show a slight translucent appearance. Unlike fruit collapse disease, pink disease does not show any soft rot symptoms.

Management: The biocontrol agent *Bacillus gordonae*has been reported as an effective method for reducing the incidence of the disease (Cha *et al.*, 1994). As vectors are presumed to aggravate disease, insecticide applications can be done. The development of resistant varieties by transfer of pink disease resistant genes is an effective strategy.

Marbling

Causal organism: *Pantoea ananatis* and *Acetobacter* sp

Etiology: It is gram negative bacteria which multiply by binary fission.

Symptoms: The flesh of the fruit becomes woody in texture and reddish brown in color. Marbling is not so serious and often occurs sporadically but becomes serious in the countries where pineapple fruit mature under lowland, tropical conditions.

Disease cycle and epidemiology: The pathogen enters through the flowers. Disease development is initiated and when fruit maturity takes place under warm, wet conditions.

Management: Proper sanitation practices must be followed. The smooth cayenne variety is moderately resistant to the disease. The damaged fruits must be rejected by processing.

Viral Diseases

Mealybug wilt

Causal organism: Pineapple mealybug wilt-associated virus-1 (PMWaV-1), 2(PMWaV-2), 3(PMWaV-3)

Etiology: It belongs to the family *Clostero viridae* and genus *Ampelo virus.* It is a long, flexuous, linear positive sense single stranded RNA containing virus (Carrasco-Lozano *et al.*, 2023). This virus is vectored by the mealybug *Dysmicoccus brevipes.*

Symptoms: The above ground symptoms are manifested by an initial reddish coloration development in the inner leaves followed by inward curling of the leaf margins (Dey *et al.*, 2018). The wilted leaves then show a change in color from red to pink coloration. Further the leaves turn glossy with leaf tips becoming necrotic and desiccating. The root system collapses leading to wilting of the plant. Sometimes plants remain asymptomatic with normal appearance, but fruits produced are drastically smaller in size than the heathy plant. During winter when plant growth and vigor are reduced the symptoms are more prominent.

Disease cycle and epidemiology: Planting material is sought to harbor the virus asymptomatically. Once the virus gets established, it is transmitted through the mealy bugs feeding on young leaves. The incidence is variable depending upon the age of the plants and prevailing weather conditions which determine the ant and mealy bug population. Younger plants when exposed to mealy bug produce symptoms within two to three months, while older plants may take up to 12 months to develop symptoms. There is a direct correlation between the amount of wilt in a field with the number of mealy bugs present.

Management: Clean planting material free from mealybug infestation must be used. Measures must be taken to control the mealybug population by using systemic insecticides. If more than 10% of plants show wilt symptoms, the field should not be used as a source of planting material.

Yellow Spot

Causal Agent: *Capsicum chlorosis virus*, *Tomato Spotted wilt virus*

Etiology: It belongs to the Family *Tospoviridae* and has a negative sense single stranded RNA genome enveloped with a lipid layer. It is transmitted by thrips.

Symptoms: Infection occurs during the first few months of planting when the plants are still in their young stage. The development of small round yellow spots takes place on the upper surface of the leaves which coalesce to form streaks and eventually become brown and die. This disease causes sideward bending of the plant due to spread of the pathogen from the leaves to the plant heart. Eventually the infection spread throughout the plant and kills it. When crown is bearing fruit is attacked, dying of the fruit occurs from top to downwards.

Disease Cycle: The virus is acquired by the vector during its 1st and 2nd instar. Once inside the virus multiplies inside the thrips and transmits the disease in a persistent propagative manner. Infection occurs mostly on plants during early growth, and crowns on developing fruit are getting infected.

Management: The vector population should be controlled by spraying systemic insecticides. As tospoviruses have wide host range, weeds must be remove from the field which may potentially harbour this virus.

References

Baiswar, P., Borah, T. R. and Singh, A. R. 2021. Postharvest Diseases of Pineapple and Their Management. Postharvest Handling and Diseases of Horticultural Produce, pp:287-294. CRC Press.

Barral, B., Chillet, M., Doizy, A., Grassi, M., Ragot, L., Léchaudel, M., Durand, N., Rose, L.J., Viljoen, A. and Schorr-Galindo, S. 2020. Diversity and Toxigenicity of Fungi that Cause Pineapple Fruitlet Core Rot. Toxins, 12(5): 339. https://doi.org/10.3390/toxins12050339.

Carrasco-Lozano, E. C., Carrillo-Ordóñez, G. A. and Gamarra-Gamarra, D.2023.Pineapple mealybug wilt-associated viruses 1, 2, and 3 are associated with mealybug wilt disease of pineapple in Peru. Journal of Plant Pathology, 105: 581–586.https://doi.org/10.1007/s42161-023-01327-y

Cha, J. S., Ducusin, A. R., Macion, E. A., Lucas, L. N., Hubbard, C. H., and Kado, C. I. 1994. Biological control of the pink disease of pineapple using bacterial epistasis.Molecular Ecology, 3: 609.

Cha, J.S., Pujol, C., and Kado, C. I. 1997. Identification and characterization of a Pantoeacitrea gene encoding glucose dehydrogenase that is essential for causing pink disease of pineapple. Applied and Environmental Microbiology, 63:71-76.

Devi, S. P., Thangam, M., Ladaniya, M. S. and Singh, N. P. 2013. Pineapple-a profitable fruit crop for Goa. Technical Bulletin No. 35, ICAR (RC), Goa.

Dey, K. K., Green, J. C., Melzer, M., Borth, W., and Hu, J. S. 2018. Mealybug Wilt of Pineapple and Associated Viruses. Horticulturae, 4:52. https://doi.org/10.3390/horticulturae4040052

FAO. 2022. Food and Agriculture Organization of the United Nations. FAOSTAT statistics database. https://doi.org/10.17660/ActaHortic.2006.702.17

Hardham, A. R., and Blackman, L. M. 2018. Phytophthora cinnamomi. Molecular Plant Pathology, 19(2): 260–285. https://doi.org/10.1111/mpp.12568

Husin, N., and Sapak, Z. 2022. Bacillus cereus for Controlling Bacterial Heart Rot in Pineapple var. MD2. Tropical Life Sciences Research, 33(1): 77–89.

Jacobs, A., Van Wyk, P. S., Marasas, W. F., Wingfield, B. D., Wingfield, M. J., and Coutinho, T. A. 2010. Fusarium ananatum sp. nov. in the Gibberellafujikuroi species complex from pineapples with fruit rot in South Africa. Fungal Biology, 114(7): 515–527. https://doi.org/10.1016/j.funbio.2010.03.013

Joy, P. P. and Sindhu, G. 2012. Diseases of Pineapple(Ananas comosus): Pathogen, Symptoms, Infection, Spread and Management. Kerala, India: Pineapple Research Station (Kerala Agricultural University). https://kau.in/sites/default/files/documents/diseases_of_pineapple.pdf

Kado, C. I. 2003. Pink Disease of Pineapple. Saint Paul: The American Phytopathological Society. doi: 10.1094/APSnetFeature-2003-0303

Kuruppu, M., Siddiqui, Y., Ahmad, K. and Ali, A. 2020. First report of postharvest stem end rot disease on MD2 pineapple fruits caused by Neoscytalidiumdimidiatum in Malaysia. Plant Disease, 105: 1564. https://doi.org/10.1094/PDIS-10-20-2318-PDN

Luo, Z. W., He, F., Fan, H. Y., Wang, X. H., Hua, M., Hu, F. C., Li, X. H., Liu, Z. X. and Yu, N. T. 2012. First report of leaf spot disease caused by Exserohilumrostratum on pineapple in Hainan Province, China. Plant Disease, 96(3): 458-459. https://doi.org/10.1094/PDIS-11-11-0979

Panabières, F., Ali, G. S., Allagui, M. B., Dalio, R. J. D., Gudmestad, N. C., Kuhn, M. L., Guha Roy, S., Schena, L., and Zampounis, A. 2016. Phytophthoranicotianae diseases worldwide: New knowledge of a long-recognised pathogen. Phytopathologia Mediterranea, 55: 20-40. DOI: 10.14601/Phytopathol_Mediterr-16423.

Petty, G. J., Tusdisease, A. and Dicks, H. M. 2006. Control of black spot disease / fruitlet core rot in queen pineapple with integrated mealybug, pineapple fruit mite and fungus control programmes. Acta Horticulturae, 702:143-149. DOI: 10.17660/ActaHortic.2006.702.17.

Rohrbach, K. G. and Apt, W. J. 1986 Nematode and disease problems of pineapple. Plant Disease, 70:81–87.

Rohrbach, K. G. and Johnson, M. W. 2003. Pests, diseases and weeds. In The pineapple: botany, production and uses. Wallingford UK: CABI publishing. pp: 203-251.

Rohrbach, K. G., and Schenck, S. 1985. Control of pineapple heart rot, caused by Phytophthora parasitica and P.cinnamomi, with metalaxyl, fosetyl Al, and phosphorous acid. PlantDisease, 69: 320-323.

Sapak, Z., Mahadeven, A. M. F., Farhana, M. H. N., Norsahira, S. and Zafri, A. W. M. 2021. A review of common diseases of pineapple: the causal pathogens, disease symptoms, and available control measures. Food Research, 5(S4): 1–14. https://doi.org/10.26656/fr.2017.5(s4).004

Varilla, C., Marcone, M., Paiva, L. and Baptista, J. 2021. Bromelain, a Group of Pineapple Proteolytic Complex Enzymes (Ananas comosus) and Their Possible Therapeutic and Clinical Effects. A Summary. Foods (Basel, Switzerland), 10(10): 2249. https://doi.org/10.3390/foods10102249

Vignassa, M., Meile, J. C., Chiroleu, F., Soria, C., Leneveu-Jenvrin, C., Schorr-Galindo, S. andChillet, M. 2021. Pineapple Mycobiome Related to Fruitlet Core Rot Occurrence and the Influence of Fungal Species Dispersion Patterns. Journal of Fungi (Basel, Switzerland), 7(3): 175. https://doi.org/10.3390/jof7030175.

Zhong, L., Ai, Y., Chun, R. and Yi, Y. 2016. Identification of Curvulariaclavata causing leaf spot on pineapple (Ananas comosus) in China. Canadian Journal of Plant Pathology, 38(2):250-253.

23

Existing Marketing System of Pineapple in West Bengal - A Look

[1]Banani Das and [2]Kalyan Kanti Das

[1]*Department of Agricultural Economics, College of Agriculture Lembucherra, Tripura, India*
[2]*Department of Agricultural Economics, Uttar Banga Krishi Viswavidyalaya Pundibari, Coochbehar, West Bengal, India*

Horticultural sector, which is contributing significantly for food and nutrition security, employment opportunity and poverty alleviation, got a real boost only in the early 1990s which coincided with liberalization of Indian economy. National area and production of horticultural crops are showing increasing trend over the years and production of horticultural crops has outpaced the production of food grain since 2012-13(Press Information Bureau, Govt. of India, Ministry of Finance, 2016). According to Economic Survey 2015-16, the share of horticultural output in agriculture is more than 33% over last decade and the sector contributes 30.4% to GDP of Agricultural sector. Horticulture sector accounts for 37% of total export of agricultural commodities at present and the exports have recorded sustained rising trend (Jha *et al.*, 2019). The export status of APEDA (Agricultural and Processed Food Products Export Development Authority) products in the recent years shows 28.18% growth in value of horticultural products over the previous year. The investment in horticultural sector with focused attention has resulted in spectacular change in the horticultural scenario and production and productivity of horticultural crops.

Of the total fruit area in the country, mango, banana & citrus shares the major part. Pineapple, the second most important tropical fruit in the world, shares about 1.6 per cent of total fruit area in India (Horticultural Statistics at a Glance 2018, Government of India). In respect of trade in pineapple, India exports both fresh pineapple and also processed pineapple products. Nepal, U.A.E, Maldives, Soudi Arabia etc. are major importing countries for Indian pineapple. However, over the years, export quantity (and also value) of fresh pineapple is showing fluctuating trend (Horticultural Statistics at a Glance 2018, Government of India). Production of pineapple is concentrated chiefly in eastern and north-eastern states like, West Bengal, Assam, Manipur, Meghalaya, Nagaland etc. in India. Pineapple happens to be the major fruit (in terms of area allocation) in two North Eastern states namely Meghalaya (37.7%) and Manipur (29.74%). But more than 45% of pineapple production comes from West Bengal (20.23%), Assam (16.97%) and Karnataka (9.59%). With about 11.41 thousand hectares under pineapple cultivation, West Bengal shares 10.40% of national pineapple acreage during 2017-18. In West Bengal, pineapple is cultivated almost in all the districts, but Darjeeling and North Dinajpur covers the maximum pineapple area and production in the state. Considering the demand in the international market and concentration of pineapple growers in the districts of Darjeeling, Jalpaiguri, and Cooch Behar, the first Agri-export zone was announced in west Bengal in 2001 by APEDA (Deb *et al.*, 2010).

A study (unpublished thesis by Banani Das, 2012) in West Bengal reveals that average size of pineapple orchard is around 0.50 ha in Darjeeling-North Dinajpur, where concentration of pineapple growers is most in the state. Three distinct typologies viz., sole pineapple growers, pineapple growers having pineapple along with other crops excluding tea, and pineapple growers having pineapple along with other crops including tea do exist in these areas. Pineapple orchard is maintained mostly in a single plot irrespective of category of farming. On an average, the orchards are 7-10 years of age and are connected with good road facility and are having the meager distance of 2.0 km. from state/national highway (National Highway No. 34). It is found that an approximate sum of Rs. 51,000/- to Rs. 73,000/- is required for carrying out pineapple production from an acre of orchard. The pineapple growers of this area use the famous variety of Giant Kew which weighs 1.5-2.5 kg, and is oblong in shape, slightly tapering towards the crown. Eyes are broad and shallow, making fruits more suitable for canning. The fruit is yellow when fully ripe and flesh is light yellow, almost fibreless, and very juicy with 0.6-1.2 percent acid, and its total soluble solids content varies from 12° to 16° Brix. In this region, average productivity of pineapple hovers around 9500 numbers of fruits (≈ 15.16 mt) per acre in Uttar Dinajpur to around 15,000 numbers of fruits (≈ 23.41 mt) per acre in Darjeeling area.

About 86-88% of the pineapple is sold as raw fruit and the per acre gross margin hovers around Rs. 98,000/- (in North Dinajpur area) to Rs. 1,70,000/- (in Darjeeling area). Thus, the avocation with cost-benefit ratio of 1:2.92 (in Uttar Dinajpur) to 1:3.32 (in Darjeeling) is found to be a remunerative one from any standard. Again, as the farmer do cultivate pineapple with their own planting material, the return prospect may be escalated further because the cost of seedling material contributes about 37-40 per cent of total cost and the cost-benefit ratio, then, becomes 1:3.55 (in Uttar Dinajpur)to 1:4.32 (in Darjeeling).

Bidhannagar, a place very near to Bagdogra airport (Siliguri, West Bengal) is reported to be the chief market hub for pineapple produced in the surrounding areas (Darjeeling, Jalpaiguri, North Dinajpur etc.). Mainly, raw fruits are transacted in this market. A large part of the supplied fruits is directed to markets of important cities like Kanpur, Lucknow, Delhi, Mumbai, Jamshedpur and Ranchi (Chadha, 1998) and even to the outside countries like Nepal, Bangladesh etc.

Two distinct type of marketing system prevail for disposal of produced pineapple in the study area. In the first one, the produced fruits are sold at the orchard field itself. The pineapple cultivators sell their produce through trade agents (locally called *'paikars'*) at village level. Generally, the small and marginal farmers or farmers who do not have sufficient family manpower to take the produce to the market are involved in this system. In the second one, pineapple fruits are brought to the nearby Bidhannagar market where various market functionaries do take part in marketing of the fruit. Similar to the observation made by Chadha, (1998), various market functionaries / middlemenlike village agents/ Trade agent, commission agents, wholesalers, retailers etc. take part in the marketing transaction of pineapple of this area. A thorough interaction with the pineapple growers and many other market functionaries reveals that there exists six (6) distinct nos. of major marketing channels in this area through which the pineapple fruits are transacted from producers to consumers (Das *et al.,* 2016).

Among these six marketing channels, four channels are incomplete (or open) at one end as these are extended towards the outside market for locations like Kolkata (W.B), Bihar, Delhi, Uttar Pradesh, Punjab, Lucknow, Haryana etc. The marketing channel comprising of producer, trade agents, wholesalers, outside market agents (outside the state, mainly) and consumer is found to be more prevalent marketing channel in that area. The next most extensively used channel comprises of producers, wholesalers, market functionaries and consumer, which incidentally is the shortest one too. In this channel, the produced fruits are brought directly to the wholesaler's godown by the producers with his own cost (cost of transportation & unloading). Then the fruits are graded in three groups according to their size and vigour (i.e., (i) grade I, <1.0 kg, (ii) grade II, 1.0-1.5 kg & (iii) grade III, >1.50 kg) by mere visual observation before being sending it to the outside market(s) through local agents. Mostly, the products are dispatched by trucks. When the goods are sent to the retail shops, it is transported through local vehicles like trolly or van-rickshaw etc. The retail shop owners sell the products in two basic forms – as raw fruit or as raw pineapple juice. Thus, the whole system of marketing in this area can be represented in the figure.

Marketing of fresh pineapple poses serious problem due to its highly perishable nature, especially that of mature pineapple fruit. Major part of loss (1.40 percent) occurs while transporting the fruits from producer to market through Trade agents in case of raw fruits. But in case of ripen fruits major part of damage loss (13 percent) occurs while fruits are stored in wholesaler's store. The fruits are kept for around 3 days in raw form in wholesaler stores and in ripen form they are stored for only one day. On an average, the producers receive about 67% -69% of consumer's price and it is found to be lower for the pineapple growers of Uttar Dinajpur than their counterparts in Darjeeling area (Das *et al.*, 2016). About 3-4 chief market intermediaries like paikars/ trade agent, wholesalers, retailers are found to be taking part in the marketing system and out of them, trade agents or wholesalers get the lion's share (70-80 per cent) of total profit in the system.

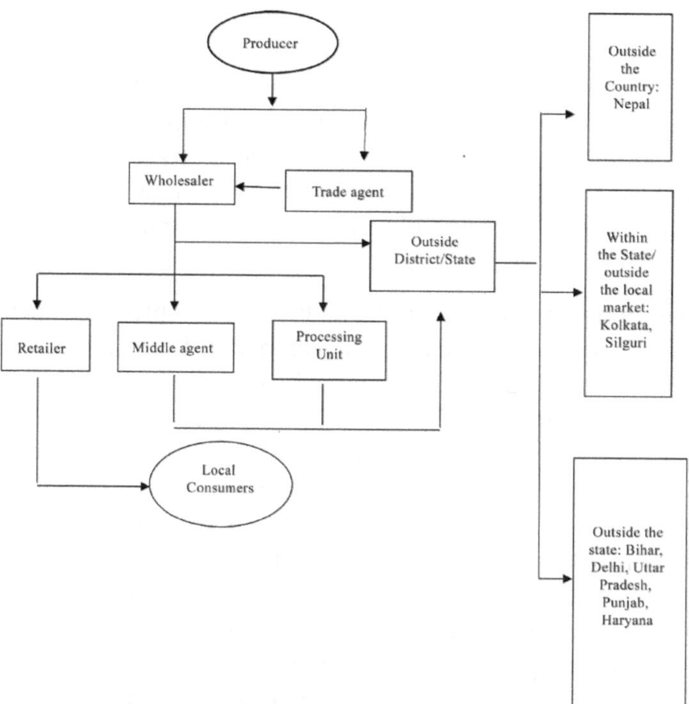

Fig. Schematic presentation of marketing system of raw fruit in the study area

Pineapple is a highly perishable fruit, and thus the mechanism of fruit preservation finds its importance to a greater extent. Moreover, the household consumer's market is rapidly changing in terms of taste and preferences of the consumers towards the ready to cook or ready to eat food (Vichare, 2008). Family consumption expenditure of processed products were found to be highly income elastic (study conducted by Vichare, 2008 on mango). It was estimated that nearly 97 percent of the world's pineapple is utilized by processing industry. Unlike the major pineapple producing countries in India the proportion used for processing is less than 10 percent. But pineapple products take second position next only to mango. Share of pineapple slices is nearly 56%, of juice and jam it is 15% & 10% respectively (Chadha *et al.*, 1998). Mainly three types of processed products viz., pineapple juice or concentrates, pineapple jam, tit-bits/canned pineappleare prepared (though, in negligible part) in the area. In the absence of organized pineapple processing units/set ups in the area, these processed products are prepared locally. Wastes, which come mainly from top and bottom ends of fruits and shell, represents around 35 percent of the total weight of the fruit at modern canneries and upto 50 percent at less equipped canneries (Py *et al.*, 1987). In this area, wastes material accounts for almost 50 percent of the fruit weight of the processed products. The common practice followed there is decomposition of the waste product and after drying it made into organic manure which costs around Rs. 2-2.5 per kg.

Bidhannagar, considered as one of the major hubs of pineapple production in India has as many as 70,000 pineapple growers. The annual production of pineapple amounts to around 0.6 million tonnes in the area from 20,000 hectares land. The region contributes to around 80 per cent of West Bengal's total output. But, following constraints are seriously hampering the growth and prosperity of the industry

- At least 26 processing units were there in the region till the 1990s, but all were shut down over the next two decades. The last one downed its shutters in 2017. Many reasons (socio-political) can be put forward on that; but, inability of the local pineapple growers to meet the minimum requirement of the raw product – both in terms of quality as well as quantity for these processing units was one of the major concerns. Pineapple growers' awareness /consciousness about the global demand (not only from quantity point of view but on quality parameter as well) is one of the prime concerns. The farming community needs to be oriented towards the selection of variety, use of planting material, application of growth hormone, protection chemicals keeping strictly in mind the consumer's preference across the country as well as in the outside markets.

- Marketing of raw fruits largely depends upon demand from outside (within the country) pockets which has been reduced to one-fifth during covid-19 pandemic situation. Distress sale with abysmally low price (in the local market) compels many pineapple growers to leave the occupation. Want of arrangement of minimum support price (MSP) or similar type of initiative(s) aggravates the problem further.
- Existing marketing system of pineapple, in general, is governed by a few market intermediaries (wholesalers/trade agents) due to which producers' share gets affected badly. Effective functioning of pineapple growers' association or a kind of multi-stakeholders platform is desperately needed.
- Though a multi-purpose market for pineapple farmers in Bidhannagar had been set up having cold storage, rooms for the buyer and other facilities long back, it remains nonfunctioning till date. Amicable political solution is warranted to revamp the pineapple marketing in the region.
- Some of the sample farmers in both the district reported that the bushy and thorny outgrowth of this pineapple crop poses a serious problem for its proper maintenance. The farmers feel that the bushy crop having hard and sword like thorny leaf creates a barrier for the labourers to do intercultural operation, especially, when the plant attains full maturity level. This nature of the crop creates the problem of labour unavailability which is the biggest problem of this crop enterprise
- Besides all these, pineapple avocation faces stiff competition from tea industry in the region which looming large on its 'future' too.

References

Anonymous. 1977. Pineapple Cultivation. Extension Bulletin No. 6. Indian Institute of Horticultural research, Bangalore, Bangalore. 1977.

Anonymous. 1977. The Pineapple in West Bengal1976. Directorate of Agriculture, Govt of West Bengal, pp, 6.

Anonymous. 1997. Production yearbook.1992 and 1995. Food and Agriculture Organization (https://www.fao.org/3/i2050e/)

Anonymous. 2003. Annual Report on social and economic impact of pineapple crop 2002. Food and Agriculture Organization (https://www.fao.org/3/i2050e/)

Anonymous. 2004. West Bengal Human Development Report 2004, Dr. Jayati Ghosh, United Nations Development Programme (https://hdr.undp.org/content/west-bengal-human-development-report-2004)

Anonymous. 2008. District Statistical Handbook. 2007. Beauro of Applied Economics and Statistics, Govt. of W.B. (http://wbpspm.gov.in/publications/)

Anonymous. 2011. Annual Report 2009 - 2010: National Horticultural Board, Govt. of India, pp, 1-57.

Anonymous. 2011. Indian Horticultural Database 2010. Annual report of National Horticultural Board, pp, 204-232.

Anonymous. 2012. Annual Report 2011-12, APEDA, Govt. of India (https://apeda.gov.in/Annual_Reports/)

Anonymous. 2015. A Strategy paper on augmenting export of value added products from India, Submitted to Agricultural and Processed food Product Export Development Authority (APEDA), Ministry of Commerce, Govt. of India. pp, 1 – 160.

Anonymous. 2016. Cash sale of agricultural produce by cultivators/producers 2016-2017, Press Information Bureau, Govt. of India, Ministry of Finance. (https://pib.gov.in/TagbasedSearch.)

Anonymous. 2019. Horticultural Statistics at a Glance 2018, National Horticulture Board, Government of India. (https://nhb.gov.in/statistics/Publication/Horticulture Glance-2018.pdf)

Chadda, K. L., Reddy, B. M. C. and Shikhamany, S. D. 1998. Pineapple. Indian Council of Agricultural Research, New Delhi, pp. 27.

Chadda, K. L., Shikhamany, S. D. and Melanta, K. R. 1977a. Correlation of growth characters with yield and quality of pineapples (*Ananas comosus* (L.) Merr.) Indian Journal of Horticulture, 34: 107-12

Das, B., Das, K. K. and Roy, T. N. 2016. Marketing and Value Addition of Pineapple Fruit in West Bengal. Agricultural Economics Research Review, 29 (2): 279-285

Deb, P., Paul, P. K., Bhowmick, N. and Suresh C. P. 2010. Pineapple under agri-export zone of India. Souvenir of 7th International Pineapple Symposium 2010. 13-15 July, 2010, Persada Johor International Convention Centre, Johor Bahru, Johor, Malaysia. pp, 126.

Jha, G. K., Suresh, A., Purnea, B. and Supriya, P. 2019. Growth of Horticulture Sector in India: Trends and Prospects. Indian Journal of Agricultural Sciences, 89 (2): 314–321

Py, C., Lacoecilhe, J. J. and Teisson, C. 1987. The Pineapple: Cultivation and Uses. Maisonneuve Publishing Co., Paris. Techniques Agricoles Et Production Tropicals. XXXIII

Vichare, A. A. 2008. Consumers Preferences for Processed Products of Mango (India). Journal of Agricultural Development and Policy, 18 :93-101.

Vyas, V. S. 1994. Agricultural policies for the nineties, issues and approaches. Economic and Political Weekly, 29(26): A54 – A63.

24

Post Harvest Management, Processing and Value Addition of Pineapple

[1]Arghya Mani, [2]V.S.S.V. Prasanna and [3]Surajit Mitra

[1]Assistant Director of Horticulture Department of Food Processing Industries and Horticulture, Govt. of West Bengal
[2]Principal Research Engineer, Linkoping University, Sweden
[3]Department of Post Harvest technology, Bidhan Chandra Krishi Viswavidyalaya, Mohanpur, Nadia, West Bengal, India

Pineapple is a tropical plant with edible multiple fruit consisting of coalesced berries. It is named after pine cone due its resemblance with its structure. It is the most economically important plant in the family of Bromeliaceae. It is widely grown throughout the world for its dual-purpose use in the form of fresh fruit and many processed products. It is a tropical fruit having excellent taste, juiciness, aroma and flavour which is preferred by many people. It is also rich source of fibre, potassium and vitamin C, vitamin B_1 and vitamin B6. One of the important constituents in the fruit is proteolytic enzyme bromelein which is known to have significant use in treating conditions like inflammation, sinusitis, arthritis and early recovery from surgery (Joy, 2010). Bromelein is a digestive enzyme also used as meat tenderizer. Pineapple is cultivated in an area of 106.22 thousand ha and total production is 1,527.93 thousand tons (APEDA). Major producing states of pineapple were West Bengal, Assam and Karnataka. Pineapple is highly perishable fruit which transpires heavily and lose moisture thus have high susceptibility to spoilage microorganism (Hossain and Bepary, 2015). There are often several difficulties in production of pineapple with its physiologically active state even after harvest and non-climacteric nature. Even though production is more, there are enormous post-harvest losses for the fruit. The overall losses being encountered constitute 32.12 percent (6.82% at farm level, 3.09% at wholesale market and transportation, 5.44 % at retail market level, 5.64% during storage, 4.31% in processing unit and 6.82% at consumer level respectively) as reported by Ningobam, (2019). The major reasons for post-harvest losses may be attributed to production area, transportation, storability, packing and number of damaged fruits. Post-harvest loss of pineapple can be minimized by adopting various suitable post-harvest handling practices that are presently in practice all over the world to extend its shelf life. To minimize the post-harvest losses occurring at various levels, efficient post-harvest system is mandatory. Starting from the harvest of the fruit, necessary care should be taken at every operation. With proper management from initial harvest to the transport of the produce shelf life of the fruits can be extended and losses can be minimized. Temperature and relative humidity regulation are important aspect that needs to be controlled. Selection of packing material must be done based on the need. Fruits for local market can be immediately transported whereas fruits for distant market should be preserved in low temperature storage. In many areas refrigerated transport and storage are not available, to minimize losses, there need to be need for refrigeration, which delays the ripening process thus helps in minimizing losses. Apart from controlled post-harvest operations, losses can be prevented by proper utilization of the produce with the help of value addition. Value added products can also be utilized throughout the year. Pineapples can be processed into different value-added product viz., juice, pineapple slices, RTS, jam, frozen slices, canned pineapple and pineapple squash (Mirza *et al.*, 2016).

Harvesting of Pineapple

Maturity indices: General indication for harvesting pineapple fruits is when fruits are firm and mature. As the fruit falls under the non-climacteric category, picking at optimum stage is crucial as sugar content will not be improved after harvest. Pineapple fruits are judged as matured according to their size attained and colour that varies from one variety to other. Some of the external and internal fruit characteristics are used for judging maturity. External characteristics include outer surface colour of the fruit, flatness of the eyes i.e., fruitlet flatness and fruit size. Internal indices include percent total soluble solids and appearance. Surface colour of the fruit naturally changes from green to reddish brown during maturity. Usually during maturation, change in the colour of the shell and flesh are not but, at the time ripening begins, the chlorophyll content of the fruit begins to degrade which results in the yellowing of the fruits (Py *et al.*, 1987); Hassan *et al.*, 2011). Fruits left on the plant until the complete yellow stage results in detoriation of the fruit quality by loss of flavour, excess sugar content and reduced acidity. Generally, acidity in the range of 0.9-1.3% is the minimum value that is preferable by consumers (Soler, 1992).After the harvest, the change in the yellow colour of the fruit does not act as an indicator of fruit ripeness or quality. The actual stage of maturity during the time of the harvest can be decided based on the storage period required, shelf life of the fruit and the mode of transport. The degree of the yellow coloration of the pineapple fruit judges its maturity. Colour stages are shown in Fig 1.

Table 1: Colour stages are categorized in to following classification according to Joy (2016)

Colour stage (CS)	Appearance of fruit
CS1:	all eyes green, no traces of yellow;
CS2:	5 to 20% of the eyes yellow;
CS3:	20 to 40% of the eyes yellow;
CS4:	40 to 80% of the eyes yellow;
CS5:	90% of eyes yellow, 5 to 20% reddishbrown;CS6: 20 to 100% of eyes reddish brown.

Fig. 1: Image *Source*: (KAU Agri InfoTech portal)

Pineapple fruit maturity starts gradually from the fruitlets in the bottom part to the fruitlets in the upper part. With the increase in maturity, fruitlets mature and get flatter. So, the extent of fruit flatness is a good visual sign for maturity of the fruit. Also fruit maturity is highly correlated with the total soluble solids in the fruit. Minimum total soluble solids required in the bottom part and top parts of the fruit during maturity are 12% and 10% respectively. According to Adikaram and Abayasekara, (2012) harvesting for distant market in pineapple should be done when the fruit has 10-20% yellow stage. Maturity index can also be calculated approximately from the days after flowering but they vary according to the variety and place. Usually the time from induction of flowering to the ripening of the fruit ranges from 140-220 days (Thompson, 2002) .On the report of Hassan *et al*., 2011, in lower altitude and equatorial region, pineapple cultivar 'Smooth Cayenne' takes shorter maturation period of normally ranging from 100-150 days. Py*et al*., 1987 reported that Pernambuco cultivar takes longer time for maturation compared with other varieties while Queen and Red Spanish cultivars usually have shorter maturation period. Fruits grown in subtropical areas usually take more than 200 days for fruit development On the report of Deka *et al.,* 2005 , Days after flowering (146-150), specific gravity (0.93-0.98), respiration rate (7-10 mg/kg/hr.), TSS (18-19 °Brix), acidity (0.77-0.83%) and TSS: acid ratio (23-25) are the ideal indices for harvesting of the Kew pineapple variety for local as well as distant markets. For export through sea shipment which requires 7-14 days, fruits are harvested at CS1 stage where the fruit eyes show no yellow colour development. Harvesting can be done at stage CS2 TO CS3 for air freighted shipments. Harvesting in the advanced stages can lead to mechanical damage to the fruit due to over-ripeness. In the same way harvesting of the immature fruit leads to production of the flavourless fruits with low brix and fruits that are easily prone to chilling injury after harvest (Paull and Rohrbach, 1985).

Harvesting techniques: First crop of the pineapple is called as plant crop and it nearly takes 16-20 months from planting to harvesting. The next crop which is called as ratoon crop takes nearly 15 months for harvest (Maria and Elhadi, 2017). Fruits should be harvested in the early morning when temperatures are lower or in the late evenings when temperatures are low. High temperatures can reduce the shelf life of fruits. Harvest fruit with a sharp knife leaving more than 2.5 cm of stalk attached to the fruit. For fresh consumption the fruits can be harvested when one eye or two eyes in the bottom are in ripe stage. However, the harvesting stage varies from cultivar to cultivar and also based on market destination. Usually 127 days after flower induction, fruits attain 25% of the peel colour i.e., one- fourth of the fruit shows yellow- orange colour. Harvesting is done generally within 7-14 days after fruit yellowing starts. Fruits are usually harvested with a long knife known as 'parang' or fruit is snapped or twisted from the stalk using hand in the downward motion (Shown in Fig 2). When harvesting is done with hands, pickers generally use thick gloves or suitable clothing to protect themselves from the spiny leaves. Fruits are placed in basket that is carried on the back of the harvester and later stored in field crates and stored in shade places. Fruit is harvested along with the peduncle leaving 3-6cm of it attached on the plant after harvesting. The stalk should not be completely removed from fruit at the basal part which leads to disease infection and reduction of shelf life of the fruit. This disease condition is more common in the Smooth Cayenne cultivar. Usually fruits that need to be transported for fresh market are harvested with peduncle length of 5-10 cm.

Fig. 2: Harvesting of the fruit

Fruit should be harvested with utmost care any damage or bruises can lead to shortage of storage life, affecting the quality of the produce and leading to infection by microbes. Out grading of the fruits should be done. Any diseased or abnormal fruits are separated from the fresh fruits. Crowns of the fruits sometimes are detached from the fruits and are left behind in the field storing in bags which can be used further as planting material. The fruits that are exported are always left with the crown. In the packing section, the stems and crowns of the fruits are trimmed to 2cm and 10cm respectively. In case of fruits selected for export trade, certain tests are necessary at the field level while selecting the fruits for harvesting.The test equipment required during export are given in Table 2.

Table 2: Equipment needed for various post harvest operations

TSS of the fruit	Refractometer
Firmness of the fruit	Penetrometer
pH of the fruit	pH meter
Pulp temperature	Probe thermometer
Weight	Weighing balance
Fruit size	Plastic rings of different size

Handling of the fruits: Pineapple fruits come under category of very perishable. Fruit quality has to be maintained at various levels of growers, purchasers, during storage till produce reach the consumers. Rough and improper handling may damage the produce and reduce the final quality. For the produce to get the maximum price various operations are mandatory before reaching to the consumers. Fruits grading and sorting are indispensable operations. Precooling should be the initial step after harvesting the produce to remove the field heat. Pineapple fruits should be harvested with care to avoid mechanical injuries. Fruits are packed in baskets made of bamboo strips or plastic crates. During packing operation, crowns can be used as natural cushioning material to prevent bruises by placing the fruits in upside down position. Temperature regulation is a key step to be regulated during storage.

Pre- Harvest Practices in Pineapple

Pre harvest management practices are the operations that need to be carried out prior to the harvesting. These are the management practices that are managed in field conditions such as soil management, fertigation, water management, and weed control, pest and disease management. Governing these conditions will directly influence the shelf life of the fruits. Hence it is indispensable to collect prior information about all the practices implemented at the farm. Prior to the harvesting, fruits may be de greened by the use of ethephon. This type of treatment may increase the process of de-greening of the shell but generally will reduce the shelf life of the fruit. (Manzar and Rejaul, 2015).

Pre harvest practices are subjected to following provisions

- Fruit should be full with crown
- Crown of the fruit should be free from dead and diseased leaves
- Fruit should be free from rotting and detoriation and should be fit for consumption.
- Fruits should be clean and free from foreign material
- Fruits should be free from blemishes
- Fruits should be healthy free from infection from pests and diseases
- Fruits should be free from internal browning
- Peduncle length should be maintained in accordance with the need
- Fruits should be devoid of low temperature or high temperature injury in the field conditions.
- Peduncle cut should be straight and clean

Post-Harvest Operations

Post-harvest handling is the manner in which fresh produce is treated during all operations between harvest and consumption. The customary sequential steps in a post – harvest handling system for pineapples are outlined in following flowchart. On farm handling describes any post-harvest practices that are carried out on the farm in order to prepare the harvested produce for sale. These may include trimming, washing and cleaning, grading, storage and transport.

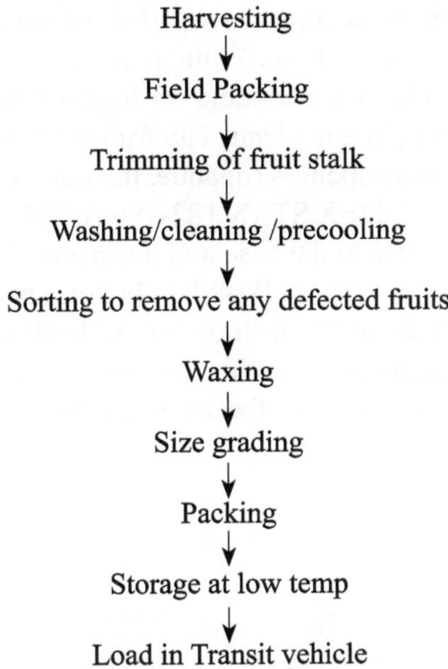

Sequential Flowchart for Steps in Post-Harvest Handling System of Pineapple Fruits

Precooling: It is the first step that needs to be performed after harvesting the produce mainly for perishable produce like pineapple. This step should be followed immediately after harvest to reduce the temperature of the fruit thus reducing the spoilage. Pineapple comes under category of temperature sensitive. Usually this step should be performed with in 10 hrs. after the harvesting operation. Precooling temperature for $1/8^{th}$ matured fruit is usually 13-15° C for 6-8 hrs. (www.kau.edu/prsvkm).

Field packing: After harvesting, the fruits are placed in plastic crates and crates are padded with foam to protect from bruising. Mechanical damage to the fruits should be avoided as it leads to microbial infections and also fruit softening. Polythene bags should not be used for pineapple packing as they are not strong enough for the weight of the fruit and cause mechanical damage. Fruits should be stored in cool place where temperature is low as it allows the fruit in removing field heat and also decrease the respiration rate which help in increase in shelf life of the fruit.it would be better if packing operation is done in a packing facility. Different varieties should be packed separately in different crates and they should also be labelled accordingly. Preliminary grading should be done priorly and the diseased, damaged, undersize, oversize, over ripened fruits should be discarded from the lot.

Trimming of the fruit stalk and crown of the fruit: Fruit stalk should be cut generally 2.5 cm from the base. Fruits that are needed to be exported for international trade shouldhave peduncle no longer than 2.0 cm, and the cut must be transversal, straight and clean. The fruit must be physiologically ripe, i.e., without evidence of unripeness (opaque, flavourless, exceedingly porous1 flesh) or over ripeness (CODEX STAN 182-1993). The stem of the fruit should be cut to a length very close to the base and the crown should be around 10 cm in length (Joy and Rashida, 2016). Based on buyers specifications, sometimes crown can be trimmed in rosette fashion, or cut back completely or left as it is(Shown in Fig 3). Dried leaves in the crown must be removed. The ratio of size of the crown to the length of the fruit can be decided based on market requirements.

Fig. 3: Caribbean Agricultural Research and Development Institute (CARDI)

Washing and cleaning of the fruit: Fruits must be cleaned in water with disinfectent like sodium hypochlorite 100ppm – 200 ppm or 150 ppm free chlorine and pH of the water should maintained around 6-7 (Manzar and Rejaul, 2015).Water used for cleaning should not be hard with heavy metal elements or of any chemicals.In addition to prevent post harvest diseses, fungicides like benomyl, thiobenzidine, or bavistin in required concentrations should be added to wash solution and fruits should be placed for about 3-5 minutes based On the size of the fruit. Later the fruit should be air dried.

Waxing of the fruits: After washing and drying of the fruit, fruits can be waxed for increase in shelf life of the fruits as pineapple fruits readily soften with in short time after harvest. Generally food grade waxes like paraffin wax or carnuaba wax can be used.In reference with study by Xueping *et al.*, (2017) wax concentration of 65 g L^{-1} for 1 min can delay the change in the colour of the fruit and also extended shelf life for nearly about 10 days It can be applied in the form of spraying or by method of dipping. As reported by Mandal *et al.*, (2015) pineapple fruits treated with GA_3 at 100 mg L^{-1} showed delayed response of ripening and high shelf life (19.05 days) during storagecompared to the wax treatment. Wax application should be limited to the fruit surface but not to te crown portion as it injures the crown part.Some of the benfits of the waxing are

- Prevention in loss of moisture from the fruit
- Glossy appearance to the fruit
- Physiological disorders like internal browning can be mimimized
- Shelf life of the fruit can be extended

According to Huigang *et al.*, (2011) waxing of pineapple fruits can have some desirable qualities like increase in total sugar co tent and ascorbic acid content in the fruit.Also waxing can alleviate the chilling injury suring cold storage thus improve the final quality of the fruit

Application of edible coatings: Edible coatings is a promising substitute in improvizing the quality and extend shelf life of produce.Edible coatings are the materials used for extension of shelf life of fruits and vegetables. They can be safely consumed and they do not add unwanted flavour to the food stuff. Recently along with edible herbs and antimicrobial componds, many coatings have been diccovered to increase the shelf life of the fruits (Pramod *et al.*, 2016). Chitosan is widely used edible coating material. In a study conducted by Ibrahim *et al.* (2014). Irradiation of pineapple fruits with 15 kGy chitosan radiation showed no fungal growth compared with untreated fruit which showed fungal growth after 18 days.Pineapples showed best qualities during storage period with minimum loss of moisture, shrivel, increased ascorbic acid content and able to maintain quality and sensory characteristics.As reported by Nimitkeatkai *et al.* (2006) pineapple fruits treated withSemperfresh (sucrose-fatty acid ester) showed extended storage life up to 5 weeks compared with the untreayed fruit which had storage storage life of only 28 days. Coating with 2% Semperfresh 7055 reduced internal browning and maintained higher ascorbic acid content when stored at 10ºC, 90-95% RH.

Sorting and grading: To meet the consumer requirements the ultimately, produce should be of high quality. Fruits should be sorted according to the size, appearance, external colour, and based on decay and damage. Damaged fruits should be rejected.Fruits should be sorted out based on the market requirements. For local markets sorting and grading operations are carried out based on visual observation and colour. Fruits that need to be exported should be graded according to the export requirements.Grading of the fruits should be done based on size, weight and degree of ripeness. Grading is also an essential operation during pineapple export process. Most importers will stipulate the expected weight per box, or the size of the box will dictate the weight of fruit that the box can accommodate. Weighing scale should be used to accommodate weight rangethat is required by the importer. The net weight of packed fruits in the carton should be around 14 – 15 kg (Robin *et al.*, 2011). According to Fruits and vegetables grading and marking rules (2004) issued by Ministry of Agriculture Department of Agriculture and Cooperation the criteria for grade designation with TSS level of 12°B is formulated as follows:

a) Extra class:1500 gm and above

b) I class:1100 – 1500 gm

c) II class:800 – 1100 gm

d) III class:< 800 gm

Source: (www.iica.int)

Table 3: Provisions concerning sizing in Pineapple fruits as per AGMARK standard

Size Code	Average Weight (+/-12%) (in grams)	
	With Crown	Without Crown
A	2750	2280
B	2300	1910
C	1900	1580
D	1600	1330
E	1400	1160
F	1200	1000
G	1000	830
H	700	560
I	<700	<560

Packing: Pineapples must be packed properly such that the produce is protected to full extent and avoid produce damaging losses. In addition to minimize the losses retailers can enhance the sales. Pineapples shall be packed in each container in compliance with the Recommended International Code of Practice for Packaging and Transport of Fresh Fruits and Vegetables (CAC/ RCP 44-1995. Some of the commonly used packing materials for pineapple in most developing countries are wooden crates, cardboard boxes, plastic crates, jute sacks and polythene bags (Idah *et al.*, 2007). Plastic cartons can have internal dividers to prevent bruising. Shredded paper should be added around the fruits and also all over the top and bottom of the crates to prevent abrasion and damage during transport. Cartons should be strengthened by taping the edges, thereby preventing fruits from accidentally dropping out of the box. Proper ventilation must be provided to the fruits. Pineapples are generally packed along with the crowns to avoid infection at crown attachment point and also to increase the shelf life of the produce (Joy and Rashida, 2016).

Table 4: Grade designation and sizing of pineapple as per AGMARK standard

Grade designated	**Requirements of the grade**	**Minimum total soluble solids required (in brix)**	**Tolerences of the grade**
Extra class	Pineapples must be of superior quality. They must be characteristic of variety and/ or commercial type. They must be free of defects. Very slight superficial defects may be there, provided these do not affect general appearance of the produce, the quality, the keeping quality and presentation in the package. The crown, if present, shall be simple and straight with no sprouts and shall be between 50 and 150% of the length of the fruit with trimmed or untrimmed crowns.	12	5% by number or weight of Pineapples not satisfying the requirements of the grade, but meeting of those of Class I grade or, exceptionally coming within the tolerances of that grade.
Class I	Pineapples must be of good quality. They must be characteristics of the variety and/or commercial type. - slight defects in shape and colouring and some skin defects such as scars are allowed - defects should not affect pulp of the product. The crown, if present shall be simple and straight and slightly curved with no sprouts, and shall be between 50 and 150% of the length of the fruit for pineapples with trimmed or untrimmed crowns.	12	10% by number or weight of Pineapples not satisfying the requirements of the grade, but meeting those of class II grade or exceptionally, coming within the tolerance of that grade.

Grade designated	Requirements of the grade	Minimum total soluble solids required (in brix)	Tolerences of the grade
Class II	Pineapples which do not qualify for inclusion in the higher grades but satisfy the minimum requirements. - Defects in colouring; including sunspots; - Skin defects(scratches, scars, bruises and blemishes) should not exceed 8% of the total surface area.defects must not affect the inside pulp in any case. The crown, if present, shall be simple or double and straight or slightly curved, with no sprouts.	12	10% by number or weight of Pineapples not satisfying the requirements of the grade but meeting the minimum requirements.

For local markets fruits are arranged in baskets weighing nearly 25 kg and the baskets are lined with the paddy straw and fruits are allowed to stand on their stumps. Second layer of fruits in the basket are arranged on the crowns of the first layer which act as cushioning materials to the fruits. For export purpose each carton should be labelled properly and necessary information such as date of packing, variety, grade, origin of the produce, name of the exporter, number of fruits per box, net weight of the fruit and recommended storage temperature and relative humidity should be mentioned.

Fig. 4: Pineapple packing

Storage: Pineapple being a perishable fruit cannot have more shelf life. So the fruits need to be transported to desired destinations as quick as possible. Fruits attached with the crown can be stored for long time without damage for nearly 10-15 days from the day of harvest. Facilities used for temporary storage of fruits should be designed for easy cleaning and sanitisation with sloping floors. For fruits that need to be transported to long distances, refrigerated storage is recommended to increase the shelf life and delay the ripening process. Pineapples can be stored for a period of 20 days when stored around 10-13 °C. The optimun storage conditions are storage is at 7.2° C and 80 or 90% relative humidity (Shweta *et al.,* 2017). Pre-export storage can be used and a suitable storage temperature should be 7.5 °C and 85-95% relative humidity. Mature green pineapples should be stored at a temperature of 8-10° C and at a relative humidity of 85-90%. Under these conditions, pineapple should have a storage life of 2-3 weeks (Robin *et al*., 2011). Storage for extended periods below this temperature will result in chilling injury. The shelf life of the pineapple fruits was prolonged by the refridgerated condition up to the 33rd day of storage in a study conducted by Joseph *et al*., 2010. During controlled atmospheric storage, oxygen level should be decreased to 2-5 % and higher carbondioxide level around 5-10% must be maintained. CA storage also improved the appearance and marketability of the fruit by decreasing disease infestation such as superficial mould growth. The symptoms of endogenous brown spot, which are externally invisible, was not affected by CA storage (Akamine and Theodore, 1971).Modified atmospheric packaging and low temperature storage also slows down the physiological process in fresh cut pineapple thus increasing the shelf life (Montero, 2008).

Transport: Quick and cheap transportation is the need of the hour. The farmers of a certain area should pool their resources at some assembling point so that they can get the benefits of bulk transportation. After harvesting, most of the farmers transport the fruits to the market in trucks and tractors. While loading the fruits in trucks, they are directly dumped into lorry without any packaging. At the time of unloading the fruits in the wholesale market, the damaged, over ripened, mechanical damage which includes loading and unloading are separated out. Losses in the wholesale market including transportation were quantified in wholesale markets. The total quantity transported was 3,385 quintals, out of which 104.65 quintal (3.09 %) was the loss [Table-2]. The major loss of 42.30 quintal (1.25%) was due to mechanical damage, followed by over ripening,36.20 quintal (1.07%) and spoilage 26.15 quintal accounting to 0.77% percent(Ningombam *et al*., (2019).

Processing and Value Addition of Pineapple

Pineapple is a very delicious fruit and is a preferred fruit due to its optimum flavour and ideal sour-sweet blend. The juicy and crispy pulp of the fruit is preferred by people all over the tropics and subtropics. The fruit is having significant amount of acidity in it and also possesses desirable therapeutic value. The only constrain about pineapple is that the fruit is very hard to eat type fruit. The crown is needed to be removed, peeling and removal of core is also needed to be done before consuming. This is the reason why the fruit is of better demand when it is processed or value added into different processed products. Processed products can effectively solve the year-round demand of pineapple. Various processed products like RTS, squash, nectar, cordial, leather, toffee, jam, jelly can be prepared from pineapple which is of enormous demand in both global and local market (Joy, 2010). Several value-added products can be prepared from pineapples which are discussed in upcoming pages. In total only 20% of pineapple is consumed fresh other 80% is processed into various products. 38 per cent is used for the preparation of juice or beverages, 30 per cent of the fruits are canned and 32 per cent fruits are processed into other products like jam, jelly, candy, leather, etc. (Saad, 2004). Fruit core can be also processed into candies and toffees.

Pineapple juice: Pineapple juice is another delicious processed product which can be obtained by crushing the fruit pieces followed by physical separation of solids and fibers. After extraction of the juice it must be properly pasteurized. Preservative might be added to the juice to enhance its storability. It can also be stored for longer duration in refrigerated storage condition. This could help in enhancement of the shelf life of the juice to considerable extent. Proper packaging of fruit juice is another important aspect which should of priority for the processing industry. Juice which is fermented or having deadly microorganism might cause health hazard, hence special care is needed to be sought for ensuring proper processing of juices. Pineapple juice is slightly acidic and hence packaging in hard plastic bottles, Plastic bags, coated cans, multilayered containers ortetra packs can be utilized. Since pineapple is acidic fruit the pH of its juice is bound to be acidic. Special attention regarding pH and acidity of pineapple juice is needed to be sought. The pH should remain agreeable for human consumption. Blending of pineapple juice with some other fruit juice can also be done so that its taste and nutritional quality is enhanced (Jan and Masih, 2012). Colour of juice can be improved by blending with coloured fruits like grape or dragon fruit

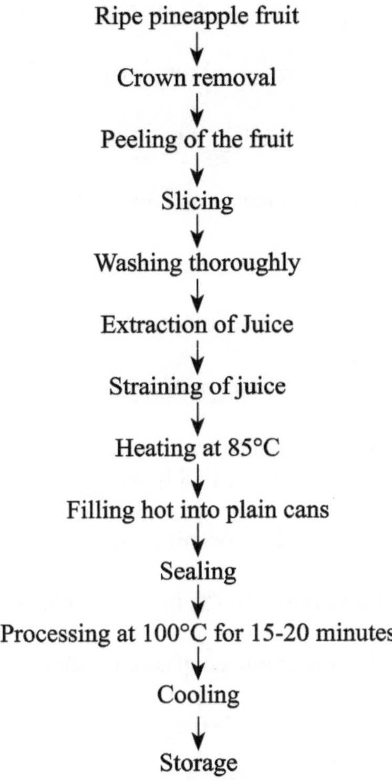

Flowchart for production of pineapple juice (Singh, 2009)

Pineapple RTS: Pineapple RTS is another important beverage which is liked by consumer of all ages and groups. Pineapple RTS is easy to be consumed and is usually ready-made. There is no need for dilution or mixing. It is preferred because it is easy to carry and consume. According to FSSAI and FPO specification and guidelines, RTS should have 10°B TSS and have minimum 10% fruit juice.At first pineapple juice is extracted, the TSS is adjusted to near about 10°B, the acidity is maintained at 0.3%. Required amount of sugar is also added to the juice to maintain the TSS of 10°B and assure sweet sour blend (Sindumathi, and Premalatha, 2013). Citric acid and clean water are also added at desirable quantity ensuring balance of acidity, taste, consistency and standards. To add consumer acceptability edible colour and artificial flavour can also be added. After filling of the beverage, a head space of 2.5 to 3.0 cm is needed to be kept. Before that the bottle and the cap should be sterilized in water at 85°C for 15-20 minutes. Hot filling of the RTS in glass bottle is done followed by air cooling (Oyeleke *et al.*, 2013).

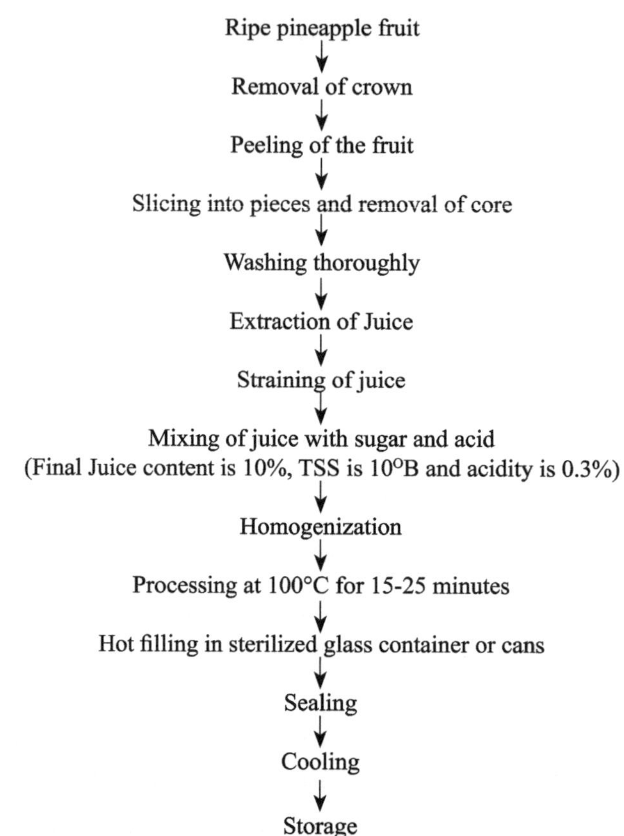

Flowchart for Production of Pineapple Juice (Singh, 2009)

Pineapple Squash: Pineapple squash is another important beverage which can be prepared from pineapple juice. Pineapple squash is prepared from fully matured and properly ripened fruit (Christina *et al.*, 2017). Pineapple squash should have 40°B TSS and have minimum 25% fruit juice. Acid content of squash generally remain between 1-1.5% but shall not exceed 3.5%. Class II preservative should not be more than 350 ppm SO_2 or 600 ppm benzoic acid. Pineapple squash is generally diluted at a ratio of 1:3 with water before serving. The process flowchart for pineapple squash is given below.

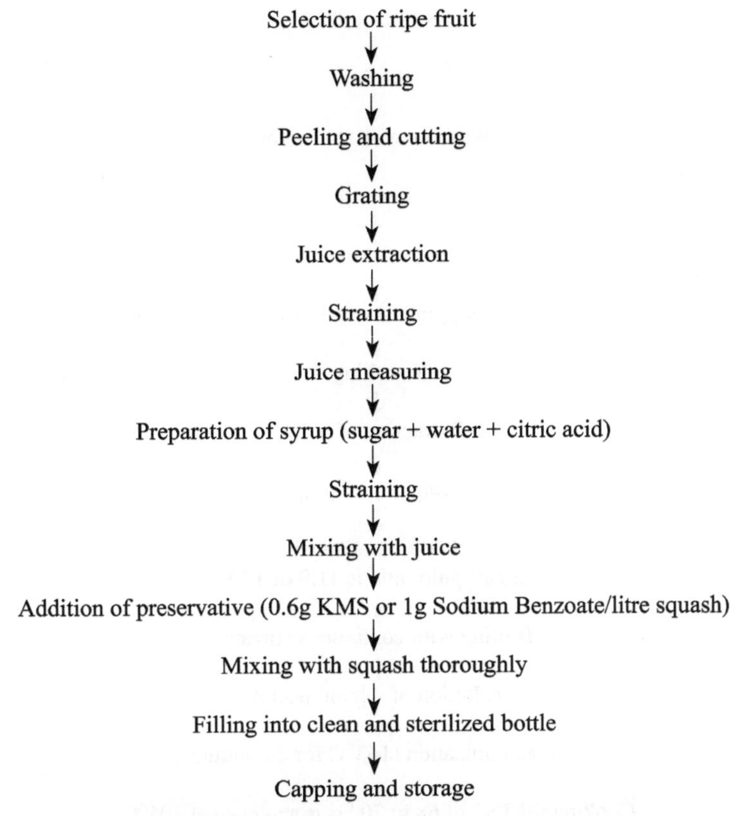

Flowchart for production of pineapple squash (Singh, 2009)

Pineapple Jam: Pineapple jam is prepared from extracted fruit pulp of perfectly ripened pineapple. Pineapple is selected at ripe stage. Unripe or over ripe pineapple is not used for jam preparation. The fruit pulp is boiled with sufficient quantity of sugar and other ingredients to a reasonable thick consistency. Sufficient amount of pectin is needed to be added in the jam to ensure that the jam is not sedimented. The jam should be firm enough to hold the fruit tissues in position. Over acidic and unripe fruits should not be used for jam preparation. According to standard guidelines, pineapple jam should contain minimum 45% fruit. 65° Brix TSS and acid 0.5%. Maximum preservative limit for pineapple jam is 40ppm SO_2 or 200ppm benzoic acid per kg of final product.

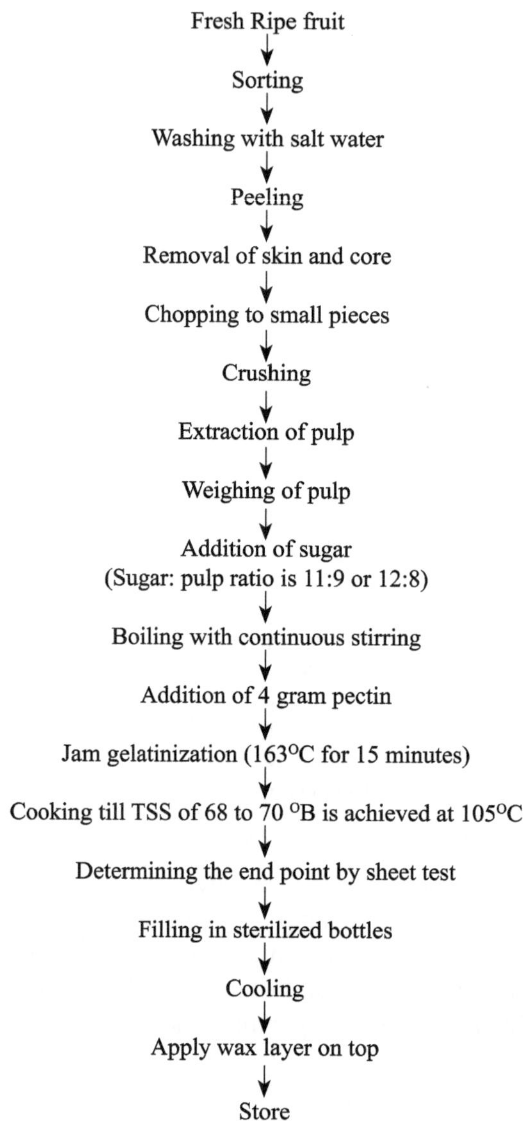

Flowchart for production of pineapple jam (Singh, 2009)

Pineapple Jelly: Pineapple jelly is another delicious processed product of pineapple which is prepared by boiling strained pineapple juice with sugar and pectin to sufficient consistency. 1 kg of grated pineapple pulp is mixed with 2.5 litre water, 10 g citric acid and 2g pectin. The mixture is boiled for 30-40 minutes and kept for cooling. The fruit pulp is pressed to separate the fiber and flesh from the liquid juicy part. The juice thus obtained is filtered and sieved in a jar and further tested for pectin quantity. Formation of a single clot indicates optimum pectin content. The product is cooled and boiled with sugar and pectin so that sufficient consistency is obtained. The final product must have a TSS of 65°B, minimum 45% fruit and acid 0.5%. After hot filling in sterilized container, a thin layer of melted wax is put and the lid is closed.

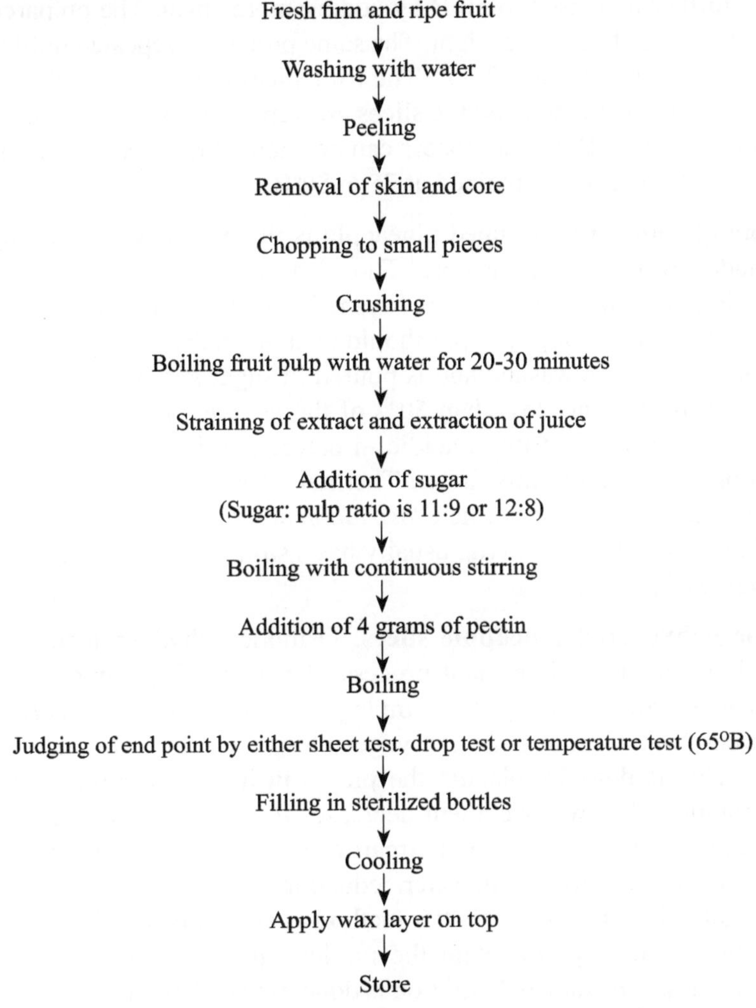

Flowchart for production of pineapple jelly (*Source*: Singh, 2009)

Pineapple preserve and candy: Pineapple preserve or candy is a very long lasting and organoleptically appealing processed product. Pineapple preserve is prepared by cutting rectangular or square shaped pineapple slices from matured ripe fruit. The fruit is peeled, core and eyes are removed. The fruits are sliced in proper shape. The fruits are kept dipped in 1.5 to 2% lime water for 3-4 hours to ensure hardness. The Calcium ion present in the lime water improves membrane integrity and retains the slice firmness. The mixture is drained and the fruit slices are washed with clean water. Around 400 to 500 g of sugar is dissolved in 600 ml hot water. The fruit slices are boiled in the sugar syrup and is kept dipped overnight. In the next day, the TSS of sugar syrup is further increased to 50°B by adding more sugar. The prepared sugar syrup is boiled and keep overnight. The same process is repeated until the TSS of the sugar syrup reaches 70-75°. The fruit slices are kept dipped for 7 to 10 days. The syrup is drained and the slices are kept in glass jar with sugar syrup concentration of 70°B. Same process can be adopted for preparation of candy. In this case the TSS is maintained at 75 to 80°B.

Canning of pineapple: Canned pineapple is the most luxury and appealing value-added product of pineapple. The ideal variety of pineapple suitable for canning is Kew. The fruits are washed, cleaned; the eyes and core are removed. The slicing of pineapple should be done in the form of circular ring. The pineapple ring is washed and is poured in sugar syrup. The drain weight of the fruit must be not less than 50% of the net weight. 1.5 cm head space is kept in the can. The fruit is acidic in nature and hence processing of the can is done considering this factor. Thermal processing is done for specific duration to destroy pathogen like *Clostridium botulinum* after exhausting. The canned pineapple in sugar syrup usually has a storage life of more than 1 year (Roodagi *et al*., 2012).

Osmotic dehydrated pineapple slices: Osmotic dehydration of fruits and vegetable pieces is an important process of preservation without impairing with the final product quality. The simple process of osmosis is utilized to pull out solvent (water) from fruit and vegetable pieces and dehydrate the final product. This is done by placing the pieces in high concentration of sugar or salt solution. The water content decreases in the fruit piece and increases in the solution side (Chavan and Amarowicz, 2012). The steps involved in osmotic dehydration are loss in water, reduction in weight and increase in solid gain.Osmotic dehydration is a better method as compared to other mechanical dehydration as it helps to retain the product quality, aroma, nutrients and wholesomeness. Osmotic dehydration is done in two steps, at first the removal of moisture using the process of osmosis and then subsequent dehydration in a dryer (Chaudhary *et al*., 2018).

According to Laura *et al*. (2005) the osmotic dehydration of 0.6 mm thick pineapple slices in sucrose solution (60% w/w) at three different temperature of 30°C, 40°C and 50°C. With increase in temperature from 30°C to 50°C, the sucrose diffusivities and apparent moisture (Ds and DW) augmented 2.8 and 3.8 times. Thus it can be said that the dehydration efficiency index (Dw/Ds) also increased with increasing temperature.Jose *et al.*, (2011) investigated on the osmotic dehydration of pineapple in a sucrose solution with respect to the effects of temperature (25-45°C) and citric acid concentration (0.5 -2.5% w/w). The results exhibited after 4 to 6 h of the process, where weight loss, moisture loss and solute gain can reach 42.62%, 36.54% and 292.16% respectively. 100% sensory acceptance and reductions in microbial counts were also observed. Rahman and Lamb (1990) described that temperature more than 50°C might not have encouraging effect on solute gain during osmotic dehydration of pineapple dipped in sucrose solution. At comparatively higher temperature, sucrose was not proficientenough to get distribute all over the fruit pieces and remove solvent through the cell membrane. Chaudhari *et al.* (2015) studied on the efficacy of different pretreatments on further drying process of pineapple cubes. Effects of osmotic dehydration on weight reduction and mass transfer during osmotic dehydration of pineapple cubes were investigated. Water loss, weight reduction and solids gain increased with increasing treatment time. Saini and Sharma (2016) studied on the consequence of storage on the colour and quality parameters of uncoated and pectin coated dehydrated pineapple slices. Higher amount of browning was evident in uncoated samples than coated samples. The browning was more evident during the later part of the storage. With increasing storage, the moisture content increased and the rehydration ratio decreased.

Pineapple powder: Pineapple powder is a commercially important product which is prepared either by dehydration of pineapple juice or dehydration of pineapple pomace. The best thing about pineapple powder is that it can be stored at ambient temperature for a considerable duration, it is convenient to use. Beside that the manufacturer also need very less transportation charge. Pineapple powder is desired by consumer as it can be consumed as an instant juice powder or a flavoring agent instantly. Its quality, flavour and texture is just like original pineapple juice (Gabas *et al.*, 2007). Usually the drying of pineapple fruit juice is done in a drum drier, spray dryer and Foam mat dryer. Spray dryer ensured the best quality of final product whereas Foam mat dryer ensure low cost drying. A temperature of not more than 100°C is desirable while spray drying of pineapple juice (Adhikari *et al.*, 2004). According to Jori *et al.*, (2013), the best quality of pineapple powder was obtained when 15% maltodextrin (MD) was added to pineapple juice, the drying temperature was 150°C and the feed released was 0.022 liter per minute.

Pineapple Leather: Pineapple leather is another dehydrated value-added product of pineapple which is obtained after dehydration of pulp or puree of pineapple. Pineapple fruit is cut and the eyes, core and peel are removed. The pulp is extracted and is homogenized properly. For improvement of quality and taste, other ingredients, other fruits or herbs can be also added to the leather. Pineapple pulp is initially cooked with high concentration of sugar till a high TSS of more than 65°B is achieved (Phimpharian *et al.*, 2011). Then the cooked puree is spread in either metal plates or in cloth for drying or dehydration. Natural sun drying is a slow process whereas dehydration in cabinet dryer is a comparatively rapid process. Beside that leather forming machine can also be used for preparation of pineapple leather. Since pineapple fruit is devoid of pectin, addition of 1% pectin can is very much needed for strong set of leather. The drying temperature for leather preparation is 55-60°C.

Pineapple wine: Pineapple wine is a popular fermented beverage which is produced by fermentation of pineapple juices in presence of desirable yeast. The juice is fermented at 5% pure yeast followed by siphoning, filtration, bottling and aging. In the process of pineapple wine preparation, pineapple juice is inoculated with 5% (v/v) active yeast and held at 20°C for 7 days. In the process of fermentation, the total sugar content is reduced whereas the acidity increases. The alcoholic strength increases with increasing duration of fermentation (Roodagi *et al.*, 2012).

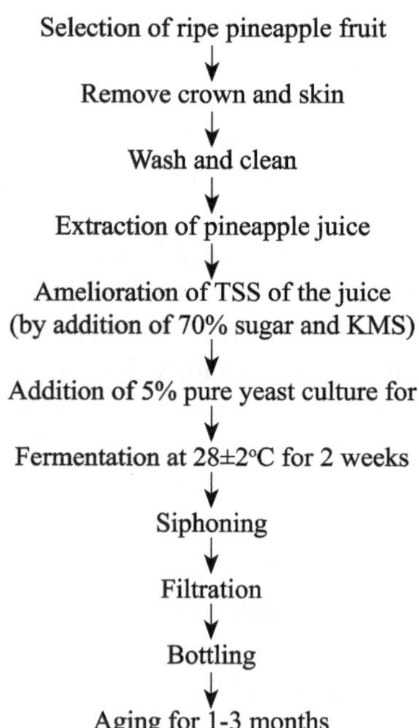

Flowchart for the preparation of pineapple wine (Ningli *et al.*, 2017)

Dehydrated pineapple: Preservation by drying and dehydration is most primitive way of preservation. The fundamental principle is to reduce the moisture content of the product, thus minimizing the amount of free water present for microbial growth. In the process of dehydration, the free water from the fruit slices or chunks are removed by a single method or combination of multiple techniques. The final moisture is reduced to 5-10%. The dried fruit slice has a comparatively higher shelf life. Drying in cabinet or tray dryer followed by osmotic dehydration of the slices.

Pineapple vinegar: Pineapple vinegar is slowly gaining popularity and is also an important product of export. Vinegar from pineapple can be extracted from over ripened fruits, core, peels, blemished and pomace. This is obtained by alcoholic fermentation followed by acetic acid fermentation of the initial product (Sossou *et al.*, 2009).

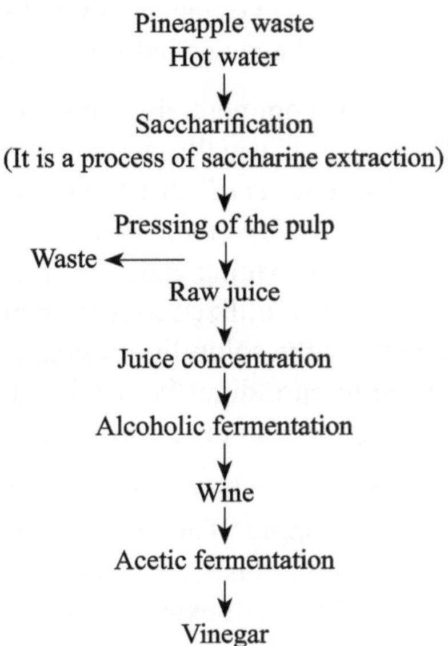

Flowchart for the production of pineapple vinegar (Arianna *et al.*, 2017)

Pineapple fiber: Fibre extracted from pineapple leaf is a matter of research in the last few decades. Pina cloth is prepared from pineapple leaves. The leaves yield a strong, smooth, glossy white coloured silk like fibre. The fibre is medium in length but is having high tensile strength. The best quality about this fibre is that it can absorb colour and retain it for a considerable duration. For obtaining fibre, the plant is not allowed to fruit. Interestingly, the fruits are removed from the plant at a very young stage so that the plant can utilize the energy and carbohydrate for its vegetative growth only. Some specific variety like

'Perolera' is especially suited for fibre extraction. For fibre extraction, outer, long leaves are preferred. Fibre is extracted either manually or mechanically. In the manual process, the leaves are first dipped in water, bitten with stick and then allowed to ret in water for some days. In this process of retting 0.5% urea or diammonium phosphate (DAP) is added to accelerate the total process and trigger the population of microorganisms needed for retting. This digest the unwanted tissue and pure fibre is left (Asim *et al.*, 2015). After retting process is completed, the fibre are cleaned in water, sun dried and combed. In mechanical extraction method the entire process is undertaken in machine. The quantity of fibre extracted is low but quality is better than manual separation.

Animal feed: Pineapple skin, core, peel, pomace and the leaves are rich in nutrients and fibre which are a favourite of ruminants. Even the cattle preferred highly acidic fermented pineapple waste along with fresh waste or waste from farm. The dried pomace or dried leaves can be a good source of roughage and can effectively replace 50% roughage needed for the cattle (Anis *et al.,* 2016)

Bromelain: Bromelain is a commercially important enzyme which is proteolytic in nature. Bromelain is usually derived from pomace, juice, core, peel or stem of pineapple (Ketnawa *et al.*, 2012). Highest amount of bromelain in obtained from crown followed by stem. Bromelain have wide utilization in food industry. It is utilized in baking industry, leather industry and meat tenderization. Beside that it is also utilized to reduce browning of apple juice, chill proofing beer and increase the solubility of gelatin (Gautam *et al*, 2010). In medical sector, it is used to cure digestive problems, reduce inflammation post surgery, reduce swelling and also in cancer treatment (Tochi *et al.*, 2008).

Ferulic acid: Ferulic acid is an abundant hydroxycinnamic acid which is present in cell wall of pineapple. The phenolic antioxidant has diverse application in cosmetic and food industry. Pineapple peel is highest source of ferulic acid which is obtained by alkali extraction (Rudra *et al.*, 2015).

Post Harvest Diseases and Pests of Pineapple

Black rot: Black rot of pineapple is due to facultative parasitic fungus *Ceratocystis paradoxa* or *Thielaviopsis paradoxa* or *Chalara paradoxa.* This fungal pathogen causes black rot in pineapple fruits where the broken peduncle remains wet; such fruits are also referred as leakers (Reyes, 2004). The pulp becomes rotted and become translucent. This disease is also known as water blister, soft rot or stem end rot as it is characterized by a soft watery rot of the flesh. The severity of disease and virulence of the pathogen depends on initial inoculum concentration, presence of any bruise or injury and the temperature the fruit is stored in. Generally, the rotting begins with the stem end and progress downwards. The fungal mycelium appears dark black in colour. Maximum fungal growth is in temperature between 21-32°C (Droby *et al.*, 1991). Hot water treatment (54°C) for 3 minutes proved to be effective in controlling black rot disease up to 21 days when stored at 10°C and for 48 hours. Fruits which are injured are more susceptible to black rot. Postharvest sanitation with 150 ppm chlorine along with fungicide treatment can be effective. Some effective fungicides for controlling black rot are benomyl, thiabendazole and captan. Besides that storing the fruits at 7°C can inhibit growth of the fungus (Marta, 2007).

Fruitlet core rot or black spot or fruitlet brown rot: Fruitlet Core Rot or black spot or fruitlet brown rot is a post-harvest disease which is due to fungal yeast and mite complex. It is due to complex action of *Fusarium moniliforme* var. *subglutinas, Penicillium funiculosum*, round yeast *Candida guilliermondi*, fruit mite *Steneo tarsonemus* and red mite *Dolichotetranychus floridanus*. The symptom is the presence of brown or black coloured spots in the centre portion of the fruitlets. The severity of the malady reduces with increasing acidity of the fruits. More the acid content lesser is the chance of infestation. Since mite is responsible for the spread of the disease, spraying of insecticide like acaricide that kills mite can be very effective to reduce the useful.

Yeasty fermentation: There is abundant presence of biological yeasts and bacteria in the skin and pulp of pineapple fruits. The bacteria or yeast can be viable only in certain stage of growth or maturity. At certain stage, these bacteria can start growing or new microbes can also enter the fruit through some cracks or holes. *Saccharomyces* is the yeast which is primarily responsible for the fermentation of pineapple flesh. Due to fermentation, juices or bubbles of gases starts escaping through the micro pores in skin and eyes. The skin eventually develops a dark yellow to brown colour. The skin becomes leathery and the flesh becomes spongy. The pulp turns acidic and the flesh develops a bright yellow colour. Minimizing mechanical damages, ensuring proper aeration during handling, packaging and storage can reduce the fermentation. Maintaining optimum temperature and relative humidity can also help to retard the fermentation process. Application of fungicide thiabendazole (TBZ) can reduce the fermentation process substantially.

Post-Harvest Insects

Mealy bug (*Dysmicoccusbrevipes*) is another serious pest of pineapple. It usually enters the fruits in the pre harvest stage and hides in the fruit cervices, between the fruitlets, below the crown leaves or inside crown whorl. After harvesting the bugs feed on the pulp and damage the entire fruit affecting the marketability.

Red mite (*Dolichotetranychus floridanus)* is another serious pest of pineapple which not only affects the fruit quality but is also responsible for the post harvest disease known as Fruitlet Core Rot.

Post Harvest Physiological Disorders of Pineapple

Internal Browning: Internal browning is a very common occurring postharvest physiological disorder of pineapples. The symptom can be marked with development of small greyish translucent areas at the base of the fruit near the core. These greyish portion starts darkening and ultimately the entire portion of the inner flesh becomes brownish-black in colour. This disorder is more prone in cultivars with low acid content. Cultivar like Smooth Cayenne is susceptible. It is prone in fruits which faces a very low night temperature followed by high day temperature during maturity and ripening. The fruits should be stored at temperature more than 7°C. This help to reduce the incidence of internal browning.

Flesh translucency: This disorder is more prone in those fruits which have been harvested at immature stage. The symptom of this disorder can be marked as development of water soaking symptoms at the skin of the fruits (Chun *et al*., 2016). The internal flesh becomes shiny and translucent in appearance (Py, 1987). Usually, the cultivars which have a small crown are more prone to the malady. Bruise or mechanical injury aggravates the disorder symptoms. At extreme stage, intracellular sap leakage takes place which led to total failure of marketability.

Sun burn: Sunburn of fruits is a post-harvest disorder which takes place if harvested fruits are kept in scorching sun for quite a significant duration. Due to exposure to solar radiation, the fruit surface skin shows a bleached appearance. The skin turns yellowish white in colour. The flesh below the bleached area gets damaged. The sun burnt area and affected pulp becomes susceptible to decay and rot diseases. It can be avoided by keeping the fruits in shaded area after harvest.

Chilling injury: Chilling injury is a storage disorder of pineapple which is stored at a temperature less than 7°C. The symptoms of chilling injury include the curling, discolouration and wilting of the crown. The fruits fail to change its colour from green to yellowish – red. The fruits become brownish in colour with poor lustre. The internal flesh also starts browning (Lim, 1985). Several small lesions are initially developed on the fruit skin which eventually coalesce and form bigger lesions. The situation further worsens leading to softening of the fruits and off-flavour development. The longer the duration of exposure to chilling temperature, the more is the extent of damage. Storage at temperature more than 10°C can minimise the disorder. Waxing and proper handling also help to reduce the intensity of the malady.

References

Adhikari, B., Howes, T., Bhandari, B. R. and Troung V. 2004. Effect of Addition of Maltodextrin on Drying Kinetics and Stickiness of Sugar and Acid-rich Foods During Convective Drying Experiments and Modeling. Journal of Food Engineering, 62:53-68.

Adikaram, N. and Abayasekara, C. 2012. Pineapple, In: Debbie Rees, Graham Farrell And John Orchard (eds.), crop post-harvest: science and technology: perishables, blackwell publishing ltd., West Sussex, UK, pp,143- 158.

Akamine, E. K. and Theodore, G. 1971. Controlled Atmosphere Storage of Fresh Pineapple (Ananas comosus (L) Merr. 'Smooth Cayenne'. Research Bulletin 152 Hawaii Agricultural Experiment Station University of Hawaii.

Anis, A. M., Senthilkumar, S. and Shailesh, K. S. 2016. A review on trends and approaches in post-harvest handling of pineapple. Journal of Chemical and Pharmaceutical Sciences, 9(4): 2885-2889.

Arianna, R. Luigi, L. Fabrizio, T. Roberta, D. Dante, M. F. and Milenal, L. 2017. Metabolite profiling and volatiles of pineapple wine and vinegar obtained from pineapple waste. Food Chemistry, 229:734-742.

Asim, M. Abdan, K. Jawaid, M. Nasir, M. Dashtizadeh, Z. Ishak, M. R. and Enamul, H. M. 2015. A review on pineapple leaves fibre and its composites. International Journal of Polymer Science, 2(1): 11-13.

Chaudhari, A. P., Dhake, K. P., and Manisha, B. 2015. Osmotic dehydration of pineapple. International Journal in IT and Engineering, 3(4):2321-1776.

Chaudhary, V., Kumar, V., Singh, G. R., Singh, J., Chauhan, N., and Kumar, K. 2018. To study the osmotic dehydration characteristics of pineapple (Ananas comosus) slices. International Journal of Chemical Studies, 6(5):1081-1083.

Chavan, U. D. and Amarowicz, R. 2012. Osmotic Dehydration process for preservation of fruits and vegetables. Journal of Food Research, 1(2)

Christina, M. Henry, L. Bendantunguka, T. Simeon, W. Daphrose, G. and Bellancile, U. 2017. Evaluation of storage stability of pineapple products processed by small and medium scale processing enterprises in Rawanda. African Journal of Food Science, 11(6):146-159.

Chun, O. U., Yu, Liu., Wang, W. and Dong, D. 2016. Integration of UV-C with antagonistic yeast treatment for controlling post-harvest disease and maintaining fruit quality of Ananas comosus. Bio Control, 2(2) 87-93.

Deka, B. C., Saikia, J. and Sharma, S. 2005. Standardization of maturity indices of 'Kew' Pineapple Proc . 5th Int. Postharvest Symp. Acta Horticulturae, 682:2215-2220.

Droby, S., Chalutz, E., Wilson, C. L. 1991. Antagonistic microorganisms as biological control agents of postharvest diseases of fruits and vegetables. Postharvest News Info, 2: 169-173.

Gabas, A. L., Telis, V. R. N., Sobral, P. J. A. and Telis-Romero, J. 2007. Effect of Maltodextrin and arabic gum in water vapor sorption thermodynamic properties of vacuum dried pineapple pulp Powder. Journal of Food Engineering, 82:246-252.

Gautam, S. S., Mishra, S. K., Dash, V., Goyal, A. K., and Rath, G. 2010. Comparative study of extraction, purification and estimation of bromelain from stem and fruit of pineapple plant. Thai Journal of Pharmaceutical Science, 34: 67-76.

Gregory, R., Ronald, P., Sharon, J. and Dorian, E. 2011.Caribbean Pineapple Production and Post-Harvest Manual, FAO Trust Fund for Food Security and Food Safety – Government of Italy Contribution.

Hassan, A., Othman, Z., and Sirivanich, J. 2011. Pineapple Ananas comosus L. (Merr). In: Post Harvest Biology and Technology of tropical and subtropical fruits: Mangosteen to white Sapotae. Cambridge Wood Head Publishing. 4: 194-217.

Hossain, M. and Bepary, R. H. 2015. Post harvest handling of pineapples: A key role to minimize the post-harvest loss. International Journal of Recent Scientific Research Research, 6:6069-6075.

http://apeda.in/agriexchange/Market%20Profile/one/PINEAPPLE.aspx

Hu, Huigang., Li, Xueping., Dong, Chen and Chen, W. 2011. Effects of wax treatment on quality and postharvest physiology of pineapple fruit in cold storage. African Journal of Biotechnology, 10: 7592-7603.

Ibrahim, S., Nahar, S., Islam, J. M. M., Islam, M., Huque, R. and Khan, M. 2014. Effect of low molecular weight chitosan coating on physico-chemical properties and shelf life extension of pineapple (Ananas sativus). Journal of Forest Products & Industries, 3: 161-166.

Idah, P. A., Ajisegiri, E. S. A. and Yisa, M. G. 2007. Fruits and vegetables handling and transportation in Nigeria. AU Journal of Technology, 10(3):175–183.

Jan, A. and Masih, D. 2012.Development and quality evaluation of pineapple juice blend with carrot and orange juice. International Journal of Scientific and Research Publications, 2(8):1-8.

Jori, D., Ladole, M., Gore, A. and Bhand, B. 2013. Study on Effect of Carbonation on Storage and Stability of Pineapple Fruit Juice. International Journal of Engineering Research & Technology, 2(12) ISSN: 2278-01

Jose, E., Zapata, M., Johan, M., Arias, A., Gelmy, L. and Ciro, G. 2011. Optimization of osmotic dehydration of pineapple using the response surface methodology optimization osmotic pineapple (Ananas cosmoses.). Aplicando Superficies De Respuesta AgronColombia, 29.

Joseph, A., Tunrayo., Okelana, M. A. and Adekoya, I. A. 2010. Storage of pineapple fruits under different conditions: implication on shelf life. Nigerian Journal of Horticultural Science, 14(1). doi: 10.4314/njhs.v14i1.62161

Joy, P. P. and Rashida, R. T. A. 2016. Harvesting and post-harvest handling of pineapple. Pineapple Research Station (Kerala Agricultural University).

Joy P. P. Benefits and uses of pineapple. Pineapple Research Station, Kerala Agricultural University, Kerala, India, 2010. [http://www.kau.edu/prsvkm/Html/BenefitsofPA.html. (DOI: 10.13140/RG.2.1.2782.4888.)

Ketnawa, S. Chaiwut, P. and Rawdkuen, S. 2012. Extraction of bromelain from pineapple peels. Food Science and Technology, 4(17):395-402.

Laura, A. and Ramallo, R. H. M. 2005. Rate of Water Loss and Sugar Uptake during the Osmotic Dehydration of Pineapple. Brazilian archives of biology and technology in International journal, 48(5):761-770.

Lim, W. H. 1985. Diseases and disorders of pineapple in peninsular Malaysia, MARDI Report No. 97, Kuala Lumpur, Malaysia, pp, 53.

Mandal, D., Lalremruata B., Hazarika, T. K. and Nautiyal, B. P.2015. Effect of post-harvest treatments on quality and shelf life of pineapple (Ananas comosus [L.] Merr. 'Giant Kew') Fruits at Ambient Storage Condition. International Journal of Bio-Resource and Stress Management, 6(4): 490-496.

Manzar, H. and Rejaul, H. B. 2015. Post-harvest handling of pineapples: a key role to minimize the post-harvest loss. International Journal of Recent Scientific Research, 6(9): 6069-6075.

Maria, G. L. and Elhadi, Y. 2017. Biology and post-harvest physiology of pineapple. Handbook of pineapple technology, production, post-harvest science, processing and nutrition. Published by John Wiley and Sons Ltd.

Marta, M. C., Maria, A. R. G. and Olga, M. B. 2007. Effect of packaging conditions on quality and shelf-life of fresh-cut pineapple (Ananas comosus). Postharvest Biology Technology, 50(2-3):182-189.

Montero, C. M., Rojas-Grau, M. A., Martín-Belloso, O. 2008. Effect of packaging conditions on quality and shelf-life of freshcutpineapple (Ananas comosus). Postharvest Biology and Technology. 50(2-3): 182–189.

Nimitkeatkai, H., Srilaong, V. and Kanlavanarat, S. 2006. Effect of edible coating on pineapple fruit quality during cold storage. Acta Horticulturae, 712: 643-648.

Ningli, Q. Lina, M. Liuji, L. Xiao, G. and Jianzhi, Y. 2017. Production and quality evaluation of pineapple fruit wine. International Global on Renewable Energy and Development, 2(7):455-460.

Ningombam, S., Noel, A. S. and Singh , J. 2019. Post-harvest losses of pineapple at various stages of handling from the farm level up to the consumer in Manipur. International Journal of Agriculture Sciences, 11(22) : 9235-9237.

Oyeleke, G. O. Ojo, A. Ajao, F. D. and Adetoro, R. O. 2013. Development and analysis of blended pineapple, watermelon ready to drink (RTD) juice. Journal of Environmental Science, Toxicology and Food Technology, 4(6): 22-24.

Paull, R. E. and Rohrbach, K.G. 1985. Symptom development of chilling injury in pineapple fruit. Journal of American Society for Horticultural Science, 110:100-105.

Phimpharian, C., Jangchud, A., Jangchud, K., Therdthai, N. and Prinyawiwatkul, W. 2011. Physicochemical characteristics and sensory optimization of pineapple leather snack as affected by glucose syrup and pectin concentrations. International Journal of Food Science and Technology. 46(5):972-981.

Prakash, A., Baskaran, R. and Vadivel, V. 2019. Citralnanoemulsion incorporated edible coating to extend the shelf life of fresh cut pineapples. LWT. 118. 108851. 10.1016/j.lwt.2019.108851.

Pramod, K. R., Nidhi, A. and Mitu, S. 2016. Edible coating of fruits and vegetables: a review. International Journal of Scientific Research and Modern Education, 1(1): 2455 – 5630.

Py, C., Lacoeuilhe, J. J. and Teison, C. 1987. The pineapple, cultivation and uses. G.P. Publisher: Paris: Editions Maisonneuve & Larose, 1987.

Rahman, M. S. and Lamb, L. 1990. Osmotic dehydration of pineapple. Journal of Food Science and Technology, 27:50-152.

Reyes, M. E. Q., Kenneth, G. R. and Robert, E. P. 2004. Microbial antagonists control postharvest black rot of pineapple fruit.Biology and Technology, 33: 193–203.

Roodagi, M. B. Gokhale, N. B. Tippanagoudar, P. G. Jagtap, J. and Salvi, V. G. 2012. Effect of different levels of TSS on the quality of pineapple (Ananas comosus L.) wine. International Journal of Processing and Post-Harvest Technology, 3(1):77-80

Rudra, S. G. Nishad, J. Jakhar, N. and Kaur, C. 2015. Food industry waste: mine of nutraceuticals. International Journal of Science, Environment and Technology, 4(1): 205-229.

Saad, S. 2004. The world pineapple market; when growth goes hand in hand with diversity. http//:www.tanusi.com/

Saini, C. S. and Sharma, H. K. 2016. Effect of pectin coating on colour and quality of dehydrated pineapple during storage. Asian Journal Dairy & Food Reserearch, 35(2):120-129.

Shweta, S., Sindhu, K. C. and Soumitra, T. 2017. Pineapple production and processing in north eastern India. Journal of Pharmacognosy and Phytochemistry, SP1: 665-672.

Sindumathi, G. and Premalatha, M. R. 2015. Development and storage studies of naturally flavored papaya-pineapple blended ready-to-serve (RTS) beverages. International Journal of Science and Research, 4(2): 856-860.

Singh, I. S. 2009. Post-harvest handling and processing of fruits and vegetables. West ville publishing house. pp, 96-166.

Soler, A. 1992. Pineapple. Paris: CIRAD-IRFA. pp, 48.

Sossou, S. K. Ameyapoh, Y. Karou, S. D. and Souza, C. 2009. Study of pineapple peelings processing into vinegar by biotechnology. Journal of Biological Science, 12(11):859-865.

Thompson, A. K., 2003. Fruit and vegetables: harvesting, handling, and storage. pp, 306-308.

Tochi, B. N., Zhang, W., Ying, X. and Wenbin, Z. 2008. Therapeutic application of pineapple protease (Bromelain): A review. Pakistan Journal of Nutrition, 7(4): 513-520.

Xueping, L., Xiaoyang, Z., Hailan, W., Xuefen, L., Hongwei, L. and Weixin, C. 2017. Postharvest application of wax controls pineapple fruit ripening and improves fruit quality. Postharvest Biology and Technology, 136 (2018): 99–110.

Index

A

Amritha 94, 107, 108, 128
Animal feed 22, 336
Artificial flower induction 114, 231
Auxin 83, 171, 224, 233

B

Baruipur Local 107, 108
Black spot 337
Bracts 13, 102, 105, 112, 139, 230, 234, 298
Bromelain 5, 7, 8, 9, 10, 18, 19, 23, 27, 291, 336
Butts 168, 169

C

Calcium carbide 67, 189, 224, 232, 235, 236, 239, 240
CAM 1, 139, 162, 239, 240, 259
Chilling injury 249, 250, 254, 316, 321, 325, 338
Chitosan 66, 321
CODEX 320
Core 8, 10, 122, 138, 168, 184, 199, 230, 236, 247, 252, 260, 262, 263, 292, 295, 296, 326, 328, 330, 332, 334, 335, 336, 337, 338
Crown 8, 59, 61, 62, 63, 65, 67, 75, 102, 104, 105, 107, 114, 122, 127, 130, 168, 169, 172, 182, 184, 185, 188, 202, 225, 226, 235, 249, 251, 252, 253, 259, 260, 261, 267, 270, 272, 283, 294, 297, 303, 306, 317, 318, 320, 321, 322, 323, 324, 325, 326, 327, 328, 334, 336, 338, 339
Crown removal

D

D-leaf 114, 188, 191, 235
Dehydrated 332, 333, 334, 335
Double row 61, 62, 70, 183, 185, 188, 215
Drip irrigation 70, 219, 220

E

Edible coating 250, 321
Ethephon 60, 223, 224, 332, 234, 235, 238, 239, 240, 318
Ethrel 67, 234, 235, 236, 239, 240, 247
Ethylene 111, 113, 154, 224, 225, 229, 230, 231, 232, 233, 234, 235, 236, 237, 238, 239, 240, 246, 250
Export 27, 28, 34, 37, 69, 70, 74, 75, 76, 77, 83, 84, 94, 97, 105, 106, 107, 155, 194, 213, 252, 253, 255, 305, 306, 316, 317, 322, 324, 325, 335
Eye 59, 122, 268, 283, 295, 296, 316

F

Fasciation 260, 261
Fibre 2, 3, 4, 6, 7, 8, 9, 17, 18, 23, 27, 41, 66, 77, 100, 102, 103, 105, 130, 313, 335, 336
Fingerprinting 148, 149
Flesh translucency 338
Flower induction 114, 160, 219, 231, 233, 236, 247, 297, 316
Fruitlet core rot 292, 295, 296, 337, 338
Genetic engineering 137, 138, 147, 148, 231, 263
Giant Kew 34, 53, 59, 67, 71, 82, 103, 104, 171, 214, 306
Gibberellin 112, 113, 237
Grade 83, 181, 224, 249, 251, 252, 307, 320, 322, 323, 324

H

Haricharanvita 82, 83, 107
Heart rot 121, 135, 163, 221, 292, 293, 299, 300
Heterozygosity 114, 115, 131, 147, 168
High density planting 60, 68, 75, 170, 215

I

Industrial 10, 17, 19, 21, 22, 23, 72, 111

Inheritance 121
Integrated nutrient management 68, 192, 193
Internal browning 109, 129, 138, 246, 247, 252, 262, 296, 318, 321, 338

J

Jaldhup 83, 106
Jam 3, 20, 31, 44, 146, 309, 314, 326, 329, 330
Jelly 3, 326, 331
Juice 5, 7, 8, 21, 27, 28, 31, 40, 41, 42, 66, 68, 69, 71, 102, 199, 239, 300, 308, 309, 314, 326, 327, 328, 329, 331, 333, 334, 335, 336

K

Kew 31, 34, 53, 58, 59, 60, 66, 67, 69, 71, 75, 77, 82, 83, 94, 95, 96, 103, 104, 107, 124, 125, 126, 128, 132, 169, 171, 189, 214, 236, 246, 248, 249, 252, 260, 291, 306, 316, 332

L

Lakhat 106
Leaf pruning 226, 227
Leather 195, 326, 334, 336

M

Maturity indices 247, 248, 249, 314
MD-2 3, 94, 101, 107, 155, 167, 168, 186, 261
Mealy bug 103, 104, 120, 121, 137, 182, 252, 265, 266, 275, 276, 277, 278, 279, 292, 302, 338
Micro-propagation 145, 147, 169, 170
Monocarpic 1, 111
Mutation 120, 124, 132, 133, 145, 146, 151, 261

N

NAA 83, 111, 125, 133, 146, 171, 173, 174, 175, 177, 224, 229, 233, 235, 236, 237, 238, 240, 247, 250

O

Organic 5, 17, 34, 39, 40, 43, 44, 61, 62, 64, 65, 66, 68, 72, 74, 75, 77, 78, 83, 85, 162, 163, 191, 192, 193, 194, 195, 196, 200, 202, 213, 214, 215, 216, 221, 222, 229, 230, 233, 309
Ornamental 10, 13
Osmotic dehydration 332, 333, 335

P

Packaging 21, 66, 246, 252, 253, 254, 322, 325, 326, 327, 337
Peduncle 44, 104, 105, 107, 122, 129, 168, 169, 181, 225, 230, 235, 237, 252, 261, 285, 295, 316, 318, 320, 337
Pina cloth 2, 335
Pinching 61, 189
Planting system 170, 181, 182, 183, 184, 185, 187, 262
Pollination 112, 115, 125, 127, 167
Processing 17, 20, 29, 31, 40, 47, 68, 69, 70, 71, 72, 73, 74, 75, 76, 77, 78, 94, 97, 100, 105, 124, 167, 215, 226, 246, 255, 262, 302, 309, 309, 313, 326, 327, 328, 332
Propagation 63, 65, 66, 100, 120, 123, 124, 136, 145, 147, 167, 168, 169, 170, 171, 176, 233, 259

Q

Queen 1, 2, 3, 42, 44, 53, 57, 58, 59, 60, 69, 71, 75, 77, 82, 83, 91, 94, 101, 103, 105, 106, 107, 108, 115, 123, 124, 125, 126, 128, 130, 131, 136, 151, 156, 171, 172, 184, 189, 214, 225, 260, 291, 315

R

Red heart 230
Ripening 59, 60, 121, 122, 129, 149, 150, 160, 200, 222, 231, 234, 235, 246, 247, 248, 250, 254, 255, 270, 296, 299, 300, 314, 315, 321, 325, 326, 338
RTS 314, 326, 327, 328

S

Slice 122, 332, 335
Slip 59, 65, 83, 104, 107, 122, 168, 171, 184, 225, 235, 239, 261, 294
Smooth Cayenne 2, 34, 91, 101, 103, 104, 107, 108, 111, 114, 115, 123, 125, 126, 128, 129, 131, 132, 133, 136, 137, 138, 146, 154, 161, 168, 171, 172, 176, 225, 232, 238, 239, 247, 260, 261, 294, 296,

302, 316, 338
Sorosis 230
Stumps 168, 169, 324
Sucker 63, 65, 67, 122, 170, 184, 225, 226, 239, 267, 277
Sun burn 62, 338
Sun scald 160, 187
Synthetic seed 176

T
Taxonomy 102
Tebuconazole 238
Tissue Culture 40, 71, 139, 145, 146, 169, 170, 233
Trimming 318, 319, 320

U
Uniconazole 238
Value addition 40, 68, 77, 78, 313, 314, 326
Vazhakulam Pineapple 94
Vermicompost 193, 195, 196
Vinegar 20, 23, 335

W
Waste 7, 9, 10, 21, 22, 41, 309, 335, 336
Waxing 250, 319, 320, 321, 339
Weed control 75, 138, 220, 318
Wine 1, 2, 66, 68, 100, 334, 335

X
Xerophytic 1

Y
Year round 60, 75, 94, 97, 189, 224
Yeasty disease 292, 299
Yeasty fermentation 292, 337

Z
Zeatin 235